GATTAATTAACCCA... ...GATTCATCATGGGTTAA
CGGAGATACCTGGA... ...TATGTATTGTCCTTTTT
AGAAAAGCATTACA... ...TTAGTAATTAAATTATA
CACGAACTTCTTCTT... ...ATTTGATTGTGGTTTCAATAATATTTAAA
CATCTTCGATTCCCAACTGGGAACTCGCCGGCTTTGCGACTGCCAATCTAGATTATCCTTTCAG
TAATTCACCAATTGCAACTGGATAATGACATATCCTGCCCATATGCTGAGTATATCCACTCCAT
TGCCTGCGTCCTTGTTGCTTGTGGAACTGTGGAATTCCATTGGGAGTCGGCTTTAATGCATTTA
CATGCAACGGGGGCCCCCAATGCAATTAGACAAAAGCCAGATACAAACGCAGCAGCAACAACTT
GGCACTACAACTATTGCATAACAATTACGAGTGCATAGCCCCTACTACACATTCGATAACTGTA
CATATACCATTAAATAGCTCTTGAAATGCTGCTCATAACCAAGTATGCTCATAATGCGTAAGTA
AGTTGTTCAACTAATGTTGCATGATTTTCCACAAAATGCCAATAGCTGTGTTCTTATGCATATT
ACTAATCCATTATAACAAATATAGGCTTATCTAGTTTTTACCTTTTCGAAATAAAACCAAATTG
GTTTAATGGGATTTCAATCTATTTACCCGATTAAGAGCGTTCATTTGCACTTTATTATTCATTT
TTTCATTTGGTTTAAACAGCTTTTAAAAACCCAAGGACTTTTTCCATTTAGTTCATAATTTAGA
TCGAATAAATGGATCAACAAACTAAATCGAAAGGCGATCAATTGGAGTTGTTAGCATTATACTT
GCCTTTGATCTCTGAAGTGGGCCAATGTTTGCCGAACGCTTGGAAATGCCGGATGATTTTGCAA
AACCGCTGGTGCACTTTAGACAATAATCACTGTAAATGCTATTTACTCAAGTGCTCGATTTGTG
TGCTTTAAATGCGTTTGTTTGCGTTACATAGCCAGAAATTTAACGCTGTAACCACAAAAACGAA
GTCACATGTGGGCCCGAAATGCCCGTGCCCTTTTATCCTTTATTTCGGTTCTTAGCTAAGCAAA
GGCCCAAAACGCAAACAAAGTTCGAGCAACAGCAGCTGGCAAGTAAATTATTGTCATCTGCATG
TTGGGCCAGCTTTTGTGGCTGATAGCAGAATTGTATGCCATTTCAGTTGTATGCGAGCAGCAAG
TTTTAGAAATTCATACTGAAAACATTTTCATATATGTACATTTTAAATCAAATAAATGCCTAAA
GCGGCAATTTTATGTTAGCTGCAGTAATCAATATTCCTACACTGCAATTTGGTGGTTTTAATTC
TGACTTAGAATGAACTGCTAATTTGTCTTTTTCTCAAACACGTTCCAGTTTGTAGTGGTATTTT
CTAGAGCGAACCCCCTTCAACAATGCTGGAAATTGAAGGGTGCCGGAAAAGTGCTGGAAAGCTG
AATCGACAAACGGGAGGAAACAGGGGAAACGGGGAAAAGCGCGTTGCGCGACACGCATAAATTG
AGCATGTTGATGGCGGTGGCGAAACACAATTTTCCTATTACTGAAAAGGATTTTGTTGCAATTA
TGCACTCGGAAAAGGGTGGGGTTAGTTAGGAAGCATGATTGGATAGCATTGTTAGACGAAAAAC
AAAACATGTTGAAAATACATTGTTGCATTTTTCATGCTTATATATTGTATATCGCGTTATTAGT
TTGCCAACTTCGCCCTAATGAAGCGTATTAGCGAACAAATTAAGCAATAAAATATTCATGATGG
CTATGCATAATTTAGCCAATGCAAATTCACAACTGCAATTTGGACGCGCAACGAAACTTTCCAA
TTGTTTGCCTTTTTTTTTCGCGTAAACTGTCAAAAGCACAGCGGCACTTAAATAAATTCTCGTA
GATTTCAAATGCAGAGTCTGTCACTTTTTGGTCTGCGCAGAAATTCTGCAGTGGGTGTACATTT
TACTACTAAGCCAAGTACTTTGCTGCTCATAAAGCGTAATCAAGCTAATTGTTTAAAAATTAAC
GCCAGACTTTTCTTACTTGGAAAAAAAAGGATTTAAAAAAAATATATACTCAAAACTACGAAA
TAAAAATAATCAAGCTTCTTTTTACCATTCAATTTCGTGGCGACCACTTTTGTTTTTATGTCCC
AAACTCGCAGTCGCCCCATTGAAATTGGAAAAACACTTTTCCTTCGTCACCTGCCCAGAGCATAA
AAAAAAAAAAAAAAAAAAACACAAATCAAAAGCGCAAGAAAATGACAAACATACGAGAAAATTAC
CCGAGCGATATAATTCAAATCGAAAAACATTAATTAACGCAACACAATATAAAACTCGAAAGCA
AGATTAATATGCGTAATTTGCCAAACTGAAAACTGCGCAAGACACAAAAATAAACATTTGGAA
TGGTCAATTTTTTCATCATTCATTTATTTCTTGGCCGGCTTGCCTTGACTCTTGTTAACCAGA
CTGTTTACGAAAATATTTATTAGTCAGCCATTCCAACGGTCTGTTAATTAAATGTTCGGAAAA
TACTTTCGGTGTCAGGGACAAAAAACAAACATTCAAAGGAGCACCGCATCTCTATAATTGTCG
TGTGCCAGAATAAAAAAAACGAATAGAGAAATCATCGGGAATTATTGCTCTATTAATATTTTG
GCACATTGTGCTCTTTAAATTCATCAAGAATTTTTGTTGTTGTTGTTATTTTTGAAGCCAGCA
AATTGAAATATTTGCTTTGACGTTGTCAGGCTGAAAAATACGTGTGAAAGAATTCTTTTTAACC
AATGAAATAAACATTTAACAGTCACAGTATGCACATAATATTGTTGCTACTGATGATTTTGCAT
GACTATGCAGTTAATTGCAAGAATAATGAATTGATAATAGCCATTCTCAACTGAATGCAAAAC
ATTTGTTTATAACCCAGCA... ...TTCTGAGTTGACAAGAAGTCGGC
GAAAACCACAACAA... ...TTGCGATAAATTTGACCA
CACTGGTCGGCGGA... ...CCCAAGGACCCTCTGACCC
CCCACCTGCTTTCAT... ...TCGGCCAAAAGCCTTTCAA
TGCCGTTGTTAAATC... ...ATGGAAGGGGAAAAGTCGG
TCGGCGAGGGAAAAT... ...CTACGTGCATGTGTTGTG
TTCATTGCAGACAAACTCTTAAACTTTACAATAAAACAACCAAAACAACACCAAAATCTAAAAC...

'Matthew Cobb is a respected scientist and historian, and he has combined both disciplines to spectacular effect in this compelling, authoritative and insightful account of how life works at the deepest level. It's a bloody brilliant book.' **Professor Brian Cox**

'*Life's Greatest Secret* is the logical sequel to Jim Watson's *The Double Helix*. While Watson and Crick deserve their plaudits for discovering the structure of DNA, that was only part of the story. Beginning to understand how that helix works – how its DNA code is turned into bodies and behaviours – took another fifteen years of amazing work by an army of dedicated men and women. These are the unknown heroes of modern genetics, and their tale is the subject of Cobb's fascinating book.' **Jerry Coyne**, University of Chicago and author of *Why Evolution is True*

'Most people think the race to sequence the human genome culminated at the 2000 White House "Mission Accomplished" announcement. In *Life's Greatest Secret*, we learn that it was just one chapter of a far more interesting and continuing story.' **Eric Topol**, Professor of Genomics and Director, Scripps Translational Science Institute and author of *The Patient Will See You Now*

'Gripping, insightful history, often from the mouths of the participants themselves.' *Kirkus Reviews*

'Rich, thrilling and thorough, this is the definitive history of arguably the greatest of all scientific revolutions.' **Adam Rutherford**, science writer, broadcaster and author of *Creation*

'Writing with flair, charisma and authority, this is Cobb's magnum opus. But more important than that, this is humankind's magnum opus. This is the story of a great human endeavour – a global adventure spanning decades – which unravelled how life really works. No area of science is more fundamental or more important; read about it and be filled with wonder.' **Daniel M. Davis**, author of *The Compatibility Gene*

'Cobb reveals the astonishing drama of the moment genetics and information technology collided, shaping the modern world and modern thought.' **Paul Mason**, Channel 4 News

LIFE'S GREATEST SECRET

The Race to Crack the Genetic Code

Matthew Cobb

PROFILE BOOKS

First published in Great Britain in 2015 by
PROFILE BOOKS LTD
3 Holford Yard
Bevin Way
London WC1X 9HD

www.profilebooks.com

1 3 5 7 9 10 8 6 4 2

Typeset in Palatino by MacGuru Ltd
info@macguru.org.uk

Printed and bound in Great Britain by
Clays, Bungay, Suffolk

A CIP catalogue record for this book is available from the British Library.

ISBN 978 1 78125 140 9
eISBN 978 1 78283 002 3

FSC
www.fsc.org
MIX
Paper from
responsible sources
FSC® C018072

The endpapers show part of the DNA sequence of the *Drosophila* gene *dunce*, which affects learning and memory. The whole gene is 124,896 base pairs long. For more on this gene, and its significance for the author, see Chapter 15.

In memory of John Pickstone (1944–2014)
– historian, colleague, friend.

CONTENTS

SECOND LETTER

		U	C	A	G	
FIRST LETTER	**U**	Phe	Ser	Tyr	Cys	U
		Phe	Ser	Tyr	Cys	C
		Leu	Ser	STOP	STOP	A
		Leu	Ser	STOP	Trp	G
	C	Leu	Pro	His	Arg	U
		Leu	Pro	His	Arg	C
		Leu	Pro	Gln	Arg	A
		Leu	Pro	Gln	Arg	G
	A	Ile	Thr	Asn	Ser	U
		Ile	Thr	Asn	Ser	C
		Ile	Thr	Lys	Arg	A
		Met	Thr	Lys	Arg	G
	G	Val	Ala	Asp	Gly	U
		Val	Ala	Asp	Gly	C
		Val	Ala	Glu	Gly	A
		Val	Ala	Glu	Gly	G

THIRD LETTER

The RNA genetic code, as finally established in 1967. U, C, A and G are the RNA bases. The 20 naturally occurring amino acids are given in the table, as three-letter abbreviations (e.g. Phe = phenylalanine). In RNA, Uracil (U) replaces the Thymine (T) base found in DNA. AUG codes for both methionine (Met) and for the start of the message. Slight variants of this code are found in some species, and in the mitochondria that are found in our cells – see Chapter 12.

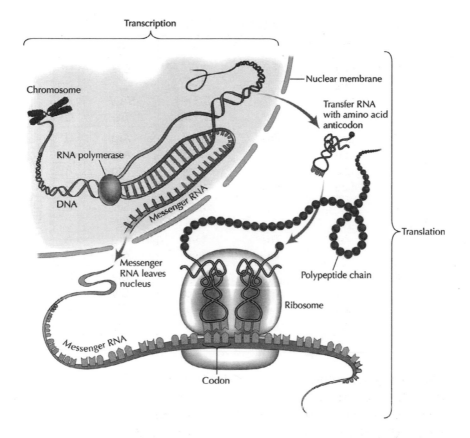

An outline of how the genetic code works during protein synthesis. A DNA double helix in the cell nucleus is partially unravelled and one strand is transcribed into RNA (mRNA). In organisms with a cell nucleus, this mRNA often contains irrelevant sequences (introns) that are spliced out to form mature mRNA which then leaves the nucleus. In the cell's cytoplasm, RNA-based ribosomes read the message, beginning at AUG. Transfer RNA (tRNA) molecules, synthesised by the cell from its DNA, carry on one side an anti-codon that binds with a particular mRNA codon and, on the other side, a binding site that links to a specific amino acid. In a process known as translation, tRNA molecules attached to an amino acid shuttle through the ribosome, bind with the mRNA codon and release their amino acid, thereby creating a protein chain.

AUG

FOREWORD

In April 1953, Jim Watson and Francis Crick published a scientific paper in the journal *Nature* in which they described the double helix structure of DNA, the stuff that genes are made of. In a second article that appeared six weeks later, Watson and Crick put forward a hypothesis with regard to the function of the 'bases' – the four kinds of molecule that are spaced along each strand of the double helix and which bind the two strands together. They wrote: 'it therefore seems likely that the precise sequence of the bases is the code which carries the genetical information.'

This phrase, which was almost certainly the work of Crick, must have seemed both utterly strange and completely familiar to those who read the article. It was strange because nothing so precise had ever been said before – no one had previously referred to 'genetical information'. This was a category that Watson and Crick had just invented. And yet it was familiar because it fitted so well with the ideas that were in the air at the time. It was adopted without debate; this new way of looking at life seemed so obvious that it was immediately accepted by scientists around the world. Today, these words, or something like them, are said every day in classrooms all over the planet as teachers explain the nature of genes and what they contain.

This book explores the surprising origin of these ideas, which can be traced back to physics and mathematics, and to wartime work on anti-aircraft guns and signals communication. It describes the way in which these concepts entered biology through the then-fashionable field of cybernetics, and how they were transformed as biologists sought to understand life's greatest secret – the nature of the genetic code. It is a story of ideas and experimentation, of ingenuity, insight and dead-ends, and of the race to make the greatest discovery of twentieth-century biology, a discovery that has opened up a brave new world for the twenty-first century.

Manchester, April 2015

– O N E –

GENES BEFORE DNA

In the early decades of the nineteenth century, the leaders of the wool industry in the central European state of Moravia were keen to improve the fleeces produced by their sheep. Half a century earlier, a British businessman farmer called Robert Bakewell had used selective breeding to increase the meat yield of his flocks; now the Moravian wool merchants wanted to emulate his success. In 1837 the Sheep Breeders' Society organised a meeting to discuss how they could produce more wool. One of the speakers was the new Abbot of the monastery at Brnö, a city that was at the heart of the country's wool production. Abbot Napp was intensely interested in the question of heredity and how it could be used to improve animal breeds, fruit crops and vines; this was not simply a hobby – the monastery was also a major landowner. At the meeting, Napp argued that the best way to increase wool production through breeding would be to address the fundamental underlying issue. As he put it impatiently: 'What we should have been dealing with is not the theory and process of breeding. But the question should be: *what is inherited and how*?'[1]

This question, which looks so straightforward to us, was at the cutting edge of human knowledge, as the words 'heredity' and 'inheritance' had only recently taken on biological meanings.[2]

Despite the centuries-old practical knowledge of animal breeders, and the popular conviction that 'like breeds like', all attempts to work out the reasons behind the various resemblances between parents and offspring had foundered when faced with the range of effects that could be seen in human families: skin colour, eye colour and sex all show different patterns of similarity across the generations. A child's skin colour tends to be a blend of the parental shades, their eye colour can sometimes be different from both parents, and in all except a handful of cases the sex of the child is the same as only one parent. These mysterious and mutually contradictory patterns – all of which were considered by the seventeenth-century physician William Harvey, one of the first people to think hard about the question – made it impossible to come up with any overall explanation using the tools of the time.[3] Because of these problems it took humanity centuries to realise that something involved in determining the characteristics of an organism was passed from parents to offspring. In the eighteenth and early nineteenth centuries, the tracing of human characteristics such as polydactyly (extra fingers) and Bakewell's selective breeding had finally convinced thinkers that there was a force at work, which was termed 'heredity'.[4] The problem was now to discover the answer to Napp's question – what is inherited and how?

Napp had not made this conceptual breakthrough alone: other thinkers such as Christian André and Count Emmerich Festetics had been exploring what Festetics called 'the genetic laws of nature'. But unlike them, Napp was able to organise and encourage a cohort of bright intellectuals in his monastery to explore the question, a bit like a modern university department focuses on a particular topic. This research programme reached its conclusion in 1865, when Napp's protégé, a monk named Gregor Mendel, gave two lectures in which he showed that, in pea plants, inheritance was based on factors that were passed down the generations. Mendel's discovery, which was published in the following year, had little impact and Mendel did no further work on the subject; Napp died shortly afterwards, and Mendel devoted all his time to running the monastery until his death in 1884. The significance of his discovery was not appreciated, and for nearly two decades his work was forgotten.[5] But in 1900 three

European scientists – Carl Correns, Hugo de Vries and Erich von Tschermak – either repeated Mendel's experiments or read his paper and publicised his findings.[6]

The century of genetics had begun.

*

The rediscovery of Mendel's work led to great excitement, because it complemented and explained some recent observations. In the 1880s, August Weismann and Hugo de Vries had suggested that, in animals, heredity was carried by what Weismann called the germ line – the sex cells, or egg and sperm. Microscopists had used newly discovered stains to reveal the presence of structures inside cells called chromosomes (the word means 'coloured body') – Theodor Boveri and Oscar Hertwig had shown that these structures copied themselves before cell division. In 1902, Walter Sutton, a PhD student at Columbia University in New York, published a paper on the grasshopper in which he used his own data and Boveri's observations to audaciously suggest that the chromosomes 'may constitute the physical basis of the Mendelian law of heredity'.[7] As he put it in a second paper, four months later: 'we should be able to find an exact correspondence between the behaviour in inheritance of any chromosome and that of the characters associated with it in the organism'.[8]

Sutton's insight – which Boveri soon claimed he had at the same time – was not immediately accepted.[9] First there was a long tussle over whether Mendel's theory applied to all patterns of heredity, and then people argued over whether there truly was a link with the behaviour of chromosomes.[10] In 1909, Wilhelm Johannsen coined the term 'gene' to refer to a factor that determines hereditary characters, but he explicitly rejected the idea that the gene was some kind of physical structure or particle. Instead he argued that some characters were determined by an organised predisposition (writing in German, he used the nearly untranslatable word *Anlagen*) contained in the egg and sperm, and that these *Anlagen* were what he called genes.[11]

One scientist who was initially hostile to the new science of what was soon known as 'genetics' was Thomas Hunt Morgan, who also worked at Columbia (by this time Sutton had returned to medical school; he never completed his PhD).[12] Morgan had obtained his PhD in marine biology, investigating the development of pycnogonids or sea spiders, but he had recently begun studying evolution, using the tiny red-eyed vinegar fly, *Drosophila*.[13] Morgan subjected his hapless insects to various environmental stresses – extreme temperatures, centrifugal force, altered lighting conditions – in the vain hope of causing a change that could be the basis of future evolution. Some minor mutations did appear in his fly stocks, but they were all difficult to observe. In 1910, Morgan was on the point of giving up when he found a white-eyed fly in his laboratory stocks. Within weeks, new mutants followed and by the summer there were six clearly defined mutations to study, a number of which, like the white-eyed mutant, seemed to be expressed more often in males than in females. Morgan's early doubts about genetics were swept away by the excitement of discovery.

By 1912, Morgan had shown that the white-eyed character was controlled by a genetic factor on the 'X' sex chromosome, thereby providing an experimental proof of the chromosomal theory of heredity. Equally importantly, he had shown that the shifting patterns of inheritance of groups of genes was related to the frequency with which pairs of chromosomes exchanged their parts ('crossing over') during the formation of egg and sperm.[14] Characters that tended to be inherited together were interpreted as being produced by genes that were physically close together on the chromosome – they were less likely to be separated during crossing over. Conversely, characters that could easily be separated when they were crossed were interpreted as being produced by genes that were further apart on the chromosome. This method enabled Morgan and his students – principally Alfred Sturtevant, Calvin Bridges and Hermann Muller – to create maps of the locations of genes on the fly's four pairs of chromosomes. These maps showed that genes are arranged linearly in a one-dimensional structure along the length of the chromosome.[15] By the 1930s, Morgan's maps had become extremely detailed, as new staining techniques revealed the presence of hundreds of bands on each chromosome. As Sutton had predicted, the patterns of these bands could be linked to the patterns with which mutations were inherited, so particular genes could be localised to minute fragments of the chromosome.

As to what genes were made of, that remained a complete mystery. In 1919, Morgan discussed two alternatives, neither of which satisfied him. A gene might be a 'chemical molecule', he wrote, in which case 'it is not evident how it could change except by altering its chemical constitution'. The other possibility was that a gene was 'a fluctuating amount of something' that differed between individuals and could change over time. Although this second model provided an explanation of both individual differences and the way in which organisms develop, the few results that were available suggested that it was not correct. Morgan's conclusion was to shrug his shoulders: 'I see at present no way of deciding', he told his readers.[16]

Even fourteen years later, in 1933, when Morgan was celebrating receiving the Nobel Prize for his work, there had been little progress. As he put it starkly in his Nobel Prize lecture: 'There is no consensus of opinion amongst geneticists as to what the genes are – whether they are real or purely fictitious.' The reason for this lack of agreement, he argued, was because 'at the level at which the genetic experiments lie, it does not make the slightest difference whether the gene is a hypothetical unit, or whether the gene is a material particle. In either case the unit is associated with a specific chromosome, and can be localized there by purely genetic analysis.'[17] It may seem strange, but for many geneticists in the 1930s, what genes were made of – if, indeed, they were made of anything at all – did not matter.

In 1926, Hermann Muller made a step towards proving that genes were indeed physical objects when he showed that X-rays could induce mutations. Although not many people believed his discovery – among the doubters was his one-time PhD supervisor Morgan, with whom he had a very prickly relationship – within a year his finding was confirmed. In 1932, Muller moved briefly to Berlin, where he worked with a Russian geneticist, Nikolai Timoféef-Ressovsky, pursuing his study of the effects of X-rays. Shortly afterwards, Timoféef-Ressovsky began a project with the radiation physicist Karl Zimmer and Max Delbrück, a young German quantum physicist who had been working with the Danish physicist Nils Bohr. The trio decided to apply 'target theory' – a central concept in the study of the effects of radiation – to genes.[18] By bombarding a cell with X-rays and seeing how often different mutations appeared as a function of the frequency and intensity of the radiation, they thought that it should be possible to deduce the physical size of the gene (the 'target'), and that measuring its sensitivity to radiation might reveal something of its composition.

The outcome of this collaboration was a joint German-language publication that appeared in 1935, called 'On the nature of gene mutation and gene structure', more generally known as the Three-Man Paper.[19] The article summarised nearly forty studies of the genetic effects of radiation and included a long theoretical section by Delbrück. The trio concluded that the gene was an indivisible phys-icochemical unit of molecular size, and proposed that a mutation involved the alteration of a chemical bond in that molecule. Despite their best efforts, however, the nature of the gene, and its exact size, remained unknown. As Delbrück explained in the paper, things were no further on from the alternatives posed by Morgan in 1919:

> We will thus leave unresolved the question of whether the individual gene has a polymeric form that arises through the repetition of identical structures of atoms, or whether it exhibits no such periodicity.[20]

The Russian geneticist Nikolai Koltsov was bolder than Delbrück or Morgan. In a discussion of the nature of 'hereditary molecules' published in 1927, Koltsov, like Delbrück, argued that the fundamental feature of genes (and therefore of chromosomes) was their ability to replicate themselves perfectly during cell division.[21] To explain this phenomenon, Koltsov proposed that each chromosome consisted of a pair of protein molecules that formed two identical strands; during cell division, each strand could be used as a template to produce another, identical, strand. Furthermore, he suggested that because these molecules were so long, the amino acid sequences along the proteins could provide massive variation that might explain the many functions of genes.[22] However perceptive this idea might look in the light of what we now know – the double helix structure of DNA and the fact that genes are composed of molecular sequences – Koltsov's argument was purely theoretical. Furthermore, it was not unique – in a lecture given in 1921, Hermann Muller picked up on a suggestion by Leonard Troland from 1917 and drew a parallel between the replication of chromosomes and the way in which crystals grow:

> each different portion of the gene structure must – like a crystal – attract to itself from the protoplasm materials of a similar kind, thus moulding next to the original gene another structure with similar parts, identically arranged, which then become bound together to form another gene, a replica of the first.[23]

In 1937, the British geneticist J. B. S. Haldane came up with a similar idea, suggesting that replication of genetic material might involve the copying of a molecule to form a 'negative' copy of the original.[24] Koltsov's views were initially published in Russian and then translated into French, but like Haldane's speculation they had no direct influence on subsequent developments.[25] Koltsov died in 1940, aged 68, having been accused of fascism because of his opposition to Stalin's favoured scientist, Trofim Lysenko, who denied the reality of genetics.[26]

Koltsov's assumption that genes were made of proteins was widely shared by scientists around the world. Proteins come in all sorts of varieties that could thereby account for the myriad ways in which genes act. Chromosomes are composed partly of proteins but mainly of a molecule that was then called nuclein – what we now call deoxyribonucleic acid, or DNA. The composition of this substance showed little variability – the leading expert on nucleic acids was the biochemist Phoebus Levene, who for over two decades explained that nucleic acids were composed of long chains of repeated blocks of four kinds of base (in DNA these were adenine, cytosine, guanine and thymine – subsequently known by their initials – A, C, G and T) which were present in equal proportions.[27] This idea, which was called the tetranucleotide hypothesis ('tetra' is from the Greek for four) dominated thinking about DNA; it suggested that these long and highly repetitive molecules probably had some structural function, unlike the minority component of chromosomes, proteins, which were good candidates for the material basis of genes simply because they were so variable. As Swedish scientist Torbjörn Caspersson put it in 1935:

> If one assumes that the genes consist of known substances, there are only the proteins to be considered, because they are the only known substances which are specific for the individual.[28]

This protein-centred view of genes was reinforced that same year when 31-year-old Wendell Stanley reported that he had crystallised a virus, and that it was a protein.[29] Stanley studied tobacco mosaic disease – a viral disease that infects the tobacco plant. Stanley took an infected plant, extracted its juice and was able to crystallise what looked like a pure protein that had the power to infect healthy plants. Although viruses were mysterious objects, in 1921 Muller had suggested that they might be genes, and that studying them could provide a route to understanding the nature of the gene.[30] Viruses, it appeared, were proteins, so presumably genes were, too. During the 1930s, many researchers, including Max Delbrück, began studying viruses, which were considered to be the simplest forms of life. Whether viruses are alive continues to divide scientists; whatever the case, this approach of studying the simplest form of biological organisation was extremely powerful. Delbrück, along with his colleague Salvador Luria, focused on bacteriophages (or 'phage') – viruses that infected bacteria, and in the 1940s an informal network of researchers called the phage group grew up around the pair as they tried to make fundamental discoveries that would also apply to complex organisms.[31]

Stanley's discovery caused great excitement in the press – for the *New York Times* it meant that 'the old distinction between life and death loses some of its validity'. Although within a few years valid doubts were expressed about Stanley's claim that he had isolated a pure protein – water and other contaminants were present and, as he admitted, it was nearly impossible to prove that a protein was pure – the overwhelming view among scientists was that genes, and viruses, were proteins.[32]

The most sophisticated attempt to link this assumption with speculation about the structure of genes was made in 1935 by the Oxford crystallographer Dorothy Wrinch. In a talk given at the University of Manchester, she suggested that the specificity of genes – their ability to carry out such a wide variety of functions – was determined by the sequence of protein molecules that were bound perpendicularly to a scaffold of nucleic acids, a bit like a piece of weaving. As she emphasised, however, 'there is an almost complete dearth of experimental and observational facts upon which the testing and further development of the hypothesis now put forward must necessarily depend.' Nevertheless, her conclusion was optimistic, as she encouraged her colleagues to explore the nature of the chromosome and of the gene:

> The chromosome is not a phenomenon belonging to a closed field. Rather it should take its place among the objects worthy of being treated with all possible subtleties and refinement of concept and technique belonging to all the sciences. A concerted attack in which the full resources of the world state of science are exploited can hardly fail.[33]

*

In the 1930s, most geneticists were not particularly concerned with finding out what genes are made of; they were more interested in discovering what genes actually do. There was a potential link between these two approaches. As the *Drosophila* geneticist Jack Schultz put it in 1935, by studying the effects of genes it should be possible 'to find out something about the nature of the gene'.[34] One of the scientists who took Schultz's suggestion very seriously was George Beadle, who had studied the genetics of eye colour in *Drosophila* in Morgan's laboratory, alongside the Franco-Russian geneticist Boris Ephrussi. When Ephrussi returned to Paris, Beadle followed him to continue their work. Their objective was to establish the biochemical basis of the mutations that changed the eye-colour of *Drosophila* flies. Beadle and Ephrussi's experiments failed: the biochemistry of their system was too complicated, and they were unable to extract the relevant chemicals from the fly's tiny eyes. They knew the genes that were involved, and they knew the effect they had on eye colour, but they did not know why.

Beadle returned to the US, determined to crack the problem of how genes could affect biochemistry, but equally certain that he had to use an organism that could be studied biochemically. He found the answer in the red bread mould *Neurospora*. This hardy fungus can survive in the near absence of an external supply of vitamins because it synthesises those it needs. To gain an insight into the genetic control of biochemical reactions, Beadle decided to create *Neurospora* mutants that could not synthesise these vitamins.

Together with microbiologist Edward Tatum, Beadle followed Muller's approach and irradiated *Neurospora* spores with X-rays in the hope of producing mutant fungi that required added vitamins to survive, thereby opening up the possibility of studying the genetics of vitamin biosynthesis. Beadle and Tatum soon found mutants that were unable to synthesise particular vitamins, and published their findings in 1941.[35] Each mutation affected a different enzymatic step in the vitamin's biosynthetic pathway – this was experimental proof of the widely held view, going back to the beginning of the century, that genes either produced enzymes or indeed simply were enzymes.[36] When Beadle presented their findings at a seminar at the California Institute of Technology (Caltech) in Pasadena, the audience was stunned. He spoke for only thirty minutes and then stopped. There was a nonplussed silence – one member of the audience recalled:

> We had never heard such experimental results before. It was the fulfilment of a dream, the demonstration that genes had an ascertainable role in biochemistry. We were all waiting – or perhaps hoping – for him to continue. When it became clear that he actually was finished, the applause was deafening.[37]

In the following year, Beadle and Tatum suggested that 'As a working hypothesis, a single gene may be considered to be concerned with the primary control of a single specific chemical reaction.'[38] A few years later, a colleague refined this to the snappier 'one gene, one enzyme hypothesis'. There was support for this view from work on human genetic diseases such alkaptonuria – in 1908 Archibald Garrod had suggested that this disease might involve defective enzyme production. But Beadle and Tatum's hypothesis met with opposition at the time, partly because it was known that genes have multiple effects, while their hypothesis – or rather, the 'one gene, one enzyme' catch-phrase by which it came to be known – seemed to suggest that each gene did only one thing: control an enzyme.[39]

*

Trinity College sits in the heart of Dublin, its grey three-storey neo-classical buildings positioned around lawns and playing fields. At the eastern end of the campus there is another grey building, built in 1905 in a rather different style. This is the Fitzgerald Building, or the Physical Laboratory as it is called in deeply engraved letters on the stone lintel. On the top floor there is a lecture theatre, and in the late afternoon of the first Friday of February 1943, around 400 people crowded onto the varnished wooden benches. According to *Time* magazine, among those lucky enough to get a seat were 'Cabinet ministers, diplomats, scholars and socialites', as well as the Irish Prime Minister, Éamon de Valera.[40] They were there to hear the Nobel Prize-winning physicist Erwin Schrödinger give a lecture with the intriguing title 'What is life?' The interest was so great that scores of people were turned away, and the lecture had to be repeated the following Monday.[41]

Schrödinger had arrived in Dublin after fleeing the Nazis – he had been working at Graz University in Austria when the Germans took over in 1938. Although he had a reputation as an opponent of Hitler, Schrödinger published an accommodating letter about the Nazi takeover, in the hope of being left alone. This tactic failed, and he had to flee the country in a hurry, leaving his gold Nobel medal behind. De Valera, who was interested in physics, offered Schrödinger a post in Dublin's new Institute for Advanced Studies, and the master of quantum mechanics found himself in Ireland.[42]

On three consecutive Fridays, 56-year-old Schrödinger walked into the Fitzgerald Building lecture theatre to give his talks, in which

he explored the relation between quantum physics and recent discoveries in biology.[43] His first topic was the way in which life seems to
contradict the second law of thermodynamics. Since the nineteenth
century it has been known that, in a closed system, energy will dissipate until it reaches a constant and even level: physicists explain this
in terms of the increasing amount of disorder, or entropy, that inevitably appears in such systems. Organisms seem to contradict this
fundamental law because we are highly ordered forms of matter that
concentrate energy in a very restricted space. Schrödinger's explanation was that life survives 'by continually sucking orderliness
from its environment' – he described order as 'negative entropy'.
This apparent breach of one of the fundamental laws of the Universe
does not cause any problems for physics, because on a cosmological
scale our existence is so brief, our physical dimensions so minute,
that the iron reality of the second law does not flutter for an instant.
Whether life exists or not, entropy increases inexorably. According to
our current models, this will continue until the ultimate heat death
of the Universe, when all matter will be evenly spaced and nothing
happens, and it carries on not happening forever.

Schrödinger encountered far greater difficulties when he came
to discuss his second topic: the nature of heredity. Like Koltsov and
Delbrück before him, Schrödinger was struck by the fact that the
chromosomes are accurately duplicated during ordinary cell division ('mitosis' – this is the way in which an organism grows) and
during the creation of the sex cells ('meiosis'). For your body to have
reached its current size there have been trillions of mitotic cell divisions and through all that copying and duplicating the code has
apparently been reliably duplicated – in general, development proceeds without any sign of a mutation or a genetic aberration. Furthermore, genes are reliably passed from one generation to another:
Schrödinger explained to his audience that a well-known characteristic such as the Hapsburg, or Habsburg, lip – the protruding lower
jaw shown by members of the House of Hapsburg – can be tracked
over hundreds of years, without apparently changing.

For biologists, this apparently unchanging character of genes
was simply a fact. However, as Schrödinger explained to his Dublin
audience, it posed a problem for physicists. Schrödinger calculated

that each gene might be composed of only a thousand atoms, in which case genes should be continuously shimmering and altering because the fundamental laws of physics and chemistry are statistical; although overall atoms tend to behave consistently, an individual atom can behave in a way that contradicts these laws.[44] For most objects that we encounter, this does not matter: things such as tables or rocks or cows are made of so many gazillions of atoms that they do not behave in unpredictable ways. A table remains a table; it does not start spontaneously turning into a rock or a cow. But if genes are made of only a few hundred atoms, they should display exactly that kind of uncertain behaviour and they should not remain constant over the generations, argued Schrödinger. And yet experiments showed that mutations occurred quite rarely, and that when they did happen they were accurately inherited. Schrödinger outlined the problem in the following terms:

> incredibly small groups of atoms much too small to display exact statistical laws ... play a dominating role in the very orderly and lawful events within a living organism. They have control of the observable large-scale features which the organism acquires in the course of its development, they determine important characteristics of its functioning; and in all this very sharp and very strict biological laws are displayed.[45]

The challenge was to explain how genes act lawfully, and cause organisms to behave lawfully, while being composed of a very small number of atoms, a significant proportion of which may be behaving unlawfully. To resolve this apparent contradiction between the principles of physics and the reality of biology, Schrödinger turned to the most sophisticated theory of the nature of the gene that existed at the time, the Three-Man Paper by Timoféef-Ressovsky, Zimmer and Delbrück.

As Schrödinger explored the nature of heredity for his audience, he was forced to come up with an explanation of what exactly a gene contained. With nothing more than logic to support his hypothesis, Schrödinger argued that chromosomes 'contain in some kind of code-script the entire pattern of the individual's future development and of its functioning in the mature state.' This was the first time that anyone had clearly suggested that genes might contain, or even could simply be, a code.

Taking his idea to its logical conclusion, Schrödinger argued that it should be possible to read the 'code-script' of an egg and know 'whether the egg would develop, under suitable conditions, into a black cock or into a speckled hen, into a fly or a maize plant, a rhododendron, a beetle, a mouse or a woman.'[46] Although this was partly an echo of the earliest ideas about how organisms develop and the old suggestion that the future organism was preformed in the egg, Schrödinger's idea was very different. He was addressing the question of *how* the future organism was represented in the egg and the means by which that representation became biological reality, and suggesting these were one and the same:

> The chromosome structures are at the same time instrumental
> in bringing about the development they foreshadow. They are
> law-code and executive power – or, to use another simile, they
> are architect's plan and builder's craft – in one.[47]

To explain how his hypothetical code-script might work – it had to be extremely complicated because it involved 'all the future development of the organism' – Schrödinger resorted to some simple mathematics to show how the variety of different molecules found in an organism could be encoded. If each biological molecule were determined by a single 25-letter word composed of five different letters, there would be 372,529,029,846,191,405 different possible combinations – far greater than the number of known types of molecule found in any organism. Having shown the potential power of even a simple code, Schrödinger concluded that 'it is no longer inconceivable that the miniature code should precisely correspond with a highly complicated and specified plan of development and should somehow contain the means to put it into operation.'[48]

Although this was the first public suggestion that a gene contained something like a code, in 1892 the scientist Fritz Miescher had come up with something vaguely similar. In a private letter, Miescher had argued that the various forms of organic molecules were sufficient for 'all the wealth and variety of hereditary trans-mission [to] find expression just as all the words and concepts of all languages can find expression in twenty-four to thirty alphabetic letters.'[49] Miescher's view can appear far-seeing, especially given that he was also the discoverer of DNA, or, as he called it, nuclein. But Miescher never argued that nuclein was the material making up these letters and his suggestion was not made public for nearly eighty years. Above all, the vague letter and word metaphor was nowhere near as precise as Schrödinger's code-script concept.

Schrödinger then explored what the gene-molecule might be made of and suggested that it was what he called a one-dimensional aperiodic crystal – a non-repetitive solid, with the lack of repetition being related to the existence of the code-script. The non-repetition provided the variety necessary to specify so many different mole-cules in an organism. Although Troland, Muller and Koltsov had all suggested two decades earlier that genes might grow like crystals, Schrödinger's idea was far more precise. His vision of gene structure was focused on the non-repetitive nature of the code-script, rather than on the relatively simple parallel between the copying of chro-mosomes and the ability of crystals to replicate their structure.[50]

*

Schrödinger's words would have had little influence had they sim-ply hovered in the Dublin air and briefly resonated in the minds of the more attentive listeners. The sole international report to describe the lectures, which appeared in *Time* magazine in April, did not refer in detail to anything that Schrödinger said, and there are no indica-tions that any of his ideas escaped to the outside world. The only detailed account appeared in *The Irish Press*, which managed to con-dense his main arguments, and included both the code-script and aperiodic crystal ideas.[51] Other newspapers found it difficult to give the story the attention it deserved; when Schrödinger gave a version of his lectures in Cork in January 1944, the local newspaper, *The Ker-ryman*, gave his talk equal coverage to the Listowel Pig Fair (there was good demand for the 126 pigs on sale, they reported).[52]

Schrödinger felt that the public would be interested in his views, and as soon as he had finished the lectures he began to turn them into a book, with the addition of a brief and deliberately controversial conclusion. Schrödinger had closed his lectures with a pious nod in the direction of his overwhelmingly Catholic audience, proclaiming that the 'aperiodic crystal forming the chromosome fibre' was 'the finest masterpiece ever achieved along the lines of the Lord's quantum mechanics'. But in a new Epilogue, written specially for publication and entitled 'Determinism and free will', Schrödinger explored his lifelong belief in the mystical Hindu philosophy of Vedanta. He argued that individual human identity was an illusion, and he criticised official Western creeds for their superstitious belief in the existence of individual souls. His point was not that there was no evidence that souls exist but rather that individual consciousness is the illusory expression of a single universal soul. He expressed this in what he admitted were 'blasphemous and lunatic' terms for the Christian tradition, but which he apparently considered to be true: 'I am God Almighty'.

The book was about to be printed by a respected Dublin publishing house when someone in the company got cold feet. In 1940s Ireland the Catholic Church retained a stranglehold over culture, and it was not possible to criticise Christian beliefs in the terms that Schrödinger had used in the final chapter. The publisher pulled out. Undeterred, Schrödinger sent the manuscript to a friend in London, and it was eventually published by Cambridge University Press in December 1944. The combination of Schrödinger's name, the intriguing title and a prestigious publisher with a global reach, coupled with the imminent end of the war, meant that the book was widely read and has remained in print ever since. Despite the commercial success of *What is Life?*, that was the end of Schrödinger's excursion into biology. He never wrote publicly on the topic again, even after the discovery of the existence of the genetic code in 1953.[53]

The immediate impact of *What is Life?* can be seen from the enthusiastic reviews it received in both the popular press and in scientific journals. There were over sixty reviews in the four years after publication, although few writers noticed what now seem to be far-seeing ideas – the aperiodic crystal and the code-script – and it was translated into German, French, Russian, Spanish and Japanese.[54]

There were two extended reviews in the leading scientific weekly *Nature,* one by the geneticist J. B. S. Haldane, the other by the plant cytologist Irene Manton. Haldane got straight to the heart of the matter, picking up on the aperiodic crystal and the code-script innovations and making a link with the work of Koltsov. Manton also noted Schrödinger's use of the term code-script, but she took it to mean 'the sum of hereditary material' rather than a particular hypothesis about gene structure and function. The *New York Times* reviewer put his finger on the central point:

> The genes and chromosomes contain what Schrödinger calls a 'code script,' that gives orders which are carried out. And because we can't read the script as yet we know virtually nothing of growth, nothing of life.

In contrast, some scientists later recalled that they had been unimpressed by the book. In the 1980s the Nobel Prize-winning chemist Linus Pauling claimed that he was 'disappointed' on reading *What Is Life?* and stated: 'It was, and still is, my opinion that Schrödinger made no contribution to our understanding of life.'[55] Also in the 1980s, another Nobel laureate, biochemist Max Perutz, wrote of Schrödinger: 'what was true in his book was not original, and most of what was original was known not to be true even when the book was written', while in 1969 the geneticist C. H. Waddington criticised Schrödinger's aperiodic crystal concept as an 'exceedingly paradoxical phrase'.[56] As well as these restrospective criticisms, some dissenting views were voiced at the time. In a review, Max Delbrück was critical, even though he had received a publicity boost from Schrödinger's espousal of his work in the Three-Man Paper. He claimed Schrödinger's term aperiodic crystal hid more than it revealed:

> genes are given this startling name rather than the current name 'complicated molecule' ... There is nothing new in this exposition, to which the larger part of the book is devoted, and biological readers will be inclined to skip it.

This was distinctly ungenerous, as Schrödinger's hypothesis was in fact quite precise and did not simply involve coining a new name. Delbrück concluded by grudgingly accepting that the book 'will have an inspiring influence by acting as a focus of attention for both physicists and biologists.' In another review, Muller said that he, too, expected that the book would act as a catalyst for 'an increasingly useful rapprochement between physics, chemistry and the genetic basis of biology'. Muller clearly felt aggrieved that Schrödinger had not cited his work, and pointed out that he had suggested the parallel between gene duplication and crystal growth in 1921 (Muller did not mention that he had taken this concept from Leonard Troland). He also dismissed the idea that there was anything novel in Schrödinger's discussion of order and negative entropy, as these were both 'quite familiar to general biologists'. Neither Delbrück nor Muller made any comment about the code-script idea.

Despite their overall scepticism, Delbrück and Muller were absolutely right: Schrödinger's book did indeed inspire a generation of young scientists. The three men who won the Nobel Prize for their work on the structure of DNA – James Watson, Francis Crick and Maurice Wilkins – all claimed that *What is Life?* played an important part in their personal journeys towards the double helix. In 1945 Wilkins was handed a copy of *What is Life?* by a friend when he was working on the atomic bomb in California. Shaken by the horror of Hiroshima and Nagasaki, Wilkins was seduced by Schrödinger's writing and decided to abandon physics and become a 'biophysicist'. Crick recalled that his 1946 reading of Schrödinger 'made it seem as if great things were just around the corner'; Watson was an undergraduate when he read *What is Life?* and as a result he shifted his attention from bird biology to genetics.[57]

Even though some of the ideas developed in *What is Life?* were visionary and the book undoubtedly inspired some individuals who played a central role in twentieth-century science, there are no direct links between Schrödinger's lectures and the experiments and theories that were part of the decades-long attempt to crack the genetic code, and historians and participants differ about the significance of Schrödinger's contribution.[58] The view of mutation put forward in the Three-Man Paper, which Schrödinger espoused so vigorously, had no effect on subsequent events, and his suggestion that new laws of physics would be discovered through the study of the material basis of heredity was completely mistaken. Even the code-script idea, which looks so prescient today, had no direct effect on how biologists looked at what was in a gene. None of the articles that later formed part of the discovery of the genetic code cited *What is Life?*, even though the scientists involved had read the book.

In fact, the meaning of Schrödinger's 'code-script' did not have the same richness as our 'genetic code'. Schrödinger did not think that there was a correspondence between each part of the gene and precise biochemical processes, which is what a code implies, nor did he address the issue of what exactly the code-script contained, beyond the vague suggestion of a plan. Ask any biologist today what the genetic code contains, and they will give you a one-word answer: information. Schrödinger did not use that powerful metaphor. It was completely absent from his vocabulary and his thinking, for the simple reason that it had not yet acquired the abstract wide-ranging meaning we now give it. 'Information' was about to enter science, but had not done so when Schrödinger gave his lectures. Without that conception of the content of the code, Schrödinger's insight was merely part of the zeitgeist, a hint of what was to come rather than a breakthrough that shaped all subsequent thinking.

TWO

INFORMATION IS EVERYWHERE

While Schrödinger was in neutral Ireland, away from the horrors of the Second World War, other scientists all over the world joined in the war effort, keen to use their skills to develop new ways of killing people on the other side, or at least to find ways of stopping people on their side from being killed. This was particularly true in the US, where in June 1940, eighteen months before the US eventually entered the war, President Roosevelt instructed the vice-president of the Massachusetts Institute of Technology (MIT), Vannevar Bush, to set up a National Research Defense Committee (NRDC) in order to develop new weapons. The NRDC went on to mobilise more than 6,000 American scientists, including those working on the ultra-secret Manhattan Project, which eventually produced the atomic bomb.[1] The scale of spending was immense: by 1944, the federal research budget was $700m per year – more than ten times the amount spent in 1938.*

One of the scientists involved in this work was a brilliant and

---

* Most of that federal money was poured into the private sector. Before the war, 70 per cent of government-sponsored research was undertaken by federal organisations, 30 per cent by companies and universities. By 1944, those figures had been reversed. There was plenty of cash to go round: both universities and the private sector were drowning in money from government contracts focused on military problems (Noble, 1986).

mercurial mathematician from MIT named Norbert Wiener (pronounced Wee-ner). In September 1940, 46-year-old Wiener – a portly, cigar-smoking vegetarian, who was short-sighted and wore a rakish van Dyke beard – wrote to Vannevar Bush offering his services: 'I hope you may find some corner of the activity in which I may be of use during the emergency.' Wiener had been a child prodigy; he later studied logic with Bertrand Russell and made important contributions to mathematics in the 1920s and 1930s.

Wiener's involvement in the preparations for war was motivated by a mixture of patriotism and deep hostility towards the Nazis – his father was a Russian Jew. However, Wiener's wife was profoundly anti-semitic and an avid supporter of Hitler, while members of her close family in Germany were Nazis. During the war, Wiener's daughter, Barbara, was punished at school for reciting passages from her mother's copy of *Mein Kampf*. All this made for an interesting home life.[2]

Vannevar Bush felt that the US was insufficiently prepared for an air attack, and that the country needed to develop 'the precise and rapid control of guns' in order to shoot down enemy planes.[3] This was the area that Wiener began working on. In autumn 1940 he showed that it was theoretically possible to develop an automatic anti-aircraft system that could destroy enemy planes with minimal human intervention. In December 1940, Wiener's proposal to turn his theoretical idea into reality was given a paltry $2,325 budget and stamped 'secret' by the newly formed section D-2 of the NRDC. Section D-2, which funded eighty projects to the value of around $10m during the course of the war, organised research on 'fire control' – systems for controlling artillery fire. The section was run by the director of the Rockefeller Institute, Warren Weaver, who two years earlier had coined the term 'molecular biology'.[4]

Existing anti-aircraft systems could involve up to fourteen men: some spotted the plane, some identified its trajectory, others rapidly calculated where the aircraft was predicted to be, while a final group cranked the gun to the appropriate orientation and elevation and then fired. But if the pilot took evasive action after the shell was fired, it would miss its target – the calculations assumed the plane was flying in a straight line. Wiener's bold idea was to find a mathematical formula for predicting where the plane would be, whatever the pilot did.

By the winter of 1941, Wiener had used his mathematical skills to predict near-random movement by a target, and then to calculate an intercept course to the most probable destination points. Julian Bigelow, a talented young ex-IBM engineer with a taste for messing about with old cars who also happened to be an amateur pilot, was assigned to work with Wiener. The pair constructed a device that simulated the movement of a target aeroplane and the response of an anti-aircraft gun crew, by projecting beams of light onto the ceiling of Room 2–244 on the MIT campus by the Charles River Basin.[5] Wiener and Bigelow also went into the field and studied how real-life gunners behaved. Here Wiener made his breakthrough, as he noticed that the soldiers would take actions designed to respond to a pattern of movement by the aircraft. The gunner used knowledge about where he expected the plane to be and attempted to compensate for that predicted movement when calculating where to fire his gun. Wiener set about trying to describe this effect in mathematical terms. The stress began to tell as Wiener gobbled amphetamines – quite legal at the time – in an attempt to meet deadlines. He became irritable and even more garrulous than usual – hardly advisable for someone working on a top-secret project – and eventually had to kick his speed habit. As he later explained: 'I had to give it up and look for a more rational way of strengthening myself to bear the burdens of war work.'[6]

Wiener realised that the way the gunner responded to the movement of the aircraft meant that he was acting as part of a feedback system – a phenomenon that was well known from acoustics and engineering. Wiener discussed this insight with a friend from his student days, a Harvard physiologist called Arturo Rosenblueth. They realised that feedback was a common feature of many systems, both technological and natural, and could be seen in the behaviour and physiology of animals. Excited by their theoretical breakthrough, the two men announced their vision at a small scientific meeting held in New York in 1942. The two-dozen strong audience was composed of an eclectic mixture of neurophysiologists and psychologists, along with the husband and wife anthropologists Gregory Bateson and Margaret Mead. Rosenblueth's speech, which described what he called 'circular causality' or feedback loops, was written up as a paper with Wiener and Bigelow and published under the title 'Behaviour, purpose and teleology' in the journal *Philosophy of Science*.[7] The use of the word teleology was deliberately provocative, as this concept explains phenomena in terms of their purpose, and purpose had been banished from polite scientific discourse for centuries. According to Aristotle, the ultimate explanation of natural phenomena was their purpose or final cause. For example, a dropped apple will fall to the ground because its final cause is to go downwards. From the seventeenth century onwards, it was increasingly realised that this approach did not explain anything, and more powerful mechanistic explanations were sought. Wiener wanted to reinstate the idea of purpose by explaining it in mathematical terms.

Wiener and Rosenblueth showed that purposeful, goal-directed behaviour can be seen in organisms and machines, and that it operates through what is known as negative feedback. Normal feedback leads to the uncontrolled amplification of the signal – this is the howl that is produced if a microphone is placed too close to a loudspeaker. Negative feedback means that a given activity ceases when a particular pre-defined state is achieved. In this way a signal can drive a machine or an organism to an end; if the goal is not attained, then continued signals will direct the behaviour towards the goal. For example, a torpedo that homes in on acoustic signals emitted by a battleship uses negative feedback to guide itself to its target – it stops altering its direction when the signal is strongest, as that indicates that the target is dead ahead.[8]

After the war, Warren Weaver finally got round to reading the Wiener, Rosenblueth and Bigelow paper. He was not impressed: 'I want to read this article but so far I have not succeeded in getting beyond the first four paragraphs', he told Wiener.[9] If Weaver had ploughed on, he might have found the rest of the document more rewarding, for it marked a shift in scientific thinking. It put all systems on the same level, be they mechanical, organic or hybrid human–machine (as in the case of the anti-aircraft guns), and suggested that behaviour could be interpreted using the same principles and analysed in terms of the same negative feedback loops. When the paper was read to the small New York audience, the effect was electric. Neurophysiologist Warren McCulloch was particularly excited, as it coincided with the models of brain function that he was developing with Walter Pitts, an odd but brilliant 20-year-old maths prodigy.[10] Even the anthropologist Margaret Mead was rapt: 'I did not notice that I had broken one of my teeth until the Conference was over', she later wrote.[11]

Although Wiener's insight excited his academic colleagues, his attempt to build an anti-aircraft device that could be engineered into a battlefield version was upstaged by a rival top-secret project, which was jointly run by MIT and a private company, Bell Laboratories. Under the deliberately misleading title the Radiation Lab (known as the Rad Lab), the project involved more than thirty scientists and in its first year alone had a budget of more than $800,000. Although it used a highly unrealistic prediction method – it assumed that the aircraft would fly in a straight line – the device made up for its lack of accuracy by firing a hail of shells around the predicted location, some of which would get lucky. In 1942 the Rad Lab project passed a practical test and more than 1,200 units were ordered by the US military. Although Wiener and Bigelow's statistical predictor was marginally more accurate than their Rad Lab rival, it soon became apparent that the improvement over the Rad Lab version would not be worth the effort and Wiener's project was cancelled in November 1943.[12] The Rad Lab system, now called the M-9, incorporated some elements of Wiener's predictive protocols and went into mass production. It eventually formed a central component of what was the first robot war – the clash between the German V-1 automatic rockets or doodlebugs and the Allies' semi-automated defence systems, in the skies over southern England in 1944–45.[13]

Wiener wrote up his method for predicting the movement of objects and filtering out noise under the daunting title 'Extrapolation, interpolation, and smoothing of stationary time series with engineering applications'. This duplicated document consisted mainly of pages and pages of Wiener's fiendishly complex mathematics and was circulated to workers on the various anti-aircraft projects. Even the briefest chapter, which was only six pages long, contained thirty-seven equations. Stamped 'Restricted', the document was printed in 300 copies and was bound in pale yellow covers that carried a declaration threatening anyone who revealed its contents with the full force of the US Espionage Act. It soon became known to its many perplexed readers as the Yellow Peril – one US engineer later said, 'copies should have been distributed to the enemy so that they would have to devote such time to it and enable us to get on with winning the war.'[14] The document was eventually declassified and has since become a classic of its kind; in 2005 a copy sold at Sotheby's for $7,200.

Wiener claimed that his method had many applications and could shed important light on the nature of communication. He argued that there was no difference between human communication and messages sent by a machine: 'the records of current and voltage kept on the instruments of an automatic substation are as truly messages as a telephone conversation.'[15] All communication possesses the same fundamental feature, argued Wiener – it has to contain what he termed variable information. One measure of that variability, he said, could be found in the mathematical theory of probability – all forms of communication could be understood in terms of a mathematical, probabilistic analysis of the information they contained. When Wiener's two fundamental insights – the importance of negative feedback in shaping behaviour, and the existence of information – were combined, it implied that the feedback loops that lay at the heart of apparently purposive behaviour were carrying information.

*

Another MIT-trained mathematician called Claude E. Shannon was working on similar problems at the same time. Shannon was a shy young man with a lifelong love of Dixieland jazz who enjoyed tinkering with electronics and spoke with a slow drawl, a bit like James Stewart. In 1938 Shannon obtained his MSc from MIT for his work on applications of Boolean logic, which played a decisive role in the development of electronics. By 1940, Shannon had completed his PhD on 'An algebra for theoretical genetics', in which he developed a mathematical way of describing how genes spread in populations. As Shannon admitted, although the proof was novel, the results were not. He was not actually interested in genes at all – according to his doctoral advisor, Vannevar Bush, 'he has only a fragmentary knowledge of this aspect of genetics'. His primary concern was with using statistics to describe the behaviour of genes in populations, not how they functioned or what they were made of.[16]

By 1942, Shannon was working for Bell Laboratories in their New York headquarters on West Street, overlooking the Hudson River. At the rear, on Washington Street, an overground subway line ran right through the building, like something out of a 1930s film of the future. Shannon was part of the cryptography group, studying the transmission of messages over the telephone. In January 1943, as Schrödinger was about to give his lectures in Dublin, the Bell Labs had a visitor from England – the mathematician and cryptographer Alan Turing, who had arrived in New York on the *Queen Elizabeth* in November. Turing began work at Bell Labs, investigating ways of setting up a securely encoded telephone link between Roosevelt and Churchill – this was later successfully implemented after Shannon's theoretical demonstration that the code could not be broken.[17]

Although Turing did not work with Shannon, the two young men regularly had tea together in the cafeteria, where they discussed Turing's ideas for a 'universal machine' that could perform any conceivable calculation. Shannon apparently surprised Turing by suggesting that such a 'Brain' would be capable of more than just doing complicated sums: 'Shannon wants to feed not just data to a Brain, but cultural things! He wants to play music to it!' Turing exclaimed in a letter.[18] More significantly, the two men also exchanged their ideas about signal transmission, how to measure the content of communication, and how to incorporate uncertainty into their mathematical procedures. Turing had developed the concept of 'decibans', which were a measure of the uncertainty contained in a message; Shannon was on the brink of defining the 'binary digit' or 'bit', which could have two states – 0 or 1 – and was at the heart of postwar computing. At the beginning of March, Turing returned to England on the hazardous north Atlantic crossing, the only civilian on a 4,300-strong troop ship. The next time the two men met was after the war, in Manchester, where Turing was working on Baby, the world's first stored-program computer.

Shannon was part of the Bell Labs team working with the MIT Rad Lab on fire control. Like Wiener, Shannon's job was to come up with a method for predicting the location of the target, and the two men discussed this question several times. Bigelow later recalled that Wiener was extremely generous, exchanging ideas with the younger man, sharing his insights. Eventually Wiener's generosity began to wear off and – perhaps because of his amphetamine abuse – he started to react in a paranoid fashion to Shannon's visits, telling close friends that Shannon was 'coming to pluck my brains', and doing his best to avoid the visitor.[19]

Although Shannon was clearly inspired by some aspects of Wiener's work in the Yellow Peril, he had already begun thinking about the nature of communication and how to describe it mathematically. He was not the first person to do this; in the late 1920s Ralph Hartley and Harry Nyquist had studied how telegraph messages were transmitted, but they did not approach the problem from a probabilistic point of view, nor did they include random variation – noise – as a factor affecting transmission accuracy. In 1945, Shannon wrote a document for the D-2 division of the NRDC, entitled 'A mathematical theory of cryptography', in which he summarised his ideas about communication and what it involved. He called the stuff that was communicated 'information', and described the nature of its fundamental unit, which he termed the 'bit'. For obvious reasons the paper was immediately stamped 'secret', but after the war a version of it was published and it was eventually declassified in 1957.[20]

The final element of the ideas about information that were coming into form around the war years was an exploration of Maxwell's Demon by the German physicist Leo Szilárd. This thought experiment was devised in 1871 by the British physicist James Clerk Maxwell, with the aim of showing how it was theoretically possible to violate the second law of thermodynamics. Maxwell imagined a demon that without effort could open a door between two chambers, allowing the more energetic molecules into one side, thereby increasing the temperature in that chamber and decreasing entropy in the system – something the second law said was impossible. Szilárd's solution, which he devised in 1929, was that the demon would have to be able to measure the speed of the molecules, and to do this would require the expenditure of energy and therefore an increase in entropy. If the demon and the chamber were taken as a whole, the entropy of the system would not decline, and the second law remained intact. Although Szilárd did not use the term information, his theoretical discussion linked entropy and measures of knowledge in a way that proved significant.

*

At the beginning of 1945, Wiener and fellow mathematician John von Neumann organised a meeting of the newly formed Teleological Society. The aim of the society was to study 'how purpose is realised in human and animal conduct and on the other hand how purpose can be imitated by mechanical and electrical means.'[21] Von Neumann was a mathematician and a pioneer of game theory – mathematical models that describe and predict simple behaviours. He played an important role in developing the models that came to dominate much of postwar economics, and which also constituted a strand in the thinking of evolutionary biologists. Above all, von Neumann played a leading role in the Manhattan Project.

Eight months after the meeting of the Teleological Society, the world changed utterly, in two terrifyingly destructive flashes of light. On 6 August 1945, the US dropped the first atomic bomb on Hiroshima, causing unimaginable devastation, instantly killing up to 80,000 people, with a similar number condemned to a slow death over the following months. Three days later, on 9 August, the city of Nagasaki was destroyed by a second bomb, which used plutonium rather than uranium and employed an implosion ignition procedure developed by von Neumann.

The Manhattan Project was a success, but many of the scientists involved were horrified at their part in the destruction. The prime reason behind the Manhattan Project – fear of being beaten to the bomb by the Nazis – had been eradicated by the surrender of Germany in May 1945. Furthermore, it soon became apparent that the Germans had not been close to success. All except the most naive or unworldly scientists came to recognise that the development and deployment of the atomic bomb showed that the Allies had something else in mind – the bomb was used to threaten the USSR. Von Neumann was quite comfortable with this. He had helped decide which two Japanese cities were to be smashed; as a committed anti-communist he accepted that the Hiroshima and Nagasaki bombs were primarily warnings to the USSR, and considered that the attendant death and destruction were quite justified.[22]

Wiener took a very different attitude. He was concerned about the moral issues raised by the use of the bomb against Japan, and by the potential for infinitely greater destruction in the future, to the extent that he considered abandoning science altogether. As he told a friend in October 1945:

> Ever since the atomic bomb fell I have been recovering from an acute attack of conscience as one of the scientists who has been doing war work and who has seen his war work a[s] part of a larger body which is being used in a way of which I do not approve and over which I have absolutely no control. ... I have seriously considered the possibility of giving up my scientific productive effort because I know no way to publish without letting my inventions go to the wrong hands.[23]

Wiener's wartime experience had convinced him that science should be as open as possible, and should not be tied to the private sector or the military. Von Neumann, in contrast, was keen to put his snout as deep as possible into the trough of the military–industrial complex that was beginning to dominate the US economy. His twin aims were to obtain funding to build the computer he had dreamt up, and to counter what he saw as the threat from the USSR.

Despite their profound differences, the two men continued to work together, most notably in the organisation of a conference that took place in March 1946 under the cumbersome title 'The Feedback Mechanisms and Circular Causal Systems in Biology and the Social Sciences Meeting' (more commonly known as the Macy conference, after the sponsors, the Josiah Macy Jr Foundation). The attendees were basically the same crowd as the people who had heard Wiener outline his negative feedback vision in 1942, with the addition of the ecologist G. Evelyn Hutchinson, the sociologist Paul Lazarsfeld and some others.[24] Wiener and von Neumann presented their project of electronic computer brains, with von Neumann drawing a parallel between the human nervous system and the digital, stored-program computer he was constructing in Princeton.

Seven months later, in October 1946, the New York Academy of Sciences held a special meeting on 'Teleological mechanisms' at which Wiener spoke, outlining the ideas in the Yellow Peril that had been withheld from public view the year before.[25] Wiener explained that underlying all examples of negative feedback control there was a single unifying idea, which he called the message – all control systems involved communication, and could be understood using the same conceptual framework. Inspired by Schrödinger's *What is Life?*, Wiener made a link between information and entropy, going even further than Szilárd's discussion of Maxwell's Demon, in that he defined entropy as 'the negative of the amount of information contained in the message'. This was 'not surprising', Wiener went on, because 'Information measures order and entropy measures disorder. It is indeed possible to conceive all order in terms of message.' The laws governing communication, he argued, were 'really identical' with the second law of thermodynamics. So, for example, once a message has been created, subsequent operations can degrade it but cannot add information. The arrow of entropy points only one way, and all that life can do is to temporarily halt the process; it cannot truly reverse it. One of the main explanatory frameworks used by postwar science – the role of information in biology – was emerging and was now connected with the fundamental measure of order on a cosmic scale.

A month later, von Neumann took a step towards linking the study of control systems with visions of how life reproduces itself. He was increasingly convinced that Wiener's focus on modelling human behaviour was a mistake: the human brain was far too complex. At the end of November 1946, von Neumann wrote a long letter to Wiener outlining a startlingly different approach, which dominated science for decades to come.[26] He began with a self-criticism, pointing out that through their shared enthusiasm for studying the human central nervous system, 'we selected … the most complicated object under the sun – literally.' But the problem was even greater than the mere complexity of the brain, argued von Neumann. He felt they had first to understand the underlying molecular mechanisms before they could hope to understand higher level activity:

> nothing that we may know or learn about the functioning of the organism can give, without 'microscopic', cytological work any clues regarding the further details of the neural mechanism.

Von Neumann's solution was radical. He concluded that they should focus on what he termed

> the less-than-cellular organisms of the virus or bacteriophage type ... They are self-reproductive and they are able to orient themselves in an unorganized milieu, to move towards food, to appropriate it and to use it. Consequently, a 'true' understanding of these organisms may be the first relevant step forward and possible the greatest step that may at all be required.

Von Neumann's grasp of virus biology was flimsy – he told Wiener that a virus was 'definitely an animal, with something like a head and a tail' (in fact viruses are not even alive by most definitions). Despite being a poor biologist, von Neumann's suggestion that simple systems can reveal principles that apply to more complex forms of organisation was absolutely right. He calculated that each virus consisted of around 6,000,000 atoms, and 'only ... a few hundred thousand "mechanical elements".' Von Neumann explained to Wiener that it should be possible to understand the interaction of these components, although he recognised that even this proposal was challenging:

> Even if the complexity of the organisms of molecular weight 10^7–10^8 is not too much for us, do we not possess such means now, can we at least conceive them, and could they be acquired by developments of which we can already foresee the character, the caliber, and the duration.

Von Neumann suggested to Wiener that they should study the 'physiology of viruses and bacteriophages, and all that is known about the gene-enzyme relationship'. His explanation for this approach was that viruses could give an insight into genes, which suggests he understood more about viruses than implied by his statement that they were animals:

> Genes are probably much like viruses and phages, except that all the evidence concerning them is indirect, and that we can neither isolate them nor multiply them at will.

Von Neumann was becoming interested in genes because one of the essential features of life that fascinated him was its ability to replicate itself. Indeed, from his previous thinking about self-reproducing automata, von Neumann now felt sure that 'self-reproductive mechanisms' in living things could be understood in terms of the framework that he and Wiener had been developing:

> I can show that they exist in this system of concepts. I think that I understand some of the main principles that are involved. ... I hope to learn various things in the course of this literary exercise, in particular the number of components required for self-reproduction. My (rather uninformed) guess is in the high ten thousands or in the hundred thousands, but this is most unsafe.

Von Neumann adopted this approach after attending the Ninth Washington Conference on Theoretical Physics, which had taken place at the end of October 1946. The subject of the small conference, 'The physics of living matter', was inspired by Schrödinger's *What is Life?* and had been chosen by an eccentric physicist and lifelong friend of Max Delbrück's, George Gamow (pronounced Gam-off).[27] The conference dealt with many of the points outlined by von Neumann in his letter to Wiener. As the rather excitable conference press release put it:

> During the past three days, a group of theoretical physicists

and biologists have been meeting at The Carnegie Institution of Washington and The George Washington University here to discuss problems relating to 'the physics of living matter.' Much of the discussion has concerned problems of heredity and the mechanisms by which the almost fantastic gene is able to imprint its characteristics on the cell constituents in a hereditary fashion. ... It was clear from the discussions this year that the borderline area between physics and biology will see a great deal of research activity during the next few years.[28]

The excitement may have been heightened by the fact that at the beginning of the meeting, it was announced that one of the thirty-six attendees, Hermann Muller, had been awarded the Nobel Prize in Physiology or Medicine for his work on genetics. Also at the meeting were two representatives of the new wave of geneticists: George Beadle and Max Delbrück. The question of the material basis of heredity – what genes are actually made of – was at the centre of everyone's attention. This interest was reinforced by a dramatic discovery that had been made more than two years previously in New York by a man who was not a geneticist and who would never have dreamt of going to any of the speculative conferences on the link between physics and biology that took place at this time.

THREE

THE TRANSFORMATION OF GENES

In December 1943, the Australian virologist Macfarlane Burnet disembarked in San Francisco after a three-week crossing of the Pacific. He was on his way to Harvard, where he had been invited to give a lecture – despite the war, academic life continued, for some. In his mid-forties, handsome and with wavy hair, Burnet had made his reputation working on influenza and other viral diseases; in 1960 he received the Nobel Prize in Physiology or Medicine for his work on the immune response to infection. After the Harvard lectures were over, Burnet travelled to Chicago and then New York, where he had an astonishing discussion with Oswald Avery – a small, bald micro-biologist in his mid-sixties, whose quiet manner impressed those who met him. Salvador Luria, a pioneer of virus genetics, recalled:

> Talking with Avery was a marvellous experience. He was a wonderful, short man. Very unpompous … He had the dignity of the nondignified people, very simple; and as he was talking he would close his eyes and rub his bald head. And always very precise.[1]

Avery had spent the whole of his academic life studying pneu-mococci – the bacteria that cause pneumonia – and had gained an international reputation for his work using immunological responses to characterise different pneumococcal strains. But the story that Avery told Burnet had nothing to do with immunology. As Burnet explained to his fiancée, Avery 'has just made an extremely exciting discovery which, put rather crudely, is nothing less than the isolation of a pure gene in the form of desoxyribonucleic acid' or DNA.[2]*

Avery's claim was amazing for several reasons. First, it was not accepted that bacteria actually had genes; second, most scientists thought that genes were probably made of proteins, not DNA; finally, Avery was not a geneticist and had no experience in the field. He was nearing retirement, and seemed an unlikely revolutionary. But revolutions can arise in many ways.

*

Oswald T. Avery – generally known as 'Fess' (short for 'Professor', although he never actually held the title) – had worked at the Rockefeller Institute Hospital in New York since 1913, apart from a brief period as a soldier during the First World War. His laboratory was on the fifth floor of the hospital; the lab had once been a hospital ward and the original partitions were still in place. The lab desks were covered with microbiological paraphernalia – Petri dishes, Bunsen burners, wooden-handled wire loops and needles, microscopes, incubators – while sinks and a fume hood were placed around the edge of the room. The whole place had the distinctive smell of a lab working on pneumonia – the microbes are bred in a blood-based broth. Avery's private lab had once been the ward kitchen; behind the swing door there was a roll-top desk that was generally crammed full of unanswered letters – Avery hated his routine to be disturbed, and even important invitations to travel to conferences would be left for weeks without reply. When he did respond, he almost always declined.

Before antibiotics became widely available in the 1940s,

---

*At the time desoxyribonucleic acid was the accepted name; this was subsequently changed to deoxyribonucleic acid. Other names used in the 1940s included thymonucleic acid and desoxypentose nucleic acid. They all referred to the same thing.

pneumonia was a major killer – in the US more than 50,000 people died each year of the disease. Physicians were powerless: treatment had little or no effect on survival rates. Some strains of the pneumonia microbe caused disease – they were 'virulent' – while others did not; Avery's approach to finding a cure was to understand why there were these differences between strains. Much of Avery's early work was carried out with his colleague, friend and flatmate, the opera-loving Alphonse Dochez. As a colleague recalled,

> not infrequently he [Dochez] returned from the Metropoli-
> tan Opera, discovered Dr. Avery, with whom he shared an
> apartment, reading quietly in bed, and then would sit down
> in full evening dress and with vast animation describe to his
> old friend some of the illuminating thoughts on the subject of
> microbiology which had occurred to him during the second
> act of *La Traviata*.[3]

Together with Dochez, Avery showed that it was possible to detect differences between types of pneumococci by injecting bacteria into a mouse and then observing the presence of specific antibodies in the animal's blood serum. Avery's technique was soon widely adopted as a way of identifying pneumococci and other infectious bacteria.

Insight into the origins of differences in virulence between strains of bacteria came in 1921, when the British microbiologist Joseph Arkwright noticed that colonies of virulent dysentery bacteria had a smooth surface, whereas non-virulent bacteria formed small colonies that appeared rough when inspected under a microscope. Rather obviously, the virulent strains were called S (smooth) and the non-virulent were called R (rough). Two years later, Fred Griffith, a medical officer with the Ministry of Health in London, showed that in pneumococci, too, virulent strains were smooth, whereas avirulent strains were rough. Avery studied the differences between the S and R strains and discovered that when pneumococci became virulent and smooth they produced a capsule that was up to four times the size of the bacterium itself. Avery showed that the capsule consisted of a layer of complex sugars or polysaccharides, which protects the bacterial cell from the body's defence mechanisms and gives the virulent colonies their smooth appearance.

Back in London, Griffith was exploring the mysterious fact that rough bacterial colonies could change into smooth colonies if they were mixed with smooth bacteria, a phenomenon that had first been described by Arkwright. Arkwright thought that this process was the outcome of competition between the two kinds of microbe; Griffith began to suspect that the avirulent R bacteria had actually changed into virulent S bacteria in a process he called transformation. Astoundingly, Griffith discovered that transformation could occur even if dead S bacteria were mixed with R colonies. Griffith injected mice with live, avirulent R bacteria together with killed S pneumococci. Some of the mice died; they were found to be full of S bacteria, even though the only living bacteria that had been injected were of the R strain. Griffith reported these and many other findings in a dense 45-page article that was published in the *Journal of Hygiene* in 1928.[4]

Griffith noted that similar effects had previously been observed in anthrax; the author of the anthrax study, O. Bail, had suggested that the effect involved 'the inheritance of the capsule-forming substance'. Griffith thought instead that the polysaccharide capsule carried by the dead S cells was being used as a kind of template by the R bacteria to make more capsules. Rather then being inherited, he argued, the 'specific protein structure' of the virulent pneumococcus was the cause.

Whatever the explanation – and there was no real evidence as to what was going on – Griffith's wealth of data and the rigour of his experiments were overwhelming. Almost straight away, Neufeld in Germany replicated the result.* Avery's group also began to study the effect, getting transformation to occur in a Petri dish rather than in a mouse. By the early 1930s they had made a breakthrough, extracting a substance from S pneumococci that could transform R bacteria. The Avery group called this substance 'the transforming principle', and the rest of Avery's working life was focused on identifying its nature.

---

*Both men were victims of the war: Griffith was killed in an air raid in 1941, while Neufeld starved to death in Berlin in 1945.

Shortly after he began studying transformation, Avery received the first of many international awards, the Paul Ehrlich Gold Medal, but severe illness prevented him from attending the ceremony in Germany. For years, Avery had been suffering from Graves' disease, or hyperthyroidism. This made his eyes bulge out, left him feeling tired and depressed, and gave him a tremor that made it difficult to carry out the delicate and precise microbiological procedures that were his stock in trade. In 1934, Avery went into hospital and had his thyroid removed. It took him months to recover, and it was more than a year before he regained the weight he had lost.

During the summer of 1934 a Canadian physician called Colin MacLeod joined the Rockefeller Institute Hospital, attached to the pneumonia service. When Avery returned from his sick leave, the pair began investigating the chemical nature of the transforming principle. A little more than a year later, Avery explained to his new colleague Rollin Hotchkiss where he thought their study might be going. Hotchkiss recalled:

> Avery outlined to me that the transforming agent could hardly
> be carbohydrate, did not match very well with protein and
> wistfully suggested that it might be a nucleic acid.[5]

There were no clear results to back up Avery's hunch, as MacLeod's work had not been conclusive. This caused a problem – the young Canadian needed to strengthen his curriculum vitae with some published articles, so he worked instead on the effectiveness of the new sulphonamide antibiotics. The Avery group did no further research on transformation until 1940.

Despite the fact that the method for separating the transforming principle from bacterial cells had been published, no scientists took up the challenge. This was not because people did not know about or appreciate the significance of pneumococcal transformation. In 1941, the leading evolutionary geneticist Theodosius Dobzhansky published the second edition of his influential book *Genetics and the Origin of Species*. In a chapter entitled 'Gene mutation', Dobzhansky described the work of Griffith and Avery and claimed that their findings were 'not unduly surprising from the standpoint of genetics', as the change from the R to the S form could be understood in terms of a mutation. More challenging was Griffith and Avery's demonstration that transformation could take place through contact with a killed sample – Dobzhansky reassured his readers that this 'extravagant' finding was 'conclusively proved'.[6] Dobzhansky emphasised that the transformed strains did not merely acquire 'a temporary polysaccharide envelope of a kind different from that which their ancestors have had, but are able to synthesize the new polysaccharide indefinitely.' Dobzhansky's conclusion was that contact with the transforming principle had somehow induced a mutation in the R bacteria, and that this could lead to the use of targeted mutation to study gene function:

> If this transformation is described as a genetic mutation – and it is difficult to avoid so describing it – we are dealing with authentic cases of induction of specific mutations by specific treatments – a feat which geneticists have vainly tried to accomplish in higher organisms … geneticists may profit by devising experiments along the lines suggested by the results of the pneumococcus studies.[7]

Dobzhansky was not claiming that the transforming principle was a gene, but the attention he paid to it showed that Avery's research was widely known and was seen as important.

*

In October 1940, MacLeod and Avery returned to the problem of identifying the nature of the transforming principle. To help with their analyses, they needed a powerful ultracentrifuge that could separate bacterial contents from the rearing medium – as the sample was spun round at high speeds, the heavier molecules sank to the bottom more quickly, concentrating compounds with a similar weight into a narrow band. The Rockefeller Institute had built some of these devices, using a design developed by the Swedish scientist Theodor 'The' Svedberg.[8] Avery's everyday needs were not so demanding – initially his group simply needed to obtain large quantities of bacteria. The solution was to adapt a kitchen cream separator made by the Sharples company. The Sharples, as it was called in the lab, consisted of a tube that was the size of a thick cucumber – about 5 cm in diameter and 25 cm long. There was one problem: the tube was not tightly sealed, and tiny gaps in the apparatus meant that every time it was used, the room became full of an invisible aerosol of potentially lethal bacteria. Sharples was therefore placed in a specially constructed containment device that could be sterilised before opening.[9] Even so, using the equipment safely was no easy matter. After centrifugation, the cake of bacteria that had accumulated at the bottom of the tube had to be removed – this was impossible to do cleanly and 'one would see small flecks of white material fly in one direction or another', recalled a lab member.[10] All of the cake was handled with towels soaked in germicide and then heated at 65°C before it was studied further, in an attempt to reduce the risk to lab members.[11] This messy and dangerous procedure so distressed the fastidious Avery that he would leave the lab when the Sharples was in action.

The group soon found that adding calcium chloride to the liquid transforming principle produced a white precipitate that contained most, if not all, of the transforming activity: adding white precipitate from smooth bacteria to a rough colony would transform it into a smooth colony. This white substance was very powerful – even at 1/1,000 dilution it could still transform a rough colony. At the beginning of 1941, MacLeod noted that the white precipitate contained both the polysaccharides typical of the smooth capsule and nucleic acids – DNA and its close relative, ribonucleic acid or RNA. When MacLeod added an enzyme that was known to destroy RNA, this had no effect on the transforming activity of the extract, strongly suggesting that RNA played no role in producing the power of the white material. In April 1941, in his six-monthly report to the Rockefeller Institute, Avery described the progress he and MacLeod had made and hinted at the potential implications:

> This study is being continued with the hope that knowledge of this important cellular mechanism may lead to a better understanding of the principles involved in certain induced variations of living cells, not only of the pneumococcus, but also those of other biological systems.[12]

*

In the summer, MacLeod left the Institute and another young physician, Maclyn McCarty, joined the Avery group. By the end of November 1941, McCarty had shown that if he used an enzyme to remove the polysaccharide, the extract nevertheless retained its transforming activity, showing that – as expected – the polysaccharide was not involved. That apparently left just two possibilities: proteins or DNA.

In December 1941 the Japanese attacked Pearl Harbor and the US entered the war. The Avery group shifted its work towards more practical aspects of pneumonia as the disease began to appear among US troops. Nevertheless, McCarty continued with his research, and in January 1942 he found that if alcohol was added to the transforming principle, a stringy white material appeared that contained 99.9 per cent of the transforming activity. It soon became apparent that this stringy stuff also contained most of the DNA that was present in the sample.

Two floors above Avery's office was the laboratory of Alfred E. Mirsky, one of the world's leading experts on nucleic acids. Mirsky gave the Avery lab some mammalian DNA extracted from the thymus gland, the traditional source of DNA, and they compared it with the white stringy stuff produced by alcohol precipitation of the transforming principle. The two substances seemed to be very similar. McCarty took an extract of transforming principle that had been treated with enzymes to remove both proteins and polysaccharides, and placed it in an ultracentrifuge. After spinning the sample for a few hours at 30,000 r.p.m., a gelatinous 'pellet' appeared at the bottom of the tube, containing the heaviest components of the extract. This contained all the transforming activity of the original solution and was apparently composed entirely of DNA.

In the summer of 1942, the suggestion that the transforming principle was made of DNA became stronger when McCarty and Avery showed that enzymes that destroyed transforming activity also affected Mirsky's DNA samples, and enzymes that had no effect on DNA did not affect the activity of the extract. This should have led to great excitement, but Mirsky was unimpressed. As he explained to the Avery group, the transforming principle could not be made of DNA because nucleic acids were all alike. As their late Rockefeller Institute colleague Phoebus Levene had argued more than three decades earlier in his tetranucleotide hypothesis, the components of nucleic acids – the two kinds of base, the purines (adenine and guanine) and the pyrimidines (cytosine and thymine; thymine is replaced by uracil in RNA) – were present at similar levels. Although DNA was known to be a component of cell nuclei, its apparently boring nature meant that it was not thought to have biological 'specificity' – the term used at the time to describe the unique effects of a particular molecule. Proteins, in contrast, were extremely varied, and could be active even at very low levels. It was quite possible that despite all the treatments to remove proteins from their extracts, minute amounts of very powerful protein molecules remained, Mirsky explained.

Although most scientists agreed that DNA did not have the necessary variability to have specificity, some were not so sure. In July 1941, at the annual Symposium on Quantitative Biology held at Cold Spring Harbor Laboratory out on Long Island, Jack Schultz pointed out that the supposed uniformity of nucleic acids was based on a single data point – all the DNA that had been studied had been taken from the thymus gland of cows. The suggestion that DNA structure was uniform could be accepted 'only as a first order approximation', he argued: 'much new data is necessary before we can exclude the possibility of specificities in the nucleic acids themselves'.[13] However, Schultz was firmly convinced that genes were made of what he called nucleoproteins – proteins that were known to be tightly associated with nucleic acids in the chromosomes.

In 1943, Mirsky underlined the growing sense of mystery surrounding nucleic acids when he wrote an article that was supposed to sum up current knowledge about nucleoproteins. Strikingly, he had little to say about proteins and instead concentrated on nucleic acids, and above all on DNA. Mirsky described how the nucleic acid component of a solution could be identified by its reaction to ultraviolet radiation; this was due to the responses of the pyrimidine and purine bases, which apparently lay in rings, perpendicular to the central axis of the molecule. Using this procedure, it was possible to show that, in animals and plants, chromosomes were largely made of DNA, and that DNA was also present in bacteria and in viruses. Finally, it seemed that nucleic acids were involved in both metabolic processes and the replication of chromosomes. Although there was no direct evidence for any link between proteins and genetic functions, Mirsky nevertheless concluded that the proteins found with DNA were at the heart of heredity:

> The great accumulation of desoxyribose nucleoproteins in the chromosome strongly suggests that these substances either are the genes themselves or are intimately related to the genes.

In retrospect, virtually all the evidence that Mirsky summarised indicated that DNA was basis of heredity, and yet – like everyone else outside the Avery lab – he argued that genes were made of proteins that were bound up with DNA. He could not see what now appears obvious because there seemed to be no way in which DNA could contain the kind of variability that was necessary to produce the wide range of genetic effects. For Mirsky, Levene's suggestion that DNA was composed of a monotonous repetition of the four bases was 'a definite restriction in possible variation among the desoxyribose nucleic acids'.[14]

Despite these arguments, Avery and McCarty were increasingly convinced that DNA was the transforming principle and therefore the main component of genes. To prove their point, production of the stuff had to be stepped up – it took 200 litres of bacteria to produce just 40 milligrams of stringy white precipitate. By this stage, McCarty had been called up to active duty by the Naval Reserve research unit. Feeling he should do something related to the war effort, McCarty asked to be put on a more practical project relating to disease treatment, but was told not to worry and to return to Avery's lab – the main difference that his call-up made was that he now went to the lab in uniform.

In April 1943, Avery's report to the Rockefeller Institute Board explicitly framed the problem of transformation in terms of genes for the first time. The transforming principle 'has been likened to a gene', Avery wrote, and the polysaccharide was like 'a gene product'. He explained:

The genetic interpretation of this phenomenon is supported by
the fact that once transformation is induced, ... both capsule
formation and the gene-like substance are reduplicated in the
daughter cells.

Nevertheless, he wrote in typically cautious style that proof was still
lacking and that all his conclusions were provisional:

If the present studies are confirmed and the biologically active
substance isolated in highly purified form as the sodium salt
of desoxyribosenucleic acid actually proves to be the transforming principle, as the available evidence now suggests,
then nucleic acids of this type must be regarded not merely
as structurally important but as functionally active in determining the biochemical activities and specific characteristics
of pneumococcal cells.[15]

Shortly afterwards, on 13 May 1943, Avery began writing a letter to his younger brother, Roy, who was professor of microbiology
at Vanderbilt University in Nashville, Tennessee. While most people called Oswald Avery 'Fess', to Roy – 15 years younger – Avery
was simply 'Brother'.[16] In the letter, written in spidery handwriting,
Fess told Roy about his plans to retire and join him in the South: 'If
this War wasn't on I tell you frankly I would liquidate my affairs
here and start for Nashville this fall', he wrote. Two weeks later, on
the night of 26 May, Avery finally got round to completing what he
called 'a rambling epistle'. In this second part, he broke with the personal and slightly weary tone of the opening section and explained
to Roy exactly what he and his group had discovered. Avery first
reminded his brother about Griffith's discovery of transformation
and the steps that his laboratory had taken in the 1930s to identify
the chemical basis of the phenomenon. Then he allowed an element
of triumph to creep into his description of the hard work that was
involved, before his habitual caution reasserted itself:

Some job – and full of heartaches and heart breaks. But at last
perhaps we have it ... the substance is highly reactive and on

elementary analysis conforms very closely to the theoretical values of pure desoxyribose nucleic acid (thymus type). Who could have guessed it? ... We have isolated highly purified substance of which as little as 0.02 of a microgram is active in inducing transformation ... this represents a dilution of 1 part in a hundred million – potent stuff that – and highly specific. This does not leave much room for impurities – but the evidence is not good enough yet.

Avery then explained the implications of his discovery, showing that he fully understood the importance of what he had found:

If we are right, and of course that's not yet proven, then it means that nucleic acids are not merely structurally important but functionally active substances in determining the biochemical activities and specific characteristics of cells – and that by means of a known chemical substance it is possible to induce predictable and hereditary changes in cells. This is something that has long been the dream of geneticists ... Sounds like a virus – may be a gene. ... Of course, the problem bristles with implications. It touches the biochemistry of the thymus type of nucleic acids which are known to constitute the major part of the chromosomes but have been thought to be alike regardless of origin and species. It touches genetics, enzyme chemistry, cell metabolism and carbohydrate synthesis etc.

But Avery would not have been Avery had he let his excitement continue, and he concluded with his usual self-deprecation:

It's lots of fun to blow bubbles – but it's wiser to prick them yourself before someone else tries to. So there's the story Roy – right or wrong it's been good fun and lots of work. ... Talk it over with [your colleague] Goodpasture but don't shout it around – until we're quite sure or at least as sure as present method permits. It's hazardous to go off half cocked – and embarrassing to have to retract later. I'm so tired and sleepy I'm afraid I have not made this very clear. But I want you to know – and sure you will see that I cannot well leave this problem until we've got convincing evidence. Then I look forward and hope we may all be together – God and the war permitting – and live out our days in peace.

And after signing off 'with heaps and heaps of love', Avery added a final postscript:

Good night – it's long after mid-night and I have a busy day ahead. God bless us, one and all. Sleepy, well and happy –[17]

*

By the autumn of 1943, Avery and McCarty were as certain as they could be that the transforming principle was composed of DNA, but Avery remained concerned that there might be some protein contaminants that were producing the effects. So he asked for advice from the protein chemists John Northrop and Wendell Stanley, and the Rockefeller biochemists Van Slyke and Max Bergmann. They all gave the same, unhelpful answer. There was no magic solution: Avery wanted to prove a negative, to show that his extract was completely free of proteins, and this was impossible. The only thing he could do was get as much evidence as possible, using a variety of techniques, and then publish. Interestingly, McCarty's notes of their meeting with Bergmann reveal that the senior biochemist had similar doubts about the allegedly uniform nature of DNA to those expressed by Jack Schultz in 1941. According to McCarty's notes, Bergmann felt that the previous certainties about nucleic acids were beginning to weaken:

In the light of present knowledge, the statement that all nucleic acids are the same regardless of the source from which they are derived is nonsense. If they are large polymeric compounds, there is an endless number of possible combinations all of which would possess the same elementary composition but would differ in chemical structure none the less. Nucleic acids hold too prominent a place in biology to be completely non-specific substances. The lack of evidence of any specificity associated with nucleic acids is only due to the fact that they have not been investigated sufficiently.[18]

For two months Avery and McCarty wrote up their findings for publication, with Avery weighing every word and often using material from his reports to the Rockefeller Institute as his starting point. From the opening sentences, the article showed the context in which the study was carried out, and gave a hint of the implications of the main findings: 'Biologists have long attempted by chemical means to induce in higher organisms predictable and specific changes which thereafter could be transmitted in series as hereditary characters.' The results they presented were detailed, and their identification of the transforming principle as DNA was based on several strands of evidence – chemical composition; inactivation of the extract by enzymes or temperatures that affected DNA; no effect of enzymes that digested proteins; absence of immune reactions typical of those produced by proteins; responses to centrifugation, electrophoresis and ultraviolet radiation were all identical to those of DNA. Every result converged on the same conclusion: the transforming principle was composed of DNA.

The discussion section of the article outlined the genetic context of their findings, using similar terms to their Rockefeller Institute report from earlier in the year:

> The inducing substance has been likened to a gene, and the capsular antigen which is produced in response to it has been regarded as a gene product.

Furthermore, the potential links of transformation with 'similar problems in the fields of genetics, virology, and cancer research' were indicated, clearly outlining the implications of their discovery. And yet, despite the overwhelming evidence, all of which suggested that the transforming principle was made of DNA and that genes might be, too, the final paragraph of the article opened with a phrase that suggested the team were not quite as confident as they ought to have been:

It is, of course, possible that the biological activity of the substance described here is not an inherent property of the nucleic
acid but is due to minute amounts of some other substance
adsorbed to it or so intimately associated with it as to escape
detection.

Despite the fact that this deflating phrase was immediately followed
by a set of counter-arguments, the tone tended to undermine the
reader's confidence. Even the final bold sentence introduced doubt
where none was needed:

If the results of the present study on the chemical nature of the
transforming principle are confirmed, then nucleic acids must
be regarded as possessing biological specificity the chemical
basis of which is as yet undetermined.[19]

On 1 November, Avery handed the manuscript to his colleague
Peyton Rous, who was the editor of the *Journal of Experimental Medicine*, which was published by the Rockefeller Institute. Rous did not
send the article out for other scientists to review before publication –
that was not the general practice at the time – but instead made some
editorial suggestions, including cutting what he considered to be a
speculative passage about the role of nucleic acids.[20] And with that
the article was accepted.

On 10 December 1943, Avery presented the findings at the regular Friday afternoon Rockefeller Institute staff meeting – the first time
in years that he had talked about the work of his group. After Avery
spoke there was a warm round of applause, followed by a deafen-
ing silence. There were no questions. Eventually Dr Heidelberger
of Columbia University rose and emphasised the many years that
Avery had been working on the problem. Then he sat down again
and another silence ensued. The meeting was then closed. Avery had
described one of the most momentous discoveries in the history of
science, and no one could think of anything to say.

*

The article appeared in the 1 February 1944 issue of the *Journal of Experimental Medicine*, but Avery's work was not over. As the concluding part of the paper indicated, Avery and McCarty feared that the evidence would not convince the majority of scientists who thought that genes were made of proteins. Even if all the protein-removing procedures were working at their best, molecules are so tiny that even in the smallest amount of 'pure' extract that worked in their system – a mere 0.003 micrograms, or 0.00000003 grams – there could still be millions of protein molecules in the sample, each of which might correspond to a gene. The biochemical and analytical techniques available at the time meant that it was not possible to confidently remove that final portion of protein, or to prove that a sample was completely protein-free. So in 1944 Avery and McCarty tried to attack the problem from the other side by showing that even minute quantities of an enzyme that attacked DNA would stop transformation. Both men began to feel the strain. Avery became increasingly withdrawn and even depressed as he tried to find evidence that would demonstrate the proof of his discoveries, but McCarty was unsympathetic. McCarty later recalled that Avery often had a 'gloomy outlook' and an 'apathetic expression', concluding with an element of self-criticism: 'I found it difficult to cope in this situation with the necessary restraint and good humour, and I'm afraid that I was not nearly as patient with Fess as I should have been.'[21]

In October 1945, Avery and McCarty submitted two more papers to the *Journal of Experimental Medicine*. These contained yet more biochemical evidence showing that the transforming principle was composed of DNA, and extended the finding to other forms of transformation in pneumonia bacteria.[22] The two articles appeared together in the January 1946 issue of the journal, and robustly addressed the potential criticism that minute amounts of protein were responsible for the effect: 'There is no evidence in favour of such a hypothesis, and it is supported chiefly by the traditional view that nucleic acids are devoid of biological specificity.'[23] Avery and McCarty made no claim for how specificity was represented in DNA. They did not use the word 'code' or anything like the concept of a code, but they clearly stated that there must be something in DNA that enabled genes to be so varied:

It remains one of the challenging problems for future research
to determine what sort of configurational or structural differences can be demonstrated between desoxyribonucleates of
separate specificities.

They explained that it was probable that only a small proportion of
the DNA they extracted was involved in transforming the bacteria
from rough to smooth. There could also be a large number of other
DNA molecules that would 'determine the structure and metabolic
activities' of both forms of pneumococcus. This was a conceptual
advance: they were arguing that all the genes possessed by these
bacteria were made of DNA.[24]

Despite the dislocation in scientific communication caused by
the war, the response to Avery's publications was immediate and
positive. In 1944, *Nature* described Avery's work in glowing terms:
'The genetic implications of this work are considerable', wrote one
scientist, and three months later another astutely suggested that
'slight differences in molecular configuration' of different forms of
DNA might explain differences in biological activity, concluding,
'this in itself must represent an entirely new and highly promising field'.[25] In October 1944, the New York Academy of Medicine
awarded Avery its Gold Medal. Although this was primarily for his
decades of work on pneumococci, the citation, which was printed
in the widely read US journal *Science*, referred to his isolation of the
transforming principle and concluded, 'this discovery has very far-
reaching implications for the general science of biology'.[26] In 1945,
the Royal Society of London followed suit and awarded Avery the
Copley Medal, again primarily for his microbiological work but
with a powerful recognition of the importance of the 1944 paper:
'the interest and importance of this work, to chemists and biologists
(and perhaps most of all to geneticists) is outstanding'.[27] In November 1945, Hermann Muller gave the prestigious Pilgrim Trust lecture to the Royal Society. His subject was 'The gene' and he focused
on recent discoveries about its physical nature. One section dealt
with 'possible roles of nucleic acids' and described the 'remarkable
experimental evidence' from Avery's group. 'If this conclusion is
accepted', said Muller, who was highly sceptical, 'their finding is

revolutionary'.[28] At around the same time, the biochemist Howard
Mueller wrote a review in which his enthusiasm was evident. He
began by summarising Avery's findings:

> a polymer of a nucleic acid may be incorporated into a living,
> degraded cell, and will endow the cell with a property never
> previously possessed ... When thus induced the function is
> permanent, and the nucleic acid itself is also reproduced in cell
> division. The importance of these observations can scarcely be
> overestimated'.[29]

Above all, some excited researchers decided to extend the Avery
group's findings. On 20 January 1945, Joshua Lederberg, a brilliant
19-year-old who had just begun a postgraduate medical degree
at Columbia University, sat down to read an article that had been
passed to him by a fellow student, Harriett Taylor. The effect was
electric. As he wrote in his diary:

> I had the evening all to myself, and particularly the excruci-
> ating pleasure of reading Avery '43 [sic] on the desoxyribose
> nucleic acid responsible for type transformation in Pneumo-
> coccus. Terrific and unlimited in its implications ... I can see
> real cause for excitement in this stuff.[30]

Inspired by Avery, Lederberg decided to turn to the study of
transformation in other bacterial systems and soon discovered that
bacteria can have a sexual phase, thereby opening the road to the
genetic study of these organisms – in 1959 he won the Nobel Prize
in Physiology or Medicine for this work. Equally importantly, Erwin
Chargaff, a 39-year-old Ukrainian-born biochemist who had fled
the Nazis and was now working at Columbia University, became
an outspoken champion of Avery's discovery. Chargaff immediately
focused his research on the chemical composition of nucleic acids,
and soon began to challenge Levene's tetranucleotide hypothesis.

In Paris, the deputy director of the Institut Pasteur, 50-year-old
André Boivin, was inspired by Avery's paper to investigate transformation in a completely different bacterium, *Escherichia coli*. Boivin

published his findings in French in November 1945. Like Avery, Boivin found that the transforming agent seemed to be 'a highly polymerised thymonucleic acid'. Bolder than Avery, Boivin explicitly argued that 'we should now look to the nucleic acid component of the giant nucleoprotein molecule that forms a gene, rather than to the protein part, to find the inductive properties of the gene.'[31] When Avery was shown Boivin's paper, he beamed and happily announced to his colleagues over lunch that they now had 'continental support'.[32] Everything seemed to be going their way.

- FOUR -

A SLOW REVOLUTION

As Europe and America emerged from the Second World War, there was a wave of research on the structure and function of nucleic acids, partly propelled by Avery's work. Between 1944 and 1947, more than 250 scientific papers were published on nucleoproteins and nucleic acids – about the same number as in the new field of antibiotics – and most of them explored the nature and function of nucleic acids rather than proteins.[1] Between 1946 and 1948, four international scientific conferences focused on the question – one in Cambridge, England (1946), two at Cold Spring Harbor Laboratory on Long Island (1947 and 1948), and one in Paris (1948). Nucleic acid structure and function was becoming one of the hottest scientific topics of the postwar world.

In July 1946, the Society for Experimental Biology held a symposium in Cambridge on nucleic acids. One of the speakers was William Astbury of the University of Leeds, a pioneer of the use of X-rays to study crystal structures. Astbury had visited Avery's laboratory in 1937 and knew all about his work on transformation. Within a few months of Avery's article appearing in 1944, Astbury told a friend that he had been 'terribly thrilled' that Avery had identified the transforming substance as DNA; he thought this was 'one of the most remarkable discoveries of our time'. Astbury wrote: 'I wish I had a thousand hands and labs with which to get down to the problem of proteins and nucleic acids. Jointly those hold the physicochemical secret of life, and quite apart from the war, we are living in a heroic age, if only more people could see it.'[2]

At the Cambridge meeting, Astbury showed X-ray images of DNA that indicated very clearly that a DNA fibre contains repeated elements, but he was unable to conclude anything about the sequence of those elements within the fibre. Astbury closed his talk by presenting the first model of DNA structure, explaining:

> A test that cannot long be dispensed with in any enquiry into the structure of a complex molecule is that of trying to build an accurate atomic model on the basis of known sizes and inter-bond angles. Chemical formulae are no more than a convenient shorthand, and it is always revealing, and often startling, to see what a molecule looks like in space.[3]

His model was a relatively uniform column. It was neither startling nor correct.

At the same meeting, Professor Masson Gulland from Nottingham summarised the work on nucleic acid structure that had been done during the war and questioned the widely held view that DNA and RNA were boring molecules: 'there is at present no indisputable evidence that any polynucleotide is composed largely, if at all, of uniform, structural tetranucleotides'.[4] There was no evidence that the four bases were repeated like beads on a string. As Gulland put it:

> there is, to choose perhaps an extreme case as illustration, no reason why four molecules of a given nucleotide should not be adjacent and be succeeded in the chain by, let us say, a group of molecules of another nucleotide.

Gulland was arguing that the sequence of bases along the DNA molecule might vary.

In another talk, Dr M. Stacey of the University of Birmingham discussed Avery's suggestion that DNA was the 'transforming principle'. Stacey accepted that DNA played an essential role but argued that it functioned as an enzyme that used tiny amounts of the poly-saccharide from rough strains as a template to begin synthesising the new kind of capsule.[5] In contrast, Edgar and Ellen Stedman, from the University of Edinburgh, were adamant that nucleic acids were merely a structural part of the chromosomes, and that protein alone could account for heredity:

> The material of which the chromosomes is composed must ... be capable of accounting in a broad manner for the heredi-tary functions of chromosomes ... The first of these require-ments can be satisfied only by one known type of compound, a protein.[6]

A few weeks earlier, Cold Spring Harbor Laboratory in the US had relaunched its annual Symposium in Quantitative Biology after a hiatus due to the war. The first two meetings were focused on top-ics that flowed directly from Avery's work: 'Heredity and variation in microorganisms' (1946) and 'Nucleoproteins and nucleic acids' (1947). At the 1946 symposium, Avery's work was cited by one-third of the speakers, and McCarty presented a paper co-authored by Avery and Harriett Taylor in which he boldly extended their view of the role of DNA in transformation to the whole of life, concluding: 'these results suggest that nucleic acids in general may be endowed with biologically specific properties not hitherto demonstrable.'[7] There was kickback from other researchers, such as Seymour Cohen, who argued in another part of the meeting that 'the data directly relating nucleic acid to specifically inheritable phenomena are very sparse indeed'.[8] Despite widespread interest, the discoveries of the Avery group were not met with unanimous approval.

An even clearer critique of Avery's interpretation appeared shortly afterwards, when Alfred Mirsky published a widely read article that he co-authored with Arthur Pollister. Mirsky and Pol-lister pointed out that 'there can be little doubt in the mind of any-one who has prepared nucleic acid that traces of protein probably remain in even the best preparations' and that 'as much as 1 or 2 per cent of protein could be present in a preparation of "pure, protein-free" nucleic acid.'[9] This protein remnant could easily account for the effects that Avery's group attributed to DNA. Mirsky focused on the quantitative estimations of protein in the extracts produced by the Avery group, ignoring the varied kinds of data that suggested that DNA was the sole active component in their extracts, such as the fact that protein-digesting enzymes had no effect on the action of the transforming principle. Mirsky's criticisms undermined confidence in Avery's claims, especially among those who were not chemists. One of the most prominent people to accept Mirsky's argument was Hermann Muller. In the written version of his 1945 Pilgrim Trust lecture to the Royal Society, Muller accepted that Avery's finding, if true, was 'revolutionary', but indicated that he was personally convinced by Mirsky's suggestion that undetected 'genetic proteins', floating free in the medium, caused Avery's results.[10]

At the June 1947 Cold Spring Harbor symposium the new abbreviation of DNA began to replace the cumbersome desoxyribonucleic acid.[11] At the meeting, which was attended by 150 people, the French DNA convert André Boivin presented the big-picture implications of Avery's findings. For two years, Boivin had been publishing evidence from *Escherichia coli* bacteria that supported Avery's conclusion that the transforming principle was composed of DNA, showing that the activity of the substance could be destroyed by the enzyme DNase but not by an equivalent enzyme that attacked RNA, RNase.[12] After summarising the evidence, Boivin presented a vision for the future of the whole of biology, speculating about the possibility of transferring genes in higher organisms much as had been done in bacteria:

> each gene can be traced back to a macromolecule of a special desoxyribonucleic acid. ... Thus, this amazing fact of the organization of an infinite variety of cellular types and living species is reduced, in the last analysis, to innumerable modifications within the molecular structure of one single fundamental chemical substance, nucleic acid ... This is the 'working hypothesis' quite logically suggested by our actual knowledge of the remarkable phenomenon of directed mutations.[13]

In the discussion of Boivin's talk, Mirsky explained why he was still not convinced. Although he agreed that there was no chemical evidence that all nucleic acids were the same, he emphasised that the only evidence for the genetic role of DNA came from bacteria. This made it difficult for Mirsky to accept what he recognised were the revolutionary implications for the whole of biology. Mirsky's critique repeated the argument he had published the year before: small amounts of protein could still be present in Avery and Boivin's extracts. 'In the present state of knowledge it would be going beyond the experimental facts to assert that the specific agent in transforming bacterial types is a desoxyribonucleic acid', Mirsky said. In response, Boivin accepted that it was impossible to be absolutely certain about the chemical composition of any extract, but he then neatly shifted the argument, underlining the varied kinds of evidence that he and the Avery group had presented: 'it seems to us that the burden of proof rests upon those who would postulate the existence of an active protein lodged in an inactive nucleic acid.'[14]

At the meeting, the chemist Erwin Chargaff also turned the tables on those like Mirsky who argued that the genetic material was a nucleoprotein, pointing out that there was no evidence that nucleoproteins were really present in cells; it was quite possible that these compounds were formed accidentally in the test tube when an extraneous protein found itself bound to the DNA while the two compounds were being isolated. Chargaff went on to praise 'the epochal experiments by Avery and his associates' and sketched out a future research programme that, although not as grandiose as Boivin's, had the great virtue of being feasible:

> If, as we may take for granted on the basis of the very convincing work of Avery and his associates, certain bacterial nucleic acids of the desoxypentose type are endowed with a specific biological activity, a quest for the chemical or physical causes of these specificities appears appropriate, though it may remain completely speculative for the time being. ... Differences in the proportions or the sequence of the several nucleotides forming the nucleic acid chain also could be responsible for specific effects.[15]

Despite Mirsky's opposition, Avery's advocates were sketching out how an apparently boring DNA molecule could be as varied as a protein, perhaps through differences in the sequence of the bases.

Masson Gulland – a cheery-looking cove with a centre parting, a moustache, full lips and laughter lines around his eyes – gave a similar talk to his presentation at Cambridge the year before, but did not fully embrace the idea that DNA had specificity: 'It seems possible that both nucleic acid and protein may contribute to the specificity, and not the protein alone as has often been thought', was as far as he would go.[16] In the discussion, Gulland was asked about the possibility that a DNA molecule might be a helix, held together by the presence of evenly spaced hydrogen bonds between the bases. Gulland, who had been studying hydrogen bonds in DNA, called this idea 'interesting and stimulating', but he did not investigate this further – he died in a train crash within weeks of returning to the UK.[17] Two other participants who had also been at the Cambridge meeting, Edgar and Ellen Stedman, continued their critique of the genetic role of DNA, even claiming that it was a moot point that nucleic acid was an integral part of the chromosomes.[18] In a similar vein, Jack Schultz complained that although the work of Jean Brachet and Torbjörn Caspersson suggested that nucleic acids played a role in controlling protein synthesis, the actual evidence was 'far from convincing'.[19] During the 1940s, Brachet and Caspersson had both found that RNA was involved in protein synthesis, although what exactly it was doing was unclear.[20]

Some scientists were more enthusiastic about the importance of nucleic acids. In an article that appeared in 1948 in a new genetics journal entitled *Heredity*, Joshua Lederberg, who had been inspired by Avery's paper to study bacterial genetics and was still only 22 years old, wrote:

> The total absence of all other components is not readily established, although none can be detected with available methods. The criticism of Mirsky and Pollister (1946) should be noted, however. The further chemical characterisation of the transforming principle is one of the most urgent problems of present-day biology, since it behaves like a gene which can be transferred by way of the medium from one cell to another.[21]

In 1948, four years after Avery's first paper appeared, Boivin was arguing that all genes are made of DNA, Chargaff had hypothesised that nucleic acid specificity might involve differences in the sequence of bases, and Lederberg was urging his colleagues that characterising the transforming principle was a central task of biology. As Astbury had put it in 1945, this was indeed a 'heroic age'.

*

Oswald Avery played no part in any of this. He left Rockefeller at the end of 1948 and moved to Nashville to live with his brother, Roy. He received no further official recognition of his part in the discovery of the genetic role of DNA; although he was nominated for a Nobel Prize, the committee apparently 'found it desirable to wait until more became known about the mechanism involved'.[22] This view may have been reinforced by the fact that the Swedish scientific community was not closely following the debate in the US and the UK about the nature of the hereditary material.[23] The brief obituary that appeared in the *New York Times* when Avery died in 1955 did not even mention DNA.[24] As science journalists began to report on the growing wave of excitement within the scientific community about DNA, history was rewritten almost as soon as it happened. In January 1949, the *New York Times* informed the world that a Rockefeller Institute researcher had discovered that 'genes consist in part of a substance called desoxyribonucleic acid'. The researcher's name was Dr Alfred Mirsky.[25]

Research into the transforming principle was continued by Rollin Hotchkiss at Rockefeller and by Harriett Taylor, who had moved to Paris, where she married the geneticist Boris Ephrussi. Taylor extended the number of bacterial characteristics that could be transformed, thereby showing the generality of transformation and its similarities to genetic factors in higher organisms, while Hotchkiss responded to Mirsky's carping criticism about the protein content of the extracts by carrying out very precise experiments.[26]

The first public presentation of Hotchkiss's new data took place in Paris at the end of June 1948. André Lwoff, a bacteriologist at the Institut Pasteur, had organised a small meeting with the rather pompous title 'Biological units endowed with genetic continuity' – these 'units' were bacteria and viruses. Hotchkiss described results from a range of techniques that were intended to eliminate proteins from the transforming principle. After treatment, at most 0.2 per cent of the extract was protein; this was within the margin of error of a result of 0.0 per cent, so it was quite possible that there was no protein at all in his samples.[27] It was impossible to be more precise. Despite this clarity, when Lwoff summed up the week's discussions, he insisted that nucleic acids 'could and should be normally combined with another constituent, most likely a protein.'[28] The old ways of thinking died hard.

Although Hotchkiss's paper was initially published only in French, news of his findings soon began circulating in the US. Another participant at the Paris meeting was the US virus biologist Max Delbrück, and in the spring of 1949 Al Hershey, a member of Delbrück's informal phage group, presented Hotchkiss's data as part of the DNA transformation story at a round-table discussion on nucleic acids organised by the Society of American Microbiologists.[29] By the end of the 1940s, it was widely known that the levels of protein in the transforming principle were effectively zero.

Boivin, who by now had moved to Strasbourg, began his talk at the Paris meeting by stating that although DNA specificity had been shown only in bacteria, the conclusion was overwhelming: 'these facts lead us to accept – until formal proof to the contrary – that this specificity is an example of a general phenomenon, which everywhere plays a major role in the biochemistry of heredity.'[30] In his conclusion, Boivin reported that in a wide range of organisms – including many animals – the nuclei of different cells in the same organism had the same amount of DNA; similarly, members of a given species all had the same amounts of DNA, whereas eggs and sperm had only half the amount found in normal cells. This was a decisive discovery. It had been known since Sutton's observations in 1902 that when most sexual species reproduced, the creation of the egg and sperm cells involved the halving of the usual double or diploid complement of chromosomes, so that one set of chromosomes went into each egg or sperm, to form what is called a haploid cell. When the egg and sperm fused to form the new organism, these two haploid components formed a new diploid set of chromosomes. Boivin's observation that the amount of DNA corresponded to the chromosome complement at different phases was what would be expected of a gene – nothing like this had been found for any protein. In bacteria, plants and animals, Boivin argued, 'each gene can, in the final analysis, be considered as a macromolecule of DNA.'[31]

This was one of Boivin's last lectures. The cancer he had been suffering from returned and he died in July 1949. In the meantime, doubts began to be raised about his reports on transformation in *E. coli* – researchers in the US were unable to replicate his findings, and his original strains were lost.[32] Despite – or perhaps because of – Boivin's bold statements and prophetic vision, an air of disbelief accumulated around his discoveries. This took years to dissipate – his findings were eventually confirmed in the 1970s, and his vision of the nature of heredity and the future of biology were also shown to be true.[33]

*

By the end of the 1940s, support for the hypothesis that DNA had a fundamental role in heredity had grown much stronger. In the summer of 1950, the cell biologist Daniel Mazia gave a lecture at Woods Hole Marine Biology Laboratory that summed up many people's thinking. Mazia could not be absolutely certain that genes were made of DNA, but it certainly looked that way: 'The "physical basis of heredity" is something in the chromosome which may or may not be DNA, but which follows DNA for all practical purposes', he said.[34] Following Boivin, Mazia outlined four criteria that had to be met by what he called the vehicle of heredity, whatever its chemical composition might be. There had to be the same amount in every diploid cell of a given species, that amount should double just before ordinary cell division, and it should be halved during the creation of haploid sex cells. It should be stable, it should be capable of specificity and it should be able to be transferred from one cell to another and act like a gene. Proteins failed at the first hurdle – there was no evidence that the levels of protein in cell nuclei were the same in all tissues of an organism. All that proteins had going for them was that they were known to be complex. Both Chargaff and Gulland had suggested that nucleic acids could vary by the sequence of the bases, perhaps providing a source of complexity. While still not excluding the possibility that proteins were involved, Mazia concluded: 'DNA is the most likely candidate so far for the role of the material basis of heredity.'[35]

A couple of months earlier, in April 1950, Mazia had chaired a session at a special conference on the biochemistry of nucleic acids, held at the US Atomic Energy Commission's Oak Ridge Laboratory in Tennessee. Among the speakers was Arthur Pollister, who with Mirsky had criticised Avery's conclusions two years earlier. Pollister was changing his tune; he enthusiastically reported the data from Boivin's laboratory that showed that the amount of DNA was constant in all diploid cells of a given species, before discussing the idea of a 'DNA-gene' and raising the possibility that the chemical structure of the gene was within reach. Nevertheless, Pollister was not completely convinced: the complexity of gene function led him to suggest that 'important genic components other than DNA remain to be discovered.'[36]

Another speaker at the Oak Ridge meeting was Erwin Chargaff, who was acquiring some of the most telling evidence in favour of Avery's conception of the role of DNA. Chargaff had been an early supporter of Avery's hypothesis, and in 1950 he presented data on the precise base composition of nucleic acids, using paper chromatography to identify the bases by weight. He found that the proportion of the different bases was constant in all tissues of any species, but differed wildly between species. As Chargaff pointed out, these data disproved the tetranucleotide hypothesis, to the extent that anyone still thought it was true.[37] DNA was clearly not boring.

The 1951 Cold Spring Harbor symposium was on the topic 'Genes and mutations', and one of the speakers was Harriett Ephrussi-Taylor, who took the opportunity to survey the seven years that had passed since Avery, MacLeod and McCarty had published their landmark paper. She was downbeat:

> Considering the interest which was aroused by the publication of the results of the chemical and biochemical study of the

capsular transforming agent of pneumococcus, it is surprising
that so few scientists are at present working in this field.[38]

As she admitted, transformation was difficult to study – for example, transformation in Boivin's *E. coli* system 'occurred only with
some irregularity' – and many researchers were not familiar with
the pneumococcal system in which transformation had first been
described. She glumly concluded that the study of transformation
remained isolated: 'as yet,' she said, 'no bridge can be seen leading
over into classical genetics'.

Ephrussi-Taylor's lament was related to what now appears to
be an odd feature of genetics research in the second half of the 1940s
– many biologists, including geneticists, simply did not 'get' Avery's
result. Not only did they not immediately accept that genes were
made of DNA, they did not even attempt to test the hypothesis in
the systems they studied. For example, Max Delbrück first heard of
Avery's breakthrough in May 1943, when his Vanderbilt colleague
Roy Avery showed him the letter from New York that announced the
discovery. Delbrück later recalled his 'total shock and surprise' at the
contents of the letter, 'which I read there standing in his office in the
spring sunshine'.[39] But despite his 'shock', Delbrück did nothing. He
did not start studying the role of DNA in viruses, nor did any of his
colleagues, even though they were all intimately aware of the results
that were coming out of Avery's laboratory. Delbrück later explained
that the suggestion that genes were made of DNA left them nonplussed. As he put it 'you really did not know what to do with it'.[40]
With the easy wisdom of hindsight, this lack of interest in what led
to the most remarkable biological discovery of the twentieth century
looks remarkably short-sighted. And at one level, it was. The phage
group did not react in the way that Lederberg, Boivin and others did.
Their diffident attitude was one component of the failure of Avery's
discovery to immediately transform biology.[41] Avery's findings now
look so obvious, and yet many scientists at the time responded to
them with hostility or bemusement.

One of the scientists who did not immediately embrace Avery's
findings was the young Gunther Stent, who worked with Delbrück.
In 1972, Stent sought to explain the lack of widespread recognition of

the importance of the Avery group's discovery by suggesting that the result was 'premature'.[42] This term does not explain anything; in fact it obscures the historical reality of how Avery's work was received, and it does not explain why some scientists accepted the finding but others rejected it. There were two valid criticisms of Avery's suggestion that DNA was the hereditary material in the transforming principle, each of which gradually became weaker. First, there was the issue of potential protein contaminants, which led the Avery group to employ increasingly precise techniques, the results of which all indicated that protein contamination was not the cause of transformation. Second, there was the conundrum of how exactly specificity might be represented in what were supposed to be boring molecules – if DNA was essentially composed of four bases, a way needed to be discovered that enabled it to bring about the almost infinitely different effects produced by genes. The leading chemists of DNA such as Gulland were happy to imagine that specificity could reside in DNA through the sequence of bases, or their proportions, but this had yet to be demonstrated. Nevertheless, as time wore on, there were fewer reasons not to accept Avery's findings.

Some scientists had strong personal reasons to reject the DNA hypothesis. Mirsky's career was based on the study of nucleoproteins and he was clearly not going to give up his world view without a fight. Through his articles, his lectures and his interventions at conferences, Mirsky sowed doubt among the undecided. Similarly, Wendell Stanley turned a blind eye to the work of the Avery group, even though he too had been familiar with it before publication. In 1936, Stanley had crystallised the tobacco mosaic virus and announced that it was a protein; this was finally shown to be wrong in 1956 – the hereditary material in this virus is in fact RNA, and small amounts of RNA in his protein extract accounted for his results. In 1946, Stanley won the Nobel Prize in Physiology or Medicine for his mistaken claim; he later said that he 'was not impressed' by Avery's discovery – otherwise he would have immediately tested tobacco mosaic virus RNA for specificity. In 1970, he concluded, somewhat shamefacedly:

I have searched my memory and have failed to find any really extenuating circumstances for my failure to recognize the full significance of the discovery of transforming DNA.[43]

The diffident response of the main members of the phage group – Delbrück, Luria and Hershey – had a rather different source, and all three of them later explained their behaviour in the same way: they were interested in genetics, not chemistry, and so they simply did not realise the potential implications. Typically robust, Delbrück said:

> And even when people began to believe it might be DNA, that wasn't really so fundamentally a new story, because it just meant that genetic specificity was carried by some goddamn other macromolecule, instead of proteins.[44]

Luria recalled: 'I don't think we attached great importance to whether the gene was protein or nucleic acid. The important thing for us was that the gene had the characteristics that it had to have.'[45] In 1994, Hershey explained that their focus was simply elsewhere – 'as long as you're thinking about inheritance, who gives a damn what the substance is – it's irrelevant.'[46] Ironically, Hershey is now best known for his attempt to resolve the issue of whether proteins or DNA are the basis of heredity, an experiment that students are now taught settled the question once and for all, even though it did not.

*

Alfred ('Al') Hershey was a tall, skinny taciturn man with a tooth-brush moustache and bad teeth. Although he was renowned for working long into the night, he was not solely focused on science – he often took afternoon naps and in the summer he would disappear for weeks on end, sailing his yacht on Lake Michigan. Like everyone else in the phage group, Hershey had followed the discussions around the chemical nature of Avery's transforming principle. In May 1949, Hotchkiss sent Hershey an update on his progress in excluding any possible protein contamination from the DNA extracts of the transforming principle; after looking at the data, Hershey

wrote to the younger man: 'The experiments are very beautiful. …
My own feeling is that you have cleared up most of the doubts.'[47] But
like Luria and Delbrück, Hershey's initial interest in Avery's experiments was unfocused – the members of the phage group could not
see how chemistry could help them understand genetics.

Nevertheless, as phage researchers tried to understand how
viruses reproduced, the question of chemistry became increasingly
pressing. By 1949, electron microscope images had shown that a
viral infection begins with the virus sitting on the outside of a cell; in
ways that were unclear, the virus then took over the cell's metabolic
system and 'lost its identity' – no viruses could be detected inside
the cell for a period, while the viral structures that were still on the
outside of the bacterium lost their infective power. It was possible to
burst viruses by subjecting them to a sudden change in the concentration of their surrounding medium; all that remained were ghost
viruses – empty protein shells that would happily adhere to the outside of a bacterium but were not infectious. Researchers had begun
to use radioactive tracers to explore this phenomenon – by growing
phage and bacteria on radioactively labelled medium, radioactive
phosphorus was taken up by nucleic acids, and radioactive sulphur
could be used to mark proteins. It was therefore possible to track the
fates of the two components of the phage virus, namely DNA and
protein, by using radioactivity.

By late 1950, several phage researchers had begun to sketch out
a hypothesis about the roles of DNA and protein in virus replication, explicitly acknowledging that Avery was right. John Northrop
of Berkeley concluded one of his articles with an outline of this
thinking:

> The nucleic acid may be the essential, autocatalytic part of the
> molecule, as in the case of the transforming principle of the
> pneumococcus (Avery, MacLeod, and McCarty, 1944), and the
> protein portion may be necessary only to allow entrance to the
> host cell.[48]

Roger Herriott of Johns Hopkins University wrote to Hershey:

I've been thinking – and perhaps you have, too – that the virus may act like a little hypodermic needle full of transforming principles; that the virus as such never enters the cell; that only the tail contacts the host and perhaps enzymatically cuts a small hole through the outer membrane and then the nucleic acid of the virus flows into the cell.[49]

Thomas Anderson later recalled:

I remember in the summer of 1950 or 1951 hanging over the slide projector table with Hershey, and possibly Herriott, in Blackford Hall at the Cold Spring Harbor Laboratory, discussing the wildly comical possibility that only the viral DNA finds its way into the host cell, acting there like a transforming principle in altering the synthetic processes of the cell.[50]

It was in this context that Al Hershey, together with his new technician, Martha Chase, decided to settle the matter. Hershey had recently moved to Cold Spring Harbor and had equipped his laboratory with the latest radioisotope technology.[51] Chase, who was only 23 years old when she joined Hershey, had a round face and short hair; generally she was as reserved as her boss, but she was nonetheless prepared to complain loudly about her low pay.[52] Their experiments, which were published in the *Journal of General Physiology* in the middle of 1952, have since taken on an iconic quality.[53] They are reproduced in textbooks and are presented as a turning point, because they are now seen as showing that genes are made of DNA. The reality is rather different.

The Hershey and Chase paper describes several experiments in which they tried to identify the functions of protein and nucleic acid in bacteriophage reproduction. First, they confirmed and extended previous findings about the function and composition of ghost phage, which they showed were made of protein, were not infectious, could still attach to bacteria, and protected their DNA contents from enzyme attack. They next demonstrated that when the phage settled onto a bacterium, it injected DNA into the cell. All this supported Herriott's hypodermic needle hypothesis, but there was no proof of what the DNA actually did, nor could they be certain that no protein entered the bacterial cell.

The final experiments are those most often taught to students today, but they are usually described inaccurately. They all involved the use of a Waring Blender, or Blendor as the Waring company trademark had it. This device was employed to agitate the viruses and their bacterial hosts, and the experiments that used it are now often known as the Blender experiments. This apparatus is often called a kitchen blender, which conjures up some kind of retro 1950s domestic device, all chrome and glass. Sadly this was not the case. Although the Waring company did make kitchen blenders, the apparatus used by Hershey and Chase was a highly specialised, unstylish bronze-coloured piece of laboratory equipment about 25 cm high that could run at speeds of up to 10,000 r.p.m. – much faster than anything you would have in your kitchen. It was not simply a centrifuge, it also produced what Hershey and Chase described as 'violent agitation', which they used to shake the protein-rich viral ghosts from the outside of the host cell. Using radioactive sulphur, they showed that they could remove up to 82 per cent of the phage protein from their preparations by separating out the ghost phage; a similar experiment with radioactive phosphorus showed that up to 85 per cent of the virus DNA was transferred into the bacterial cell.

Students are now generally taught that these experiments provided the evidence that DNA is the genetic material, but in fact they did no such thing, nor did Hershey and Chase claim that they did. The problem faced by Hershey and Chase was similar to that encountered by Avery and his colleagues, but in spades. Hotchkiss had reduced the protein component in his version of Avery's experiment to effectively zero (at most 0.02 per cent), and still people did not accept his findings; in Hershey and Chase's extracts around 20 per cent of the protein was still floating around. It was quite possible that some of this protein played a role in the reproduction of the virus. Furthermore, as Hershey and Chase put it, none of the experiments proved anything more than that DNA had 'some function' in viral reproduction. The paper concluded with Hershey's typical terseness, beginning with the question of 'adsorption', or how the virus sticks to the outside of the bacteria:

The sulfur-containing protein of resting phage particles is confined to a protective coat that is responsible for the adsorption to bacteria, and functions as an instrument for the injection of the phage DNA into the cell. This protein probably has no function in the growth of intracellular phage. The DNA has some function. Further chemical inferences should not be drawn from the experiments presented.[54]

Hershey remained troubled by his findings and later admitted, 'I wasn't too impressed by the results myself'.[55] When he first presented his experiments in a small laboratory seminar at Cold Spring Harbor, he expressed his surprise that protein apparently had no function inside the infected cell. And when he made his first public presentation of the results, at the June 1953 Cold Spring Harbor meeting, speaking after the double helix structure of DNA had been described, Hershey was still sure that DNA could not be the sole carrier of hereditary specificity. He addressed this issue head-on by summarising the evidence from Avery, Boivin, Taylor and himself as follows:

1. The amount of DNA in chromosomes is consistent in a species, not in a given kind of tissue in different species.
2. DNA can transform bacteria.
3. DNA plays some unidentified role in one kind of viral infection.

Hershey then told his audience that this evidence was not enough to support the conclusion that DNA was the hereditary material. He remained convinced that proteins must play a role:

None of these, nor all together, forms a sufficient basis for scientific judgement concerning the genetic function of DNA. The evidence for this statement is that biologists (all of whom, being human, have an opinion) are about equally divided pro and con. My own guess is that DNA will not prove to be a unique determiner of genetic specificity, but that contributions to the question will be made in the future only by persons willing to entertain the contrary view.[56]

Hershey's caution shows us the rigorous nature of his scientific thinking – strictly speaking, his interpretation was absolutely correct; he would go no further than the data allowed. It also shows the continued uncertainty about the possibility of contamination – this was an important problem in the Hershey and Chase experiment, although nobody pointed it out at the time.

Hershey later argued that the complex route from Avery's 1944 discovery to the widespread acceptance that genes were made of DNA 'shows that some redundancy of evidence was needed to be convincing and that diversity of experimental materials was often crucial to discovery'.[57] Although that is undoubtedly true, it is also the case that while some people immediately embraced Avery's discovery, others – including the phage group – were reluctant to recognise its significance. For a decade, scientists spent their time arguing over something that now seems blindingly obvious. There are many such moments in the history of science, and they can only be understood in terms of the evidence and attitudes of the time. In this case, the predominant problem was the power of the old ideas about the dominant role of protein, and the difficulty of imagining how in reality – not in theory – DNA could produce specificity.

*

While the chemists and the microbiologists battled it out over the chemical nature of the gene, there were several bold attempts to look at gene function – how genes do what they do. In 1947, Kurt Stern, a 43-year-old German biochemist who had left Nazi Germany for the US, published a highly speculative article on what he called the gene code – one of the first uses of the word code since the publication of Schrödinger's book three years earlier. In a prescient guess, Stern argued that the chemical basis of genes might take the form of helical coils – he assumed that genes were made of nucleoproteins, although he recognised that Avery could be right and DNA alone might be the genetic material.

Like Chargaff, Stern suggested that variations in the sequence of the bases in the DNA molecule could lie at the heart of gene specificity. According to Stern's theory, genes were physical modulations, much like a groove on a vinyl record. The role of the nucleoprotein, Stern argued, was to fix the DNA molecule in a particular shape; removal of the protein would return the nucleic acid to its unmodulated form. To prove his point, Stern provided photographs of physical models that he had made, showing DNA and an associated protein spiralling around each other in a double helix, although he did not use the term. Despite the rich biochemical and structural data that were at the heart of Stern's work, his model was too hypothetical to generate experiments, and his ingenious views had no influence.[58] Although Stern used the term code, he did not embrace the idea that the code might abstractly represent protein structure; his vision was that of a template – genes were the physical form on which proteins were synthesised.

At the same time, André Boivin and Roger Vendrely came up with a hypothesis about the relation between the two kinds of nucleic acid found in a cell and the enzymes that were thought to be the product of genes, building on Torbjörn Caspersson and Jean Brachet's investigations into the role of RNA in protein synthesis.[59] Boivin and Vendrely's idea was pithily expressed by the editor of the journal, *Experientia*, in an English-language summary:

> the macromolecular desoxyribonucleic acids govern the building of macro-molecular ribonucleic acids, and, in turn the production of cytoplasmic enzymes.[60]

In other words, DNA led to the production of RNA, which then led to protein synthesis and the production of enzymes. This hypothesis, which was correct, was another example of Boivin's perception.

In 1950, two Oxford chemists, P. C. Caldwell and Sir Cyril Hinshelwood, proposed a physical model to explain how protein synthesis takes place on a nucleic acid.[61] If the synthesis of an amino acid depended on the presence of pairs of the five components in a DNA molecule (the four bases, plus the phosphate backbone), they argued, 'twenty-five different internucleotide arrangements are possible'. This was enough to produce the twenty different kinds of amino acid that are found in organisms. Caldwell and Hinshelwood accepted that DNA was 'the principal seat of the unchanging hereditary characters', but this apparent enthusiasm hid doubts about the importance of genetics: Hinshelwood was a believer in the inheritance of acquired characteristics, a mode of heredity and evolution proposed by the French naturalist Lamarck in the early decades of the nineteenth century and since abandoned. Above all, like Stern, Caldwell and Hinshelwood were thinking in terms of a physical relation between the protein and the nucleic acid upon which it was synthesised. The gene acted as a template, they thought.

In 1952, Alexander Dounce, a biochemistry professor at Rochester Medical School, had a similar starting point: there must be some structural relation between the bases in the nucleic acid of the gene and a corresponding amino acid in a protein. Dounce argued that the nucleic acid was a physical template for protein synthesis. His model proposed that each amino acid was determined by a set of three bases; this is indeed the case, but not for the reasons that Dounce suggested.[62] Like Boivin, Dounce took the evidence from Brachet and Caspersson about the role of RNA and argued that protein synthesis occurred on an RNA molecule, not on DNA. Dounce described this idea in a way that is now taught to students all over the world every day: 'deoxyribonucleic acid (DNA) → ribonucleic acid (RNA) → protein'.[63] Dounce was the first person to spell out this series of links, following the work of Boivin and Vendrely five years earlier. Their names are now forgotten to all except a handful of historians and scientists.

In retrospect, what is most remarkable about these models of gene function is that they were strictly physical in nature. They were all based on three-dimensional template structures rather than the kind of abstract one-dimensional 'code-script' that Schrödinger had suggested. They were all analogue, rather than digital. For this to change, a new set of ideas, different ways of thinking, had to enter biology. That was just around the corner. The developments in mathematical thinking about information, codes and control that had occurred during the war were being popularised at the very same time as the biological community was coming to terms with the implications of Avery's discovery.

FIVE

THE AGE OF CONTROL

In his 1988 best-seller *A Brief History of Time*, the physicist Stephen Hawking recounts that his editor told him that every equation he used would halve the potential readership. Hawking obligingly included just one equation ($e = mc^2$) and the book went on to sell more than 10 million copies. Things were clearly different back in the 1940s – Norbert Wiener's 1948 popular science book *Cybernetics or Control and Communication in the Animal and the Machine* was stuffed full of hundreds of complicated equations, and yet it became a publishing sensation around the world.

With its weird title and its promise of a new theory of nearly everything, *Cybernetics* took the bookstores by storm. The first edition was sold out in six weeks and the book went through three printings in six months, jostling for top place in the best-seller lists.[1] According to a critic in the *New York Times*, *Cybernetics* was one of the 'seminal books … comparable in ultimate importance to … Galileo or Malthus or Rousseau or Mill.'[2] Wiener's book popularised the research on control systems and negative feedback that had taken place during the war, spreading this new approach throughout the scientific community – and especially into biology. It helped invent a new vocabulary of information that transformed the postwar world and shaped a radical new view of genetics. We are still living in that world.

*

Cybernetics was born in Paris. In 1947, Wiener was invited to a conference in Nancy, in eastern France. On his visit to Europe he met thinkers in England, such as Alan Turing, J. B. S. Haldane and the University of Manchester computer pioneers, 'Freddie' Williams and Tom Kilburn. During a visit to Paris, Wiener went to 'a drab little bookshop opposite the Sorbonne', where he met Enriques Freymann, the Mexican-born head of a French publishing firm.[3] Fascinated by Wiener's ideas, Freymann suggested that the American should write a book to explain them to the general public. Wiener was taken with the proposal and the deal was sealed over a *chocolat chaud*. Wiener's first task was to find an overall description of the new vision of control and information that he had been working on. He eventually came up with a new word – 'cybernetics', from the Greek word for 'steersman'.[4] All of the 'cyber' words we now use come from Wiener's coinage.

Cybernetics was published simultaneously on both sides of Atlantic at the end of October 1948. It was full of equations, which were riddled with mistakes due to a mixture of Wiener's haste and the hazards of transatlantic proof-reading. But because the maths was so complicated nobody really noticed – few readers were able to follow what *Science* magazine called the 'troublesome mathematical portions', far less spot the flaws. The algebra was in fact pretty irrelevant. As a writer in the *New York Times* explained, the success of *Cybernetics* was explained by the fact that between the equations there were 'pages of sparkling, literate and provocative prose.'[5]* Wiener put forward a theory of the role of information and control in behaviour, underlining the similarities between animals and machines, and sketched out a vision of a cybernetic future in which automation would dominate: 'the present time is the age of communication and control', he wrote.

---

*Not all the prose is sparkling. The conclusion to chapter 2 reads: 'I do not wish to close this chapter without indicating that ergodic theory is a considerably wider subject than we have indicated above. There are certain modern developments of ergodic theory in which the measure to be kept invariant under a set of transformations is defined directly by the set itself rather than assumed in advance. I refer especially to the work of Kryloff and Bogoliouboff, and to some of the work of Hurewicz and the Japanese school.'

In a postwar world marked by anxieties about the destructive power of technology, Wiener's vision was simultaneously apocalyptic and humane. He described the 'unbounded possibilities for good and for evil' contained in the increased automation of production. This was bold thinking, given that factory automation was in its infancy, and computers were only just able to store a brief program in their valve-based memories. Nevertheless, Wiener foresaw the overall tendency that would characterise production in the second half of the century. He predicted a new industrial revolution that would devalue the 'simpler and more routine' aspects of the human brain, as machines began to perform menial mental tasks. The only way to limit the destructive aspects of this development, he argued, was to create 'a society based on human values other than buying and selling.' Ultimately, Wiener was gloomy about the future. After describing the creation of a new science, he felt powerless and pessimistic: 'We can only hand it over into the world that exists about us, and this is the world of Belsen and Hiroshima.'[6]

At the heart of *Cybernetics* was Wiener's view that all control systems, and the negative feedback they embodied, were based on information flows. Information, he argued, was at the heart of all systems – mechanical, electronic or organic – and this was closely related to the physicists' concept of entropy. Five years earlier, Schrödinger had argued that life was 'negative entropy', because of its ability to temporarily resist the second law of thermodynamics. Now Wiener was extending that concept to information as a whole:

> The quantity we here define as amount of information is the negative of the quantity usually defined as entropy in similar situations.

As he explained: 'Just as the amount of information in a system is a measure of its degree of organization, so the entropy of a system is a measure of its degree of disorganization; and one is simply the negative of the other.' Information, argued Wiener, was 'negative entropy'. According to this view, life and information were intimately connected. There was therefore a continuum between the most ordered state of matter – living beings – and inanimate forms of organised matter, a continuum that could be viewed in terms of a new quality: information. As Wiener stated emphatically: 'Information is information, not matter or energy. No materialism which does not admit this can survive at the present day.'[7] While this is true, information requires a material substrate and, as Szilárd pointed out in his solution to Maxwell's Demon, energy has to be expended in order to obtain or create information.

Cybernetics was not just about information, it was fundamentally based on the concept of negative feedback that Wiener had first explored in his anti-aircraft work during the war.[8] Negative feedback had been known in antiquity and during the golden age of Islam, in the form of liquid-level regulators (like a cistern in a toilet), and in the eighteenth century the phenomenon had been used by the steam engineer James Watt, who developed a 'governor' to prevent his machines from running out of control. Although Watt's invention led to new forms of everyday language – it was the origin of terms such as 'self-regulation' and 'checks and balances' – it was not generalised into something that operated in all systems. Wiener's vision of the importance of negative feedback, and his ability to spin together threads from behaviour, physiology, sociology, electronics and automation, showed the general public how an extraordinary variety of natural and mechanical phenomena could be interpreted using the same framework. It made for a heady mixture.

The influence of *Cybernetics* extended right across the academic spectrum, just as Wiener had hoped. There were summaries in the *New York Times* and *Scientific American* and positive reviews in a wide range of academic journals, from *American Anthropologist* to *Psychiatric Quarterly*.[9] According to the *Annals of the American Academy of Political and Social Sciences* it was 'a fascinating book that is at present being read with more than just novelty or academic interest by scientists of all disciplines and philosophical camps.'[10] The US journal *Science* hailed the birth of 'a new discipline', while in *American Scientist* the French physicist Léon Brillouin described the link between information and 'negative entropy' as 'an entirely new field for investigation and a most revolutionary idea', before eventually writing his own book on the topic.[11] *Cybernetics* was read widely and was enthusiastically discussed – at the 1949 Summer School held at the world-famous Woods Hole Marine Biological Laboratory, it was a topic of debate among the students and young scientists.[12]

Cybernetics had a similar impact in France, where it attracted wide attention despite having been published in English. In December 1948 a long article appeared in *Le Monde* by the French Dominican friar and professor of philosophy Dominique Dubarle. Dubarle praised Wiener's 'extraordinary' book, which he claimed announced the birth of 'a new science'.[13] The world was being reshaped, but Wiener remained bemused by the success:

> Freymann had not rated the commercial prospects of *Cybernetics* very highly – nor, as a matter of fact, had anybody on either side of the ocean. When it became a scientific best-seller we were all astonished, not least myself.[14]

*

Wiener was not the sole creator of this revolution in thinking, nor did he claim to be. In the Introduction to *Cybernetics*, he described how this new vision of control had been developed through years of discussions with his intellectual partners, including Claude Shannon. Wiener generously explained that something like his 'statistical theory of the amount of information' had been simultaneously arrived at by Shannon in the US, by R. A. Fisher in the UK and by Andrei Kolmogoroff in the USSR. Fisher's approach to information was in fact quite different, and the works of Kolmogoroff were unobtainable, but Shannon's views were virtually identical to Wiener's, as readers with the necessary mathematical ability were soon able to appreciate.

In 1948, Shannon published two dense theoretical articles in the pages of the *Bell System Technical Journal*. These were then used as the basis of a 1949 book, *The Theory of Communication*, which also contained an explanatory essay by the Rockefeller Foundation administrator and one-time head of Section D-2, Warren Weaver. Although *The Theory of Communication* soon became well known in the scientific community, it did not have anywhere near the public success of *Cybernetics*. Shannon's arid stretches of equations were not punctuated by anything like Wiener's stylish and thought-provoking prose, and all except the most committed reader would probably have tripped up over the opening sentence: 'The recent development of various methods of modulation such as PCM and PPM which exchange bandwidth for signal-to-noise ratio has intensified the interest in a general theory of communication.'[15] In contrast, Wiener's *Cybernetics* opened with a German folk song.

Shannon was interested primarily in communication, not control. His model was highly abstract, dealing with information without reference to the content of that information. As he stated at the outset:

> The word *information*, in this theory, is used in a special sense that must not be confused with its ordinary usage. In particular, *information* must not be confused with meaning.[16]

Unlike Wiener's vision, Shannon's approach had no place for feedback – this was a linear transmission system with no connection between the reception of the message and the source. It was not a model that could explain the flow of control in a behavioural or mechanical system; it did no more than what it claimed – it explained how a single message can pass from a transmitter to a receiver despite the presence of noise in a system. For Shannon, the critical process was encoding and decoding: 'the function of the transmitter is to *encode*, and that of the receiver to *decode*, the message'.[17]

Shannon showed that information could be conceptualised mathematically, as a measure of freedom of choice between all possible messages. The simplest such choice occurs when there are two options, as in a binary system. 'This unit of information,' wrote Shannon, 'is called a "bit," this word, first suggested by John W. Tukey, being a condensation of "binary digit."' Shannon's formal mathematical description of information was essentially identical to that of Wiener, except that the two mathematicians approached the question from different sides: whereas Wiener saw information as negative entropy, for Shannon information was the same as entropy.[18]

The two men were well aware of this difference; in October 1948 Shannon wrote to Wiener:

I consider how much information is produced when a choice is made from a set – the larger the set the more information. You consider the larger uncertainty in the case of a larger set to mean less knowledge and hence less information. The difference in viewpoint is partially a mathematical pun. We would obtain the same numerical answers in any particular question.[19]

The explanation for these apparently opposite approaches, and the source of what Shannon called a pun was that whereas he was strictly focused on the relatively narrow issue of communication, Wiener wanted to create a grander, broader theory, which integrated communication, control, organisation and the nature of life itself. Wiener's model therefore had to include meaning, something that Shannon opposed. This contrast between the two formulations was therefore far more than a mathematical pun – it reflected differences in the two men's ambitions.

*

Wiener and Shannon's ideas had an important influence on the scientific community. Information came to be seen as a characteristic of matter that could be quantified, preferably in terms of binary coding, while control and negative feedback seemed to be fundamental features of organic and engineered systems. One of the ways

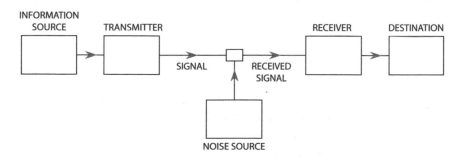

1. Claude Shannon's model of communication.
From Shannon and Weaver (1949).

in which these ideas came to influence biologists was through the promise of creating automata and thereby testing models of how organisms function and reproduce.[20] These links were explored at a symposium on Cerebral Mechanisms in Behavior that was held at the California Institute of Technology (Caltech) a month before the publication of *Cybernetics*, at the end of September 1948. The symposium was a small affair – only fourteen speakers, with a further five participants, one of whom was the Caltech chemist Linus Pauling.

Von Neumann gave the opening talk, entitled 'General and logical theory of automata', and explored one of the defining features of life: its ability to reproduce. Von Neumann's starting point was Alan Turing's prewar theory of a universal machine that carried out its operations by reading and writing on a paper tape. But this was too simple for von Neumann: he wanted to imagine 'an automaton whose output is other automata'.

Von Neumann argued that such a machine needed instructions to construct its component parts, and that these instructions would be 'roughly effecting the functions of a gene'; a change in the instruction would be like a mutation. Von Neumann explained that a real gene 'probably does not contain a complete description of the object whose construction its presence stimulates. It probably contains only general pointers, general cues.' In contrast, the 'fundamental act of reproduction, the duplication of genetic material' could be conceptualised in terms of the copying of a paper tape – von Neumann was implicitly arguing that a gene was like one of Turing's instruction-containing tapes.[21] Although von Neumann did not make the link, this was a computer version of Schrödinger's aperiodic crystal that contained a code-script.

Scientists who were not at the meeting were also becoming interested in the links between cybernetics and genetics. Wiener was in contact with the British geneticist J. B. S. Haldane, who had been following Wiener's work with close attention. In mid-November 1948, Haldane wrote to Wiener as he struggled to apply the new concepts to his field:

I am gradually learning to think in terms of messages and

noise. … I suspect that a large amount of an animal or plant is redundant because it has to take some trouble to get accurately reproduced, and there is a lot of noise around. A mutation seems to be a bit of noise which gets incorporated into a message. If I could see heredity in terms of message and noise I could get somewhere.[22]

Similar ideas were being developed by the University of Illinois physicist Sydney Dancoff, together with his colleague, the Austrian-born radiologist Henry Quastler. In July 1950, Dancoff wrote a letter to Quastler summarising the two men's thinking on the links between genetics and information. The chromosome, wrote Dancoff, could be seen as a 'linear coded tape of instructions.' He went on:

The entire thread constitutes a 'message'. This message can be broken down into sub-units which may be called 'paragraphs,' 'words,' etc. The smallest message unit is perhaps some flip-flop which can make a yes-no decision. If the result of this yes-no decision is evident in the grown organism, we can call this smallest message unit a 'gene'.[23]

Dancoff and Quastler were forcing genetics into a binary mode, viewing organisms as an example of the automata imagined by Turing and von Neumann.

In Europe, too, the scientific community discussed the new ideas about information and control, with each country taking a slightly different approach to the question. In September 1950, the Royal Society of London organised a three-day conference on 'information theory' – this was not a term used by Shannon or Wiener, although it has since become widespread. Around 130 attendees crowded into the small lecture theatre in Burlington House on Piccadilly to discuss the mathematical and electronic aspects of the field. Shannon gave three talks at the meeting, but there was little exploration of Wiener's cybernetic approach. There were no geneticists to be seen. The only person to speak about genes was Alan Turing, and he was more interested in how natural selection alters the shape of organisms than in thinking about how heredity works.[24]

The degree to which information had inserted itself into scientific thinking and ordinary language was revealed at the end of 1950, when the British zoologist J. Z. Young gave the prestigious Reith Lectures on the BBC, under the title 'Doubt and certainty in science'. The first three radio lectures were about how biologists study brain function, and they teemed with the word 'information'.[25] Listeners were presented with this new vision as though it were the only way of understanding how nervous systems work. What it really showed was that the way that biologists thought about life had been transformed.*

In France the Nobel Prize-winning physicist Louis de Broglie gave a lecture series in spring 1950 under the general title 'Cybernetics', and in 1951 a congress was held in Paris, funded by the Rockefeller Foundation, that was attended by over 300 people including Wiener and McCulloch. After the congress was over, Wiener remained in Paris to give several talks on the subject at the prestigious Collège de France. There were also articles in French journals such as *Esprit* and the *Nouvelle Revue Française* while, for the general public, science journalist Pierre de Latil wrote a lively book called *La Pensée Artificielle*, which explained the nature and genesis of cybernetics, full of useful diagrams and photos, but focusing on feedback and showing how French engineers had come up with the concept in the fifteenth and nineteenth centuries.[26] De Latil's book embodied the contrast between the French and British approaches to cybernetics: whereas the British focused on information, the French emphasised the control and robotics aspects of the subject. Strikingly, de Latil's book did not refer to Shannon at all. In the UK and the US 'information' was widely discussed in popular science magazines such as *The Times Science Review* and *Scientific American*, both as an abstract concept and in Shannon's mathematised version.[27]

Whatever the contrasts between different subject areas and different countries, in Britain, America and France, everyone in science knew that a conceptual revolution was taking place. Not everyone

---

*Intriguingly, as early as the 1930s, the pioneering Cambridge physiologist Edgar Adrian was using the terms 'information' and 'code' to describe the activity of neurons – Garson, J. 'The birth of information in the brain: Edgar Adrian and the vacuum tube', *Science in Context*, vol. 27, 2015, pp. 31–52.

was impressed, however. In 1948 Max Delbrück was invited to one of the cybernetics conferences. It was the only such meeting he attended. Not a man to mince his words, Delbrück later recalled that he found the discussion 'too diffuse for my taste. It was vacuous in the extreme and positively inane.'[28]

*

In 1950, public interest in cybernetics was cranked up even further when Wiener published a second book – this time without a single equation – in which he outlined the potential changes that society would have to face as a result of increased automation. With the unwieldy title *On the Human Use of Human Beings*, Wiener's new book explained how society should respond to the looming cultural and economic developments that would follow the introduction of automation and the development of computers in the second half of the twentieth century. It spanned a wide range of culture in a free-wheeling style, dealing with language, law and individuality, exploring their changing meaning in the context of machines that could apparently embody aspects of purposeful behaviour. Wiener was concerned that top-down social control was becoming typical of all economic and political systems across the planet. As he explained: 'I wish to devote this book to a protest against this inhuman use of human beings.'[29] Professor Cyril Joad, the BBC's favourite philosopher, hated the book. His review in the *Times Literary Supplement* was scathing, criticising Wiener's 'incoherent and slovenly language' and branding the book as 'highly dangerous'.[30]

Despite Joad's fears, Wiener's remarkable book had a lasting influence because it showed that everything – including humans – could ultimately be reduced to mere patterns of information. Wiener again emphasised the links between the latest technological developments and the way in which organisms behave and function: 'It is my thesis that the operation of the living individual and the operation of some of the newer communication machines are precisely parallel.'[31] If humans were essentially machines in both form and function, then it should be possible to define a human being in terms of the information they contained, by calculating 'the amount of hereditary information,' he claimed. If that information could be represented in some way, then it would even be possible to transmit it by electronic means, preserving the identity of the individual, argued Wiener.[32] Although he realised that such a procedure would remain in the realms of science fiction for the foreseeable future, Wiener had made his point: a human being was fundamentally no different from any other form of organised matter. In the end, it was information.

Wiener was not the only person to be thinking along these lines. In July 1949, Shannon sketched a list of different items and their 'storage capacities'. He considered that a 'phono record' contained about 300,000 bits of information, one hour of broadcast TV contained 10^{11} bits, and the 'genetic constitution of man' was a mere 80,000 bits.[33] Nothing became of these wildly inaccurate guesstimates – the sketch remained in the Shannon archives until it was recently discovered by James Gleick – but they show how the concept of information could be applied to virtually anything. In May 1952, J. B. S. Haldane wrote a letter to Wiener in which he announced that he had 'worked out the total amount of control (= information = instruction) in a fertilized egg'.[34] It is not known what number Haldane came up with, nor on what basis he made his calculation – he never published his answer to this conundrum.

Henry Quastler was bolder. In March 1952 he organised a symposium on Information Theory in Biology, which was held in his Control Systems Laboratory at the University of Illinois. The rising star of bacterial genetics, Joshua Lederberg, had been invited to attend, but he was wary because the meeting was funded by the US Office of Naval Research. Lederberg was concerned that the discussions were to be recorded and might involve matters that could be the subject of a future security classification.[35] Lederberg was not being paranoid – the McCarthyite witch-hunts were getting into their stride, leading to US academics having to swear 'loyalty oaths' or risk losing their jobs; one ill-judged comment could lead to disaster.

The speakers at the symposium showed how scientists were trying to apply the new concept of information to biology. One participant discussed Linus Pauling's models of the molecular structure of the protein keratin in terms of the information they contained, while another explored the information content of a zygote, which he argued was 'a set of instructions coded into the fertilized egg as dictated by the genetic constitution'. There were even two attempts to calculate the information contained in organisms. Henry Linschitz used molecular and energetic calculations to conclude that 'the information content of a bacterial cell' was around 10^{13} bits.[36] In a joint paper that was completed after Dancoff's untimely death, Quastler outlined what he accepted were 'crude approximations and vague hypotheses' and then calculated that a human genome contains at most 10^{10} bits of information. This calculation took as its starting point the view that each gene and its different versions, or alleles, is 'an independent source of information, with an entropy which depends on the number of allelic states, or different messages'. Quastler admitted that he knew 'neither the number of genes nor the average number of allelic states', both of which would appear to be essential for such a calculation to be valid. Unperturbed, Quastler concluded: 'this is an extremely coarse estimate, but it is better than no estimate at all.'[37] Not everyone agreed. In 1965, after the genetic code had been cracked but before it had been completely deciphered, Michael Apter and Lewis Wolpert returned to Dancoff and Quastler's figures and dismissed them as 'so arbitrary as to make the values obtained meaningless'. Their conclusion about the pointlessness of Dancoff and Quastler's calculation was cutting:

> We believe that, on the contrary, they are not better than no estimate at all, since such estimates are liable to be misleading and to breed a false confidence.[38]

Applying the new concepts of information to genetics was proving to be more difficult than many people expected.

*

It seemed that cybernetics and communication theory were going to sweep away everything in their path, changing the whole of science. But then some of its principal proponents were thrown into turmoil by personal and political events. The backdrop to the developments in cybernetics, and indeed the source of much of its funding, was the Cold War. In February 1949, the US lost its monopoly on nuclear weapons when the USSR exploded its first atom bomb. In 1950, the Cold War began to heat up as the Korean War broke out and the US fought a proxy war against the Russians and the Chinese. Shocked by these developments, the anti-communist John von Neumann pressed the US government to focus all its research effort on building a hydrogen bomb. Thanks in part to his lobbying, a major development programme began in which he was heavily involved, leaving little time for his other interests. The project culminated in the explosion of the first H-bomb in November 1952, with a yield that was nearly 1,000 times more destructive than that of Hiroshima. Nine months later, the USSR exploded its own thermonuclear device, and the arms race began in earnest. Von Neumann abandoned his interest in creating self-reproducing automata and spent much of the rest of his life until his death in 1957 working on intercontinental ballistic missiles, applying his mathematical genius to the potential destruction of the human race.[39]

In 1951 Wiener's collaboration with Pitts and McCulloch, which had been at the heart of the development of cybernetics for nearly a decade, came to an abrupt end as he broke off all contact with the two younger men. Wiener had been severely irritated by a trivial but jocular letter sent by Pitts and McCulloch, but the source of the crisis appears to have been his wife's malicious and entirely untrue suggestion that Pitts and McCulloch had seduced Wiener's 19-year-old daughter Barbara. The cybernetics group was severely weakened and never recovered from the blow.[40]

Other thinkers were taking up the banner of cybernetics. In 1950 Hans Kalmus, a geneticist based at McGill University in Montreal who worked closely with J. B. S. Haldane, published a brief article in *The Journal of Heredity* entitled 'A cybernetical aspect of genetics'. Kalmus had read *Cybernetics* and had been struck by what he called 'certain unifying principles' that shed an interesting light on heredity. Kalmus suggested that a gene 'is a message, which can survive the death of the individual and can thus be received repeatedly by several organisms of different generations.' Kalmus even claimed there was a parallel between what he called the racial memory of genes and the recently developed ability of computers to store information.[41]

Kalmus went on to argue that there was no contradiction between the widely held enzymatic view of gene function and the cybernetic vision he was putting forward. Even more ambitiously, Kalmus tried to show how cybernetic concepts of feedback could explain interactions between genes at the genomic and populational levels, and interactions between genes and environmental factors – climate, other organisms and so on. However, Kalmus had little to say by way of detail and his ideas had no discernible influence. The first person to cite his article was ... H. Kalmus, in 1962.[42]

As cybernetics and information theory became fashionable in fields far removed from automata and electronic communications, it became a target for ridicule. One example was a satirical letter to *Nature* that was cooked up over a well-oiled lunch in the Italian Alps, in September 1952.[43] Boris Ephrussi and Jim Watson were dining with one of Ephrussi's colleagues, Urs Leopold, and they decided to write a brief spoof letter, taking the mickey out of a recent review by Joshua Lederberg in which Lederberg had rather pompously suggested that several well-established terms in bacterial genetics should be replaced by fancy new words that he had invented.[44]

Ephrussi and Watson's 'joke' consisted of satirically suggesting that words such as transformation, induction, transduction and infection, all of which had recently come into currency in bacterial genetics, should be replaced by the single term 'inter-bacterial information'. Several points could have alerted the reader that this was not intended to be taken seriously. The suggested change did not make sense – 'information' could not be a grammatical substitute for 'transformation'. Furthermore, the brief letter closed by reassuring the reader that their preferred term did 'not imply necessarily the transfer of material substances' – but the only alternative to a material transfer would be something like a radio broadcast between bacteria. The final phrase was equally facetious, as it highlighted 'the possible future importance of cybernetics at the bacterial level', conjuring an apparently surreal comparison of single-celled organisms with the most complex machines on Earth at the time.[45]

The editors of *Nature* did not get the joke – to be fair, it was

Terminology in Bacterial Genetics

THE increasing complexity of bacterial genetics is illustrated by several recent letters in *Nature*[1]. What seems to us a rather chaotic growth in technical vocabulary has followed these experimental developments. This may result not infrequently in prolix and cavil publications, and important investigations may thus become unintelligible to the non-specialist. For example, the terms bacterial 'transformation', 'induction' and 'transduction' have all been used for describing aspects of a single phenomenon, namely, 'sexual recombination' in bacteria[2]. (Even the word 'infection' has found its way into reviews on this subject.) As a solution to this confusing situation, we would like to suggest the use of the term 'interbacterial information' to replace those above. It does not imply necessarily the transfer of material substances, and recognizes the possible future importance of cybernetics at the bacterial level.

BORIS EPHRUSSI

Laboratoire de Génétique,
Université de Paris.

URS LEOPOLD

Zurich.

J. D. WATSON

Clare College,
Cambridge.

J. J. WEIGLE

Institut de Physique,
Université de Genève.

[1] Lederberg, J., and Tatum, E. L., *Nature*, **158**, 558 (1946). Cavalli, L. L., and Heslot, H., *Nature*, **164**, 1058 (1949). Hayes, W., *Nature*, **169**, 118 (1952).
[2] Lindegren, C. C., *Zlb. Bakt.*, Abt. II, **92**, 40 (1935).

2. Spoof letter by Ephrussi and others to *Nature*, April 1953

not very funny – and they published the letter in the 18 April 1953 issue of the journal. At the time the letter met its deserved fate and disappeared into oblivion, apart from a couple of ironic citations from sharp-eyed bacterial geneticists who picked up on the attempt at humour.[46] Recently, some historians have taken this apparently brilliant insight seriously, hypnotised by the sudden appearance of the words 'information' and 'cybernetics' in a letter signed by Jim Watson. The main historian who has studied this period, the late Lily Kay, argued earnestly that the letter represented a 'gestalt switch' in the thinking of the scientific community.[47] Despite not 'getting it', Kay was absolutely right: the very fact that Ephrussi was poking fun indicated that concepts and words had changed. For Ephrussi and his boozy pals, information and cybernetics were now so commonplace that they could be used in an ironic spoof. The real irony was that this squib of a joke appeared in *Nature* just one week before the three articles that described the structure of DNA, and seven weeks before Watson and Crick changed our view of life by allying that structure with the term 'genetical information', this time used in a deadly serious fashion.

SIX

THE DOUBLE HELIX

On 6 August 1945, when the atomic bomb destroyed Hiroshima, Maurice Wilkins was a 28-year-old British physicist working on the Manhattan Project. Like many of his colleagues, Wilkins had begun to have doubts about the morality of building the bomb as soon as Germany surrendered in 1945. The use of the bomb against Japan was the final straw. With his recent marriage in tatters and his love of physics poisoned by the horror of Hiroshima and Nagasaki, Wilkins returned to the UK.

Schrödinger's *What is Life?* inspired Wilkins to use physics to investigate biology, so he approached his PhD supervisor, John Randall, who suggested that he should trace how the amount of DNA doubled just before a cell divided. The two men, who worked at the new Medical Research Council biophysics unit that Randall had set up at King's College, London, knew of Avery's work and felt that DNA was at the very least a vital component of nucleoproteins, if not the sole genetic material. In 1947 Wilkins met Francis Crick, whose physics PhD had been interrupted by the war. Crick had also read *What is Life?* and had also turned to biophysics. The two men became close friends, even though they had very different personalities – Crick was a noisy, brilliant magpie, with an eye for shiny new ideas, whereas Wilkins was quiet and reserved, with an odd habit of turning away from the person he was speaking to. He was also prone to suicidal thoughts and was in psychoanalysis.[1]

The friendship between Wilkins and Crick led to what is probably the most intensely studied moment in the history of twentieth century science: the discovery of the double helix structure of DNA. The events surrounding this event have been described in memoirs, biographies, exhibitions, TV programmes, countless academic articles, many inaccurate blog posts and even in a video rap contest.[2]* The story is of fundamental scientific importance and shows how science is an intensely human, collaborative and competitive enterprise in which luck, ambition and personality can play a central role. Above all, it was the advance that revealed the existence of the genetic code.

*

The King's College biophysics unit was initially a minor player in the small world of groups that were studying the structure of DNA. The most influential work was being carried out in Columbia University by Erwin Chargaff, who was making a detailed biochemical analysis of the relative proportions of the four DNA bases – adenine, cytosine, guanine and thymine. Between 1948 and 1951, Chargaff showed that the four bases were not present in equal amounts; his conclusion was confirmed by Avery's arch-critic, Alfred Mirsky, who by 1949 had become convinced that the old tetranucleotide hypothesis was 'no longer tenable'.[3]

Chargaff's insight went much further. In 1951 he summarised the results he had published over the previous three years: different tissues of the same species yielded DNA with an identical composition in terms of the proportions of the four bases, and furthermore DNA molecules showed a 'composition characteristic of the species from which they are derived'. DNA composition was constant in all tissues of a given species, but each species had its own profile. Even more importantly, he repeated a remarkable conclusion that he had come to the year before:

---

*The rap contest can be seen at http://www.youtube.com/watch?v=35FwmiPE9tI.

It seems that in most specimens examined until now, the ratios of adenine to thymine, of guanine to cytosine, and of total purines to total pyrimidines were not far from one.[4]

The ratios of the bases actually reported by Chargaff were not always as telling as he suggested: for example, in the cow the ratios ranged from 0.75 to 0.80 for C:G and were clustered at around 1.16 for A:T. Chargaff's analytical procedure recovered only 70 per cent of the bases, so the ever-sceptical Mirsky dismissed the seductive similarities as experimental errors.[5] Even Chargaff was not certain whether the ratios had any meaning:

> As the number of examples of such regularity increases, the question will become pertinent whether it is merely accidental or whether it is an expression of certain structural principles that are shared by many desoxypentose nucleic acids, despite far reaching differences in their individual composition and the absence of a recognizable periodicity in their nucleotide sequence. It is believed that the time has not yet come to attempt an answer.[6]

Chargaff was clearer about the possible significance of the nucleotide sequence, although when he gave a lecture on the subject, in the summer of 1949, he could not exclude the possibility that genes were made of nucleoproteins, not nucleic acids:

> We must realize that minute changes in the nucleic acid, e.g. the disappearance of one guanine molecule out of a hundred, could produce far-reaching changes in the geometry of the conjugated nucleoprotein; and it is not impossible that rearrangements of this type are among the causes of the occurrence of mutations.[7]

The leading researcher into the structure of DNA was Bill Astbury of the University of Leeds. In 1938, Astbury and his PhD student Florence Bell had published X-ray images of DNA and had described a model in which the bases were strung perpendicularly along the phosphate-sugar backbone, like 'a pile of pennies'. But at the Cambridge meeting of the Society for Experimental Biology in 1947, Astbury had changed his mind and had argued that the bases were in fact parallel to the phosphate-sugar backbone, as though the 'pennies' were laid flat, rotated at 90° compared to his previous view. At the same time, Masson Gulland's group in Nottingham published evidence suggesting that the bases were linked by hydrogen bonds – a particularly strong and biologically widespread form of atomic bond – but Gulland could not determine whether these were bonds between nucleotides of the same DNA chain or between different DNA chains.[8]

Meanwhile, at Birkbeck College in London, a Norwegian PhD student called Sven Furberg was struggling to analyse the organisation of each of the four bases found in DNA with the use of X-ray crystallography. This technique involved crystallising a sample of one of the bases and then bombarding it with X-rays for hours on end. A piece of photographic film captured the result – patches of light and dark on the film that had been produced by the diffraction of the X-rays by the crystal structure of the sample. With a great deal of effort and no small amount of luck, these blobs could be interpreted in terms of the molecular structure of the crystal, which had diffracted the X-rays in a consistent fashion. Furberg was using some highly complex mathematics called the Patterson function to calculate a three-dimensional molecular structure on the basis of a series of two-dimensional X-ray images taken at different orientations. In a 1950 paper, Furberg concluded that Astbury's initial insight was correct and that the bases were indeed regularly spaced perpendicular to the phosphate-sugar backbone. Furberg's model of DNA structure was a spiral or helix, with the backbone twisting around on itself. But he could not determine the exact form of the molecule: either the backbone was in the middle, with the bases sticking outwards, or it was on the outside of the helix, with the bases all converging on the centre. Although Furberg's idea was not published until 1952, a copy of his 1949 thesis ended up at nearby King's College, where Wilkins was becoming increasingly interested in the molecular structure of DNA.

In May 1950, Wilkins heard a talk by Rudolf Signer from Bern, who described a new method for creating high-quality DNA, and generously gave out samples. Wilkins initially tried flattening Signer's pure DNA into a thin gel – it had the consistency of snot, he later recalled. Then something interesting happened:

> Each time that I touched the gel with a glass rod and removed the rod, a thin and almost invisible fibre of DNA was drawn out like a filament of a spider's web. The perfection and uniformity of the fibres suggested that the molecules in them were regularly arranged.[9]

Together with his PhD student Raymond Gosling, Wilkins made a frame from a bent paperclip (it was later upgraded to fine tungsten wire), stretched a DNA fibre across the metal and then put the sample in front of an X-ray source. To reduce background scatter by the X-rays, the inside of the camera was filled with hydrogen, and the seal between the X-ray tube and the camera was bound in a condom.[10] After much fiddling about with the humidity of the sample, they produced good X-ray diffraction images on a sheet of photographic film. As with Astbury and Bell's prewar picture, the patterns produced by Wilkins and Gosling showed a set of concentric curved and vertical lines, but their image was far sharper. It showed that when it had been pulled into a fibre Signer's DNA was in a quasi-crystalline state: it was organised in a regular fashion with all or most of the molecules apparently arranged in the same orientation. Gosling later recalled his reaction; he was 'standing in the dark room outside this lead-lined room, and looking at the developer, and up through the developer tank swam this beautiful spotted photograph ... it really was the most wonderful thing. ... I went back down the tunnels over to the Physics Department, where Wilkins used to spend his life, so he was still there. I can still remember vividly the excitement of showing this thing to Wilkins and drinking his sherry by the glass ... by the gulpful.'[11]

Both Wilkins and Gosling later suggested that at this moment they thought that genes were made of DNA, and that they had therefore crystallised a gene. But none of the publications from the King's group stated matters so clearly, and even in August 1950, four months after the image was made, Wilkins expressed his uncertainty in a letter to a friend: 'What we would really like to do, of course, is to find what nucleic acid is in cells *for*.'[12] A year later, Wilkins showed the picture in a lecture but still suggested that nucleoproteins, rather than nucleic acid, were the physical basis of genes.[13] The successful use of X-ray diffraction had in fact deepened the fundamental paradox of DNA. If it were the genetic material, then it should show variability, in order for the specific effects of genes to be expressed. But the fibre diffraction images were tantalisingly suggesting some fixed, repetitive and relatively simple structure.

Soon after this, the X-ray tube broke. It was months before it could be replaced, so the King's team tried to attack the problem using other techniques. Wilkins and Gosling noticed abrupt changes in the length of the DNA fibres with increasing humidity. This was remarkably similar to an effect seen in one of the most intensely studied proteins, keratin, which switched its form as the molecule stretched. In February 1951, two papers on DNA structure were sent to *Nature* by the King's group. Wilkins, Gosling and their colleague Willy Seed described the stretching of DNA fibres as seen under polarised light, and concluded that as the molecule stretched when it dehydrated, the orientation of the bases changed, becoming more parallel to the phosphate-sugar backbone.[14] In a second paper, a King's PhD student called Bruce Fraser, together with his wife, Mary, used infrared measurements to confirm Furberg's suggestion that the bases were normally perpendicular to the phosphate-sugar backbone. Above all, the Frasers reported that their data could 'be interpreted in terms of a structure similar to that proposed by Furberg' – a helix.[15]

*

In the related but distant world of protein structure, a helix was also causing a storm. Years earlier, Astbury had studied X-ray photos of unstretched keratin (known as α-keratin – α is pronounced 'alpha') and had suggested that it had a helical structure. There followed a long and intense competition between Lawrence Bragg's group at the Cavendish Laboratory in Cambridge and Linus Pauling's group at Caltech in Pasadena, as each tried to come up with a precise description of what became known simply as 'the α-helix'. In a wave of papers that appeared between November 1950 and May 1951, Pauling presented exact models that accurately described the helical structure of keratin, and showed that the same structure could be found in a range of biological tissues.[16] Bragg and his group were devastated. They had lost a race that had lasted for years.

There were still some kinks to be ironed out, including explaining how the α-helix seemed to coil round itself, but by early 1952 these had been fixed by both Pauling and by an infuriatingly brilliant and garrulous newcomer at the Cavendish, Francis Crick. Crick was vaguely studying for his PhD on the molecular structure of haemoglobin, including the mathematical theory behind the X-ray diffraction data produced by helical molecules.[17]

Pauling's description of the structure of the α-helix was a tour de force, but it was far more impressive to chemists than it was to biologists. The structure provided no explanation of keratin function. Max Delbrück was particularly scornful, as he later recalled: 'the α-helix, even if correct, had not provided any biological insights'.[18]

✳

Randall was acutely aware that the King's group lacked the skills required to interpret X-ray crystallography data. In late 1950, he recruited a British woman researcher in Paris who was using X-ray diffraction to study the molecular structure of coals. The initial plan was that she should study the structure of proteins in solution, but then Wilkins fatally suggested to Randall it might be a good idea for her to work on the X-ray diffraction analysis of DNA. Randall agreed and sent a letter to his new researcher, 30-year-old Rosalind (pronounced 'Ros-lind'[19]) Franklin, explaining the change of plan – 'nucleic acid is an extremely important constituent of cells and it seems to us that it would very valuable if this could be followed up in detail'.[20]

Randall's letter to Franklin, which Wilkins did not see for several decades, was the source of several tragic misunderstandings between Wilkins and Franklin. In the letter, Randall stated that

Franklin would be the only researcher studying DNA with X-rays – 'as far as the experimental X-ray effort is concerned there will be at the moment only yourself and Gosling'. Franklin understandably concluded that she would have sole control of her research. But Wilkins was still interested in using X-ray diffraction on DNA and had no idea that Randall apparently wanted him to hand the project over to Franklin. When Franklin arrived in the laboratory, Wilkins was on holiday, and when he returned he discovered that his PhD student Gosling was now working with the new arrival without any explanation. This uncomfortable situation could have been solved by talking, but Wilkins and Franklin were the victims of an immediate and appalling clash of personalities. Wilkins was quiet, diffident and hated arguments; Franklin was forceful and thrived on the rough and tumble of intellectual debate. Her friend Norma Sutherland recalled: 'Her manner was brusque and at times confrontational – she aroused quite a lot of hostility among the people she talked to, and she seemed quite insensitive to this'.[21] Wilkins was intimidated by Franklin's character and bewildered by her refusal to work with him. Franklin, in contrast, was irritated by Wilkins's rather limp behaviour and by his ignorance of the fundamentals of X-ray diffraction. Their working relationship was doomed before it began.

It is not known why Randall wrote his letter to Franklin in the way that he did. Wilkins later wondered whether it was an attempt to sideline him from the project, either because Randall wanted to keep DNA to himself or because he was frustrated by Wilkins's slow progress. Perhaps Randall hoped to get the best out of both researchers by setting them up to compete with each other. Whatever the case, the inherent personality differences between Wilkins and Franklin were horribly amplified by their totally different impressions of their respective roles, and their inability to simply talk about things.[22]

*

At the end of May 1951, Randall was due to speak at a dull-sounding 'Symposium on submicroscopical morphology in protoplasm', to be held at the Naples Marine Biological Station. He was unable to attend, so he sent Wilkins in his place. In his talk, Wilkins explained their work but presented it in terms of nucleoproteins, not nucleic acids: 'when living matter is to be found in a crystalline state, the possibility is increased of molecular interpretation of biological structure and processes. In particular, the study of crystalline nucleoproteins in living cells may help one to approach more closely the problem of gene structure.'[23] At that point he showed the DNA X-ray diffraction image he had taken with Gosling. In the audience, a gangly 23-year-old American suddenly paid attention. His name was Jim Watson. He later described the scene:

> Maurice's dry English tone did not permit enthusiasm as he stated that the picture showed much more detail than previous pictures and could, in fact, be considered as arising from a crystalline substance. ... Suddenly I was excited about chemistry. Before Maurice's talk I had worried about the possibility that the gene might be fantastically irregular. Now, however, I knew that genes could crystallize; hence they must have a regular structure that could be solved in a straightforward fashion. Immediately I began to wonder whether it would be possible for me to join Wilkins in working on DNA.

At the time, Watson was based in Copenhagen, rather unsuccessfully studying phage duplication, having obtained his PhD on phage genetics with Salvador Luria a year earlier, at the remarkably young age of 22 years.[24] Watson had become obsessed with how genes copied themselves, and Wilkins's data seemed to suggest a way of attacking the problem. On his return to Copenhagen, Watson read Pauling's slew of articles on the α-helix and became determined to use X-ray crystallography to study genes. After some squabbles with his funding agency, in autumn 1951 Watson moved to Cambridge and began work in Max Perutz's group, which had world-leading expertise in X-ray crystallography. In Cambridge he shared an office with Francis Crick.[25] One of the great scientific partnerships began. Watson described his new colleague in a letter to Delbrück:

The most interesting member in the group is a research student named Francis Crick. ... He is no doubt the brightest person I have ever worked with and the nearest approach to Pauling I've ever seen. ... He never stops talking or thinking and since I spend much of my time in his house (he has a very charming French wife who is an excellent cook) I find myself in a state of suspended stimulation.[26]

Crick recalled he was 'electrified' by meeting Watson. 'It was *remarkable*', he said, because they had the same focus on understanding gene structure but they had entirely different skills – phage genetics and crystallography, respectively. In 1947, Crick had described his desire to unravel 'the chemical physics of biology'.[27] He was not initially particularly interested in genes or DNA – like most people, he assumed that the genetic material would be a protein. In 1950 he had teased Wilkins for his work on DNA, telling him, 'What you ought to do is get yourself a good protein.'[28]

*

During the summer of 1951, Wilkins became increasingly convinced that the X-ray diffraction data showed that DNA had a helical structure. When Wilkins first presented this idea, Franklin's response was to tell him to stop working on X-ray data from DNA and 'go back to your microscopes'. Wilkins, who was the assistant director of the laboratory, was incredulous that a postdoctoral researcher should speak to him in such a fashion, but typically said nothing.[29] Matters got worse when another member of the King's group, Alex Stokes, found mathematical support for Wilkins's intuition. When Wilkins told Franklin of their success, she responded furiously, saying, 'How dare you interpret my data for me?'[30]

To overcome the tension between Wilkins and Franklin, Randall adopted the simplest solution: the two researchers would be kept apart. There was a ready justification for this. Franklin had shown that DNA came in two distinct forms, named A and B, depending on the degree of humidity. The A form, which could be seen under drier conditions, gave precise but highly complex images; the B form, which occurred at high humidity levels, was more blurry and less enticing. X-ray crystallography requires precise measurements from the diffraction images; if the photo is blurred, it will be impossible to come to an accurate description. Randall decided that Franklin should study the A form DNA using Signer's samples, while Wilkins would investigate the B form using Chargaff's less pure DNA. Randall could perhaps have resolved everything by being frank with the pair of them; but he was not, and matters got no better.

Shortly afterwards, on 21 November 1951, the King's group organised a small meeting on DNA at which they all presented their latest findings. Jim Watson was in the audience and when Franklin spoke, his attention was torn between her results and idle musings about her looks. Franklin showed DNA images she had made with the lab's new fine-focus X-ray tube, which had a very narrow beam, and she described the two forms, A and B, highlighting the fact that the A form produced clearer images, showing 'evidence for spiral structure'. The next day, Watson and Crick excitedly discussed the details that Watson could recall – as was his habit, the cocky young American had taken no notes. Crick became convinced that only a few structures would fit Franklin's data and within a week they had a model for DNA. This was accompanied by a rather pompous eight-page 'memorandum' by Crick that outlined their strategy, which was above all 'to incorporate the *minimum* number of experimental facts'.[31] This was quite appropriate, as neither man had done a single experiment on DNA structure.

The first Watson and Crick model of DNA was a triple helix – there were three intertwined phosphate-sugar strands in the centre, with the bases sticking out like fingers. In triumph the pair invited Wilkins, Gosling and Franklin to come to Cambridge to view their creation. Franklin took one look at the model and dismissed it. Her X-ray data that had so entranced Watson had clearly shown that the phosphate-sugar groups were on the outside, not the inside, whereas the magnesium ions that held Watson and Crick's triple helix together would be unable to fulfil this function because they would be surrounded by water molecules. If such a structure ever existed, it would instantly fly apart. All this had been explained in her talk at the King's meeting, but Watson had not fully understood what she was saying and had not written anything down. Watson and Crick's first venture into model-building ended in embarrassing failure.

Watson's failure to pay attention was even more significant than he realised. According to Franklin's notes, when she spoke at the November meeting she described the shape of the 'unit cell' (the shape of each molecule) of the DNA crystal as 'monoclinic'. This crystallographic jargon meant that the molecule would show rotational symmetry, and that if there were chains of molecules wrapped around each other in the structure, they must run in opposite directions. This turned out to be a vital insight into the structure of DNA, but Watson did not understand enough crystallography to grasp its significance. Crick did, in an instant, when he eventually learned of it fifteen months later.

When Randall heard about the Cambridge duo's attempt to muscle in on the structure of DNA, he furiously asked the head of the Cavendish Laboratory, Sir Lawrence Bragg, to tell the two upstarts to leave DNA alone. Bragg had a low opinion of Crick and probably no opinion of Watson, who was beneath his notice, so he willingly banned them from doing any further work on DNA. Crick returned to his PhD on haemoglobin structure, and Watson began studying the nucleic acids in the tobacco mosaic virus, learning elementary X-ray crystallography. The fiasco also reinforced Franklin's prejudices against building speculative models. The data had to lead to the model, she felt. Watson and Crick, high on mathematics and fixated with Pauling the alpha-helix male, had been utterly confident that logic and 'the *minimum* number of experimental facts' would lead them to the answer. Instead it had led to ridicule.

Despite Franklin's conviction that the results would speak for themselves, her data were confusing because she was looking at the precise and detailed images produced by the A form of DNA. She apparently assumed they were so sharp because the A form was an array of crystals that were all oriented in the same direction. In fact, the A form is made up of small crystalline blocks in which the crystals within a block have the same orientation, but where different blocks have different orientations, producing an image that is both clear and complex.[32] Deducing the structure of DNA from the A form image was going to be extremely difficult. When Crick eventually saw the A form data, in 1954, he told Wilkins:

> This is the first time I have had an opportunity for a detailed study of the picture of Structure A, and I must say I am glad I didn't see it earlier, as it would have worried me considerably.[33]

Eventually, putting too much faith in the sensitivity of her apparatus, Franklin concluded that the A form was not helical and sent round a spoof death notice, edged with black, to members of the laboratory:

> It is with great regret that we have to announce the death, on Friday 18th July, 1952, of D.N.A. HELIX (crystalline).

By this time Franklin had already decided she had had enough of the terrible atmosphere at King's; she agreed with Randall that at the end of the year she would move to Birkbeck College and would abandon her studies of DNA.

Despite the triple helix fiasco, Watson and Crick did not stop thinking about the structure of DNA. Crick asked a colleague to work out the chemical bonds that could exist between the bases and was delighted when he was told that A would bind with T, and C with G. In a flash, Crick realised that this provided the clue to gene replication, through what he called complementary replication. If A on one molecule bound with T on another, you would get a kind of mirror image of the original DNA; if the same process was then repeated with that 'mirror', a new strand of DNA, identical to the original one, would have been created. If there were two molecules of DNA bound together at the outset, replication would be even more straightforward – simply copy each strand and you would duplicate the original molecule.[34]

At the end of May 1952, Chargaff visited Cambridge and had a meal with Watson and Crick at which they talked about DNA. It did not go well. Chargaff – a notoriously prickly character – poured scorn on them because of their ignorance of chemistry and of his work. He later recalled that his first impression was 'far from favorable; it was not improved by the many farcical elements that enlivened the ensuing conversation … So far as I could make out, they wanted, unencumbered by any knowledge of the chemistry involved, to fit DNA into a helix. The main reason seemed to be Pauling's alpha-helix model of a protein.'[35]

Despite his evident irritation, Chargaff told Watson and Crick of the enigma of the apparent 1:1 ratios of A:T and C:G. As Crick later recalled:

> Well, the effect was electric … I suddenly realized, by God, if you have complementary replication, you can *expect* to get one-to-one ratios.[36]

In a rare foray into the laboratory, Crick spent the next week trying to get bases to pair spontaneously in the test tube. It did not work, and Crick's flash of insight led nowhere, for the moment.

*

King's and Cambridge were not the only places where scientists were trying to understand the structure of DNA. On 28 May and 1 June 1951, Elwyn Beighton in Bill Astbury's laboratory in Leeds took some of Chargaff's DNA and made several X-ray images. They had the classic X-shape which we now know reveals the presence of a helix. Astbury was not impressed, and did not encourage Beighton to continue his work; he was not able to interpret the images correctly because Crick had not yet published his papers that described the diffraction pattern produced by a helix.[37] Astbury may have felt that the material was less pure than the extracts he had worked with before the war, or he may have merely been frustrated that the image seemed too simple at a time when all the lines of argument were suggesting that DNA was the genetic material, the carrier of specificity. At around this time, the Medical Research Council (MRC) rejected Astbury's proposal to create a new department, and the arrival on the scene of the well-funded group at King's College may also have discouraged him from pursuing the structure of DNA. Whatever the case, the Leeds images became a historical dead-end, an enigmatic curiosity, and Astbury's direct involvement in the determination of the structure of DNA was over.[38]

At about the same time, Edward Ronwin of the University of California produced a model of DNA. This was superficially similar to Astbury's 1947 model – it had the phosphate-sugar backbone in the centre, with the bases fanning out. However, Ronwin had made some basic biochemical errors and his model contained too much phosphorus. Linus Pauling was outraged, and wrote a letter to the editor of the *Journal of the American Chemical Society* criticising the 'foolishness' of publishing Ronwin's model, and complaining about 'the irresponsible publication of unsupported hypotheses.'[39]

More seriously, in the first half of 1952, John Rowen at the National Cancer Institute in Maryland studied the molecule using light-scattering electron microscopy and viscometry. In 1953 he published an article describing its configuration as 'intermediate between a rod and a coil' before concluding that one of 'its most striking properties is its tendency to spiral, twist and intertwine with neighbouring molecules'.[40]

*

Beginning in the second half of August 1952, Franklin took X-ray photos of her DNA samples every day, using a heavy disc-shaped metal camera that was about the width of an orange. Her PhD student, Raymond Gosling, recalled that she bounced ideas off him – they had what he called 'pretty hot discussions' in which she played the role of devil's advocate and which he found enormously stimulating.[41] She never had that kind of discussion with Wilkins. Together with Gosling, Franklin began to calculate the Patterson function, the difficult mathematical procedure used by Sven Furberg at Birkbeck. This involved projecting the X-ray photos in a dark room, measuring the position and intensity of the various blobs on the pictures and then spending hours doing complex calculations. In November 1952, Franklin summarised the data she had obtained with Gosling as part of a brief report by Randall for the Biophysics Committee of the MRC. There was nothing in Franklin's few paragraphs that had not been presented at the King's symposium the previous year, but this time the data – including the different sizes of the repeating patterns in the A and B forms and above all the dimensions of the monoclinic unit cell, were written out clearly and slightly more precisely.[42] In the middle of December, members of the Biophysics Committee made an official tour of the King's lab and were each given a copy of the informal document. One of the visitors was Max Perutz from Cambridge.

*

At the end of 1952, Pauling finally stirred. Even though he still assumed that proteins explained genetic specificity, Pauling had been snuffling around DNA for some time – a year earlier he had cheekily written to Randall asking to see their data; Randall had given him the brush-off.[43] In Watson's mind, Pauling had become a bogeyman, a powerful competitor who had the ability to steal the prize, should he so desire. At the end of the year, just as Watson feared, Pauling submitted an article to the US journal *Proceedings of the National Academy of Sciences*, based entirely on measurements from Astbury and Bell's data from 1938.[44] But when the manuscript arrived in Cambridge at the end of January, Watson and Crick were relieved to see that the structure was just as wrong as their own first attempt. It, too, was composed of three intertwined helices; it, too, had the bases on the outside; and to their amazement the way in which the model was built meant that the molecule was not an acid at all. As Watson later put it, 'a giant had forgotten elementary college chemistry'. Finally, the model did not explain the nature of the gene or its essential functions: replication and specificity. It was biologically mute.

A few days after reading Pauling's manuscript, Watson went to King's, where he had a brief squabble with Franklin over her apparent refusal to accept that DNA was helical. He then saw Wilkins, who took him to his office and showed him a photo of the B form that Raymond Gosling had given him a few days earlier as the pair worked on Gosling's thesis; with Franklin's departure imminent, Gosling was once again being supervised by Wilkins. This photo – 'photo 51'– had been taken in May 1952 by Gosling, but it had not been studied and had lain in a drawer for months.[45] Watson was stunned – the image was so much simpler than any he had previously seen. As he put it: 'The instant I saw the picture my mouth fell open and my pulse began to race.'[46] In the centre was an X shape; Watson was a crystallographic novice, but from his discussions with Crick, who had been working on the crystallographic interpretation of helical molecular structures, he knew that the X could only come from a helix.

The significance of photo 51 in the identification of the double helix structure of DNA has often been overstated, mainly because of the weight given to it in Watson's own world-famous account, *The Double Helix*. In reality, the insight given by the photo was extremely limited. Everyone at King's – even Franklin – now accepted that the B form was helical, and without any more precise details of the measurements of the molecule, all that happened was that Watson's preconception was confirmed – DNA had a helical structure. Furthermore, there was nothing underhand going on – Watson was shown the photo in perfectly legitimate circumstances. Raymond Gosling was absolutely certain: 'Maurice had a perfect right to that information', he said later. Whether Wilkins used that information wisely is another matter. He later regretted his action – 'I had been rather foolish to show it to Jim', he wrote in his memoirs.[47] Despite the excitement that Watson felt, all the main issues, such as the number of strands and above all the precise chemical organisation of the molecule, remained a mystery – a glance at photo 51 could not shed any light on those details. The decisive information, which was unwittingly provided by Franklin herself, came from another source.

Pauling's foray into DNA structure led Bragg to lift his injunction against Watson and Crick working on a model of DNA – he was not going to allow Pauling to repeat the coup of the keratin α-helix. Wilkins reluctantly agreed. There followed a rapidly evolving frenzy of dead ends, brilliant insights and chance encounters as Watson and Crick worked furiously to solve the problem. Most decisively, Max Perutz showed Crick the MRC report that included Franklin's brief summary of her data from fifteen months earlier. Although it contained nothing that Watson should not have noted in November 1951, it now provided Crick with the information he needed. By chance it chimed completely with what he had been working on for months: the type of monoclinic unit cell found in DNA was also present in the horse haemoglobin he had been studying. This meant that DNA was in two parts, each matching the other. As Crick later recalled, 'the chains must come in pairs rather than three in a molecule, and one must run up and the other down.'[48] Then another occupant of the Watson and Crick office, Jerry Donohue, pointed out that Watson was using the wrong forms of the bases when he was trying to build the model. With the correct structures, adenine bound with thymine, and cytosine bound with guanine, using hydrogen bonds, as Gulland had reported six years earlier.

Back at King's, Bruce Fraser, like Watson and Crick and Pauling before him, was struggling with a triple helix structure, but this time with the bases on the inside. The model led nowhere. Meanwhile, Rosalind Franklin was finishing up her work on DNA before leaving the lab, groping her way towards a solution, oblivious to what was happening in Cambridge. The progress she made on her own, increasingly isolated and without the benefit of anyone to exchange ideas with, was simply remarkable. In January 1953 she struggled with the data from the Patterson function and complained in her notebook that she could not 'reconcile nucleotide sequence with Chargaff's analysis': her data suggested a cumbersome seven-nucleotide unit that gave approximate 1:1 ratios of purines and pyrimidines, but nothing as precise as some of Chargaff's data.[49] But by 24 February, she had realised that both the A and the B forms of DNA were double helices. She also suggested that the bases on either strand were interchangeable (A with T, C with G), and above all she realised that 'an infinite variety of nucleotide sequences would be possible to explain the biological specificity of DNA'.[50] Franklin was almost there, but she did not have a chance to get any further, because Watson and Crick had already crossed the finishing line.

By the end of February 1953, Watson and Crick had agreed the basic outline of the double helix model, but this was merely a seductive concept. It needed to be turned into precise numbers, spatial relationships and chemical bonds, in the shape of a physical model. It took a week of calculation and intense work before the double helix, with complementary base pairing between A and T and between C and G, finally emerged from a tangle of precise metal templates held together by clamps. It was a molecular model informed by experimental data; it did not simply emerge from the diffraction data, as

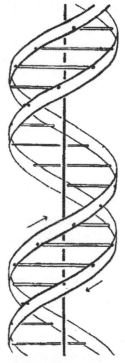

This figure is purely diagrammatic. The two ribbons symbolize the two phosphate—sugar chains, and the horizontal rods the pairs of bases holding the chains together. The vertical line marks the fibre axis

3. Double helix model of DNA from Watson and Crick (1953a)

Franklin had wanted. On 7 March, Wilkins wrote to Crick announcing that the 'dark lady' (Franklin) was leaving King's the following week, that 'much of the 3 dimensional data is in our hands' and promising to launch a 'general offensive on Nature's secret strongholds on all fronts', signing off with 'It won't be long now'. When Crick opened the letter on 12 March, the double helix model stood in front of him.[51]

Wilkins and Franklin came to Cambridge to see the model, and immediately agreed it must be right. Although Watson described Wilkins as being remarkably magnanimous at being scooped,

Wilkins recalled that he felt 'rather stunned' and bitter and that he made 'an angry outburst'.[52] Whatever the case, it was agreed that the model would be published solely as the work of Watson and Crick, while the supporting data, without which the model would not have existed, would be published by Wilkins and Franklin – separately, of course. The first public announcement of the discovery was made by Sir Lawrence Bragg in April, at a conference in Solvay, Belgium. Pauling, who had seen the model in Cambridge and given it his blessing, told the audience that the Watson and Crick model was 'very likely' to be 'essentially correct', and that his triple helix was mistaken.[53] On 25 April there was a party at King's when the three articles were published in *Nature*. Franklin did not attend. She was now at Birkbeck and had stopped working on DNA.

*

Franklin was never told the full extent to which Watson and Crick had relied on her data to make their model; if she suspected, she did not express any bitterness or frustration. In subsequent years she became very friendly with Crick and his wife, but she was never close to Wilkins or Watson, although she interacted with Watson as she worked on the structure of the tobacco mosaic virus.[54] Franklin died of ovarian cancer in 1958, four years before the Nobel Prize was awarded to Watson, Crick and Wilkins for their work on DNA structure.

In 1968 Jim Watson published *The Double Helix*, in which he gave a gripping but partial account of events and a frank description of his own bad behaviour, particularly with regard to Franklin. The epilogue to the book contains a generous and fair description of Franklin's vital contribution and a recognition of his own failures. *The Double Helix* also suggested that Max Perutz had given Watson and Crick a confidential document when he handed over the MRC report containing those vital paragraphs by Franklin. Perutz was hurt by this allegation and showed that it was not true. There can be no doubt that the data in the report provided Crick with the insight he needed to come up with the correct structure, but the document was not confidential, and above all Franklin had publicly communicated the essential results nearly eighteen months earlier. Watson had been in the audience, but he had not understood the significance of what was being said.[55]

One of the main facts that in retrospect seems so obvious, but which was not at the time, is the role of what are sometimes called the 'Chargaff rules' – the fact that the amounts of A and T and of C and G are equivalent. These ratios were not known to be so precise at the time and they were certainly not 'rules'. As Jerry Donohue, who shared Watson and Crick's room at the Cavendish laboratory, later recalled in somewhat exaggerated fashion:

> When the final model of DNA was discovered – more or less
> by accident – it wasn't Chargaff's rules that made the model,
> but the model that made the rules.[56]

The Watson and Crick paper in *Nature* concluded coyly, 'It has not escaped our notice that the pairing we have postulated immediately suggests a possible copying mechanism for the genetic material.'[57] This was the solution that Watson had been searching for – the complementary pairing of the bases gave a potential insight into gene duplication; with a single molecule it was possible to create two identical daughter molecules, simply by copying each strand using complementary pairing. Even more significant was what the three *Nature* papers did not say – none made any reference to how genes functioned, or the significance of the sequence of bases. There was still no genetic code.

SEVEN

GENETIC INFORMATION

On 19 March 1953, about two weeks after the double helix model had been completed, Francis Crick wrote a letter to his 12-year-old son, Michael, who was at boarding school. Crick told Michael what he had discovered, and included a sketch of the structure of DNA. He then went on to explain the significance of the double helix:

> It is like a code. If you are given one set of letters you can write down the others. Now we believe that the D.N.A. is a code. That is, the order of the bases (the letters) makes one gene different from another gene (just as one page of print is different from another).[1*]

Although the idea that the sequence of bases might be the source of genetic specificity had been in the air for some time, this was the first time that anyone had said that DNA contains a code, and Crick's son

---

* As Crick later recognised, from a strict cryptographical point of view, the genetic code is a cipher, not a code. Crick had in mind something like the Morse Code, which is equally not a code from a technical point of view. And as he later put it, '"genetic code" sounds a lot more intriguing than "genetic cipher"' (Crick, 1988, pp. 89–90). The term genetic code was first used in print in 1958, by Geoffrey Zubay (Zubay, 1958).

Michael was the first to read it. In 2013 the letter was sold at auction for $6m.

Crick went even further in the second paper he published in *Nature* with Watson, which appeared on 30 May 1953. Like their first publication, this article contained no data at all – it was purely theoretical. As the title explained, its aim was to explore the 'genetical implications of the structure of deoxyribonucleic acid'.[2] The article began by suggesting that DNA was 'the carrier of a part of (if not all) the genetic specificity of the chromosomes and thus of the gene itself'. This 'big picture' was absent from all three DNA papers published in April, which dealt solely with the structural chemistry of the molecule, not its function, with the exception of the coy closing phrase 'it has not escaped our notice …'. Much of Watson and Crick's second article was devoted to expanding on that cheeky insight. They described how, during gene duplication, the double helix could unwind, with each chain forming the template for the construction of a new molecule, leading to the creation of two identical daughter double helices.[3]

In the middle of this discussion there was a half-sentence, almost a throwaway remark, that echoed the terms used in Crick's letter to his son, but which expanded them to a far broader conception and propelled biology into the modern age:

> The phosphate-sugar backbone of our model is completely regular, but any sequence of pairs of bases can fit into the structure. It follows that in a long molecule many different permutations are possible, and it therefore seems likely that the precise sequence of the bases is the code which carries the genetical information.

With the exception of Ephrussi and Watson's unfunny satirical letter to *Nature* seven weeks earlier, this was the first time that the content of a gene had been described as information. It is not known where the phrase 'genetical information' came from – Watson recalls that Crick wrote most of the manuscript in less than a week at the end of April, and the form and the content are more typical of Crick's style than of Watson's.[4] Given the growing presence of 'information' in scientific articles from a wide range of disciplines, and the popular interest in communication theory and cybernetics, it seems likely that the idea just came naturally to Crick, as part of the zeitgeist. There is no indication that either Crick or Watson had read Shannon or Wiener, or that they were using the term in explicit reference to their mathematical ideas. As with the word 'code', 'information' seems to have been used as an intensely powerful metaphor rather than a precise theoretical construct.

In the past, scientists had spoken of genetic specificity, but with the introduction of the idea that the DNA sequence contained 'the code which carries the genetical information' a whole new conceptual vocabulary became available. Genes were no longer mysterious embodiments of specificity, they were information – a code – that could be transmitted (another word from the electronic age), and the central hypothesis was that the code was composed of a series of letters – A, T, C and G. How exactly that code might function, what it might represent, was not stated. Nevertheless, the words used so lightly by Crick and Watson changed the way in which scientists could speak and think about genes. Eventually this new vocabulary contributed to the development of novel parallels between genes and electronic communication and processing.

None of this was obvious at the time. Watson was not particularly keen on the article being published at all, as he explained to Delbrück in May:

> Crick was very much in favour of sending in the second *Nature* note. To preserve peace I have agreed to it and so it shall come out shortly.[5]

Wilkins was not impressed, either. He recalled: 'Some of Francis and Jim's friends, however, thought the second paper was rather "going over the top".' Among those friends was Wilkins himself.[6] Despite these doubts, the advantage of Crick's approach was that it explicitly set out the two revolutionary implications of the double helix: complementary base pairing explained gene duplication, while the sequence of bases explained genetic specificity. Here were two hypotheses that would revolutionise biology, if they were true.

Crick recognised the links between the discovery of the double helix and Schrödinger's ideas from a decade earlier. On 12 August 1953, he sent copies of the two *Nature* articles to Schrödinger, accompanied by a brief letter:

Watson and I were once discussing how we came to enter the field of molecular biology, and we discovered that we had both been influenced by your little book, 'What is Life?'

We thought you might be interested in the enclosed reprints – you will see that it looks as though your term 'aperiodic crystal' is going to be a very apt one.

*

In July 1953, Watson and Crick received a strange letter from the US. Handwritten in big letters on headed notepaper from the University of Michigan students' union, the letter was full of crossings-out and spelling mistakes and looked as though it was from a crank. In fact it was from the Russian-born cosmologist George Gamow (pronounced 'Gam-off'), who was a long-time friend of Max Delbrück and had chaired the 1946 meeting on 'The Physics of Living Matter'.[7] Although he was an expert in nuclear physics, Gamow had not passed security clearance for the Manhattan Project and had not been involved in the development of the atomic bomb at all. FBI surveillance of Gamow continued after the war, and he was interviewed by them as late as 1957, although they never found any evidence against him.[8]

Gamow was an eccentric 50-year-old who liked his whisky and had a sideline in popular science books based around an everyman character called Mr Tompkins. In his odd letter, Gamow seized upon Watson and Crick's suggestion that the sequence of bases contained a 'code', and audaciously tried to come up with ways of cracking that code. Gamow's starting point was that each organism could be characterised by 'a long number', which corresponded to the number of positions in the DNA sequence. He then dismissed decades of research in classical genetics showing that genes are located in definite positions on chromosomes, and argued that it seemed 'more

logical' if genes were instead 'determined by the different mathematical characteristics of the entire number'. In an attempt to make
his idea clearer, Gamow wrote, with his typically erratic spelling:

> the animal will be a cat if Adenine is always followed by
> cytosine in the DNA chain, and the characteristics of a hering
> is that Guanines allways appear in pairs along the chain …
> This would open a very exciting possibility of theoretical
> research based on mathematics of combinatorix and the theory
> of numbers![9]

Gamow said he would be in England in the autumn, and asked
whether they could meet. Watson and Crick were both about to
leave Cambridge and pursue their separate careers – Watson was
going to Pasadena, while Crick was headed for Brooklyn Polytech-
nic, once he had finished his PhD on the structure of haemoglobin.
So the pair simply ignored Gamow's letter.[10] Or, rather, they did not
reply to it. Crick did not ignore it: Gamow had planted an idea that
would not go away.

Gamow did not give up easily. Over the next couple of months
he worked up his ideas about the genetic code and in October he
sent a brief note to *Nature*, which was published in the following
February. He tried to publish a longer article on the same subject in
Proceedings of the National Academy of Sciences, co-authored by his fictitious character Mr Tompkins. The Editor of *PNAS* spotted the jape
and was not amused, so Gamow sent the article to the Royal Danish
Academy, with Tompkins's name excised.[11] Gamow addressed the
link between the DNA code and proteins by pointing out that the
central question was how four-digit 'numbers' in the gene (A, C, T
and G) were translated into an amino acid 'alphabet' in a protein.

Gamow's answer was ingenious and was not dissimilar to
the template idea that Caldwell and Hinshelwood had published
three years earlier. Gamow assumed that proteins were synthesised
directly on the DNA molecule, so that the shape formed by the bases
as the DNA molecule twisted round acted as a kind of template upon
which the amino acids were arranged. Because of the spiral shape of
DNA, there would be a diamond-shaped 'hole' between different

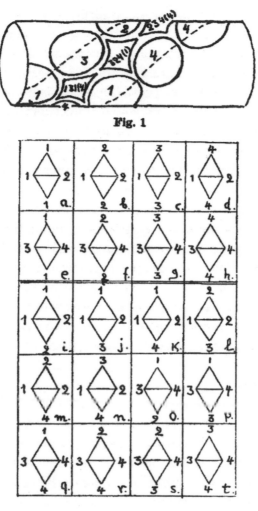

Fig. 1

1. Gamow's 'diamond' model of the genetic code, from Gamow (1954). The round structures numbered 1–4 are the bases, the diamond shapes labelled a–t are the 20 naturally occurring amino acids.

rows of bases; the four bases on each side of this diamond therefore constituted the code.

Gamow noted that there were twenty different possible kinds of 'hole' and continued, 'it is inviting to associate these "holes" with twenty different amino acids essential for living organisms.' Gamow even came up with a prediction that would test his model: because each base contributed to the shape of the 'hole' of more than one amino acid, 'there must exist a partial correlation between the neighbouring amino acids in protein molecules, since the neighbouring holes have two common nucleotides.'[12] By treating the code as a mathematical problem rather than a biological one, Gamow was opening the door to years of speculation about the nature of the genetic code. He was also committing the classic physicist's error of assuming that living systems are designed according to elegant, logical principles that can be revealed by mathematics. In fact they are historical, carrying the baggage of their evolutionary past, and have not been designed at all. They are often far from logical, nor are they generally elegant. They work, and that is enough.

Gamow sent Crick a copy of his paper and eventually the two men met in Brooklyn in December 1953. Crick's office-mate Vittorio Luzzati recalled:

> It was amazing. These two spirited men debated, argued and fought their way through the subject of the code disposing of issues, one after another, in their exuberance their voices rising to shouts.[13]

Crick was not convinced by Gamow's ideas – for a start, he did not think that protein synthesis took place on the DNA molecule, as Gamow assumed. It was by now well known that DNA was found in the nucleus, as part of the chromosomes, whereas RNA was found freely in the cell, where protein synthesis took place. Crick and Watson, following Caspersson, Brachet, Boivin, Vendrely and Dounce, considered that RNA acted as an intermediary between gene and protein. This was the meaning of the equation DNA $\rightarrow$ RNA $\rightarrow$ protein. The very starting point of the diamond code was wrong.

But Gamow had put his finger on a fundamental and seductive issue: the potential relation between the number of naturally occurring amino acids (twenty) and the number of possible combinations in the code formed by the bases A, C, T and G. As Gamow immediately realised, if the code were composed of two-letter 'words' (AA, AT, AC, AG, etc.), there would be sixteen possible combinations – not enough for each 'word' to correspond to a different amino acid. But if the code were composed of three letters (AAA, AAT, AAC, etc.), there would be sixty-four possible combinations – more than enough.

What Crick called 'the magic twenty' came to be the criterion against which all potential codes were measured.[14] What was called 'the coding problem' began to have a whiff of numerology – coding schemes were developed, with the objective always being to come up with twenty possible combinations that might correspond to the twenty widely occurring amino acids. Neither Crick nor Watson had previously thought much about the relationship between a base sequence in DNA and the amino acid sequence – they were more focused on the problem of how the double helix might unwind during gene replication. They suggested that the sequence was 'the code that carries the genetical information', but they had not thought beyond that fundamental insight. As Watson told the Cold Spring Harbor meeting in the summer of 1953, he and Crick could not explain how the gene controlled the activity of the cell.[15] Gamow's intervention brought the question of coding to the forefront of everyone's mind. For Crick, it occupied an important portion of his life for the next fifteen years.[16]

*

In the months following his letter to Watson and Crick, Gamow organised informal discussions involving biologists and a small gang of physicists and mathematicians – including Edward Teller ('the father of the H-Bomb') and the future Nobel Prize winner Dick Feynman – who were seduced by Gamow's infectious enthusiasm and by his zany and intellectually provocative correspondence, which came complete with jokes, comments and cartoons. Gamow spent his time flitting from laboratory to laboratory, writing his slightly mad letters on headed notepaper from hotels or railway companies, scattering forwarding addresses like confetti.[17] Gamow was larger than life – he was more than six feet tall, with a taste for hard liquor, practical jokes, magic tricks and women, and he spoke in a thick Russian accent that was often hard to understand.[18] Having him around could be hard work – in February 1955, Watson wrote to Crick: 'Gamow was here for 4 days – rather exhausting as I do not live on whisky.'[19] Crick recalled:

> And he was what is called good company, was Gamow. I wouldn't quite say a buffoon, but – yes, a bit of that, in the nicest possible way. You always knew, if you were going to spend the evening with Gamow that you would have a 'jolly time'. You know. And yet there was something behind it all.[20]

When it came to discussing coding, Gamow was irrepressible; as soon as one of his schemes crashed into the hard wall of biochemical fact, he simply came up with another. So when it became evident that his DNA-based diamond model had little traction, Gamow, undaunted, switched to thinking about coding in RNA. This molecule was not so amenable to the diamond model – its precise structure was unclear, but it was probably some kind of single helix, meaning that Gamow's original scheme would not work.

In March 1954, Gamow and Watson came up with the wheeze of creating an informal group in which a select few scientists – the number twenty again – could bat around ideas related to the coding problem. An entirely male body, this jokey group was known as the 'RNA Tie Club' after the hand-embroidered ties bearing the single helix of the RNA molecule they were each given.[21] Each member was also given a name corresponding to one of the twenty amino acids and a small gold tie-pin carrying the three-letter abbreviation of 'their' amino acid (so Gamow was ALA for alanine, Crick was TYR for tyrosine, etc.). The members of the 'Club' never actually met all together, but small groups of them gathered to drink and chat, with Gamow generally acting as master of ceremonies. The outstanding feature of the Club was that it allowed its members to present half-formed ideas without the pressure of publication or presentation at a conference – Crick later drew a parallel with the modern circulation of e-mails among groups of scientists.[22] Reading the slow correspondence between members of the RNA Tie Club who were scattered around the world, with letters taking days to arrive at their destination – sometimes on the other side of the planet – you cannot be sure whether they would have benefited or suffered from the rapidity of modern electronic communication.[23] The correspondence would have bounced back and forth far more quickly, but the participants would have had less time to think and develop their ideas.

More than half of the RNA Tie Club members were either physicists, chemists or mathematicians, but none of them were involved with the two main points at which mathematics and biology were intersecting – cybernetics and what had become known as information theory.[24] Although Gamow did attend a 1956 conference on information theory in biology, there were no direct interactions of any significance with scientists working in either cybernetics or information theory, with the exception of von Neumann.

Gamow's role in this phase of work on the genetic code was fundamental. He gave the project a shape – in fact, he made it a project – by pulling together a group of disparate individuals. As RNA Tie Club member Alexander Rich later recalled:

> What Gamow did was to bring a kind of enthusiasm to the problem, and an intensity and a focus.[25]

What Gamow also brought was an emphasis on the role of information. And with his showman's instinct, he made sure that his influence was not restricted to the twenty or so members of the RNA Tie Club. In October 1955 he published an article in *Scientific American* that gave the scientific community and the public a glimpse into the thinking of the elite group.

Entitled 'Information transfer in the living cell', Gamow's article began by describing the fundamental unit of life in a radically new light, written in a way that was sure to grab the reader: 'The nucleus of a living cell is a storehouse of information,' he wrote. Gamow summarised the various coding schemes that had been devised, and showed readers how the work on the genetic code was linked with the other scientific sea-change that was taking place – the development of computers and cybernetics. The cell was described as 'a self-activating transmitter which passes on very precise messages that direct the construction of identical new cells', a reference to von Neumann's work on self-replicating automata. In fact, Gamow thanked von Neumann for helping with the calculations involved as RNA

Tie Club members looked for some kind of pattern in the frequency of amino acids found in different proteins. Gamow explained that the code was involved in transmitting information, drawing a rather laboured analogy between the cell and a factory, in which genes were blueprints stored in filing cabinets (chromosomes) which were used by workers (enzymes) on the factory floor (in the cytoplasm of the cell). For Gamow, information was at the heart of the cell. What that information was, he did not say.

Over the next year, Gamow, Teller and other members of the RNA Tie Club devised a number of theoretical models of the genetic code; each time their aim was to devise a scheme that produced twenty unique combinations corresponding to the widely occurring amino acids.[26] Gamow even managed to rope in a colleague from the Los Alamos laboratory, who used precious time on one of the rare computers available for scientific work – it was called MANIAC – to crank through the myriad possible combinations.[27] There was no direct way of testing any of these models – even rudimentary DNA sequencing was two decades ahead.

After a year of excited chatter, Crick's enthusiasm for the coding mania began to wane, and in January 1955 he produced the first of the RNA Tie Club's informal documents, surveying the various ideas that had been tossed about 'to subject them to the silent scrutiny of cold print'.[28] Crick tested Gamow's model against the first protein amino acid sequence data, which had been obtained thanks to some brilliant and complex chemistry by Fred Sanger, who was also based in Cambridge. In 1951, Sanger published a paper describing the sequence of amino acids in the insulin molecule. First he described the B chain, which was thirty amino acids long; two years later, he published the sequence of the shorter, but technically more demanding, A chain (twenty-one amino acids).[29] This work proved so significant that Sanger won the Nobel Prize in Chemistry a mere seven years later.

Crick used amino acid sequence data for cow and sheep insulin to show that Gamow's diamond code could not work; the two insulin sequences differed by a single amino acid in the middle of the chain, but in the diamond code a minimum of two amino acids would be altered by any base change. The diamond code was wrong. As Crick put it acidly:

it is surprising how quickly, with a little thought, a scheme can be rejected. It is better to use one's head for a few minutes than a computing machine for a few days![30]

Despite this jibe, Crick's view of Gamow's contribution was overwhelmingly positive – 'without our President, the whole problem would have been neglected and few of us would have tried to do anything about it,' he wrote. Crick praised Gamow for his three main contributions: he introduced the idea of 'degeneracy' (several different base sequences may code for the same amino acid – this is now generally called redundancy), he suggested that the words in the code might overlap (Crick admitted that neither he nor Watson had thought of this possibility), and he focused thoughts on coding as an abstract idea, 'independent as far as possible of how things might fit together'.[31]

Crick had a better grasp than Gamow of 'how things might fit together', and realised that he had to come up with a coding mechanism that fitted biological reality. But the exact details of protein synthesis – which was what coding ultimately related to – were not yet apparent. In a flash of brilliance, Crick described an idea he had discussed with a young South African researcher and junior member of the RNA Tie Club, Sydney Brenner, which they called 'the adaptor hypothesis'. Crick argued that the physicochemical structure of nucleic acids showed that neither DNA nor RNA was used as a direct template for building proteins – there was simply no physical correspondence between nucleic acid structure and the shape of proteins. Crick therefore hypothesised the existence of small 'adaptor' molecules, each specific to a particular amino acid, which would bind onto the DNA or RNA chain on one part of its structure and to an amino acid on another. Crick recognised that there was no evidence for the existence of such molecules, but he breezily dismissed this problem by supposing that there would be very few of them in the cell, so they would be difficult to detect.

If the adaptor hypothesis were true (and it was), this meant that the role of DNA and RNA would be reduced to simply carrying genetic information: they had no direct structural role as templates; they were merely a medium. If large nucleic acid molecules did not act as a physical template for the production of proteins, it became possible to imagine more complex forms of overlapping codes in which each word included some letters from the previous word. One of the consequences of such codes was that some pairs of amino acids should be found together in a protein sequence more often than expected by chance – each word in the code was not strictly speaking independent, because it shared some letters with the words on either side in the DNA sequence. Despite – or rather because of – this rash of new codes which were marked by what Crick called 'bewildering variety', his conclusion was not optimistic: 'In the comparative isolation of Cambridge I must confess that there are times when I have no stomach for decoding', he wrote.[32]

*

Crick's appetite might have been flagging, but other thinkers from outside the RNA Tie Cub were still keen. Drew Schwartz from Oak Ridge Laboratory produced a complex model that involved tweaking Watson and Crick's DNA structure and then allowing only particular kinds of amino acid to be bound in 'holes', as in Gamow's diamond code. Schwartz's hypothesis, which he admitted was speculative, was a way of linking the idea of a code (a word that Schwartz did not use) with previous models of protein synthesis, in which the gene – whatever it was made of – was thought to determine the shape of a protein molecule, by acting as a template.[33] The fact that most people realised that proteins were not synthesised on DNA did not seem to bother Schwartz.

For his part, Gamow remained as determined as ever. Together with Alex Rich and Martynas Yčas, Gamow summarised all the various coding alternatives that had been developed and explained why they had all failed.[34] Then, in a ingenious attempt at predicting what the code might be, the trio took data on the frequency of the different amino acids found in tobacco mosaic virus proteins and predicted what the frequency of each of the sixty-four possible 'triads' (AGC, etc.) might be, given the observed proportions of each of the four bases in viral RNA. But applying this method to different viruses gave completely different results: alanine was supposedly coded either by GCU or AAG according to the tobacco mosaic virus data, but by AAC or GGC if the turnip yellow virus was used (both viruses are based on RNA, so have U in place of T). If the genetic code were universal, which everyone accepted, it seemed likely there was a problem with this clever approach. In the end, all that Gamow and his colleagues could conclude was that overlapping codes looked unlikely and that there were at least three bases in each 'word'. This was hardly a great step forward from Watson and Crick's initial vague suggestion that the sequence of bases was a code.

The final blow to the idea of overlapping codes was delivered by Sydney Brenner, who had recently returned to Witwatersrand University. In the summer of 1956, Brenner gathered all the known amino acid sequence data and tried to see whether it would fit the coding model that was most popular at the time. This code was again based on groups of three letters or triplets (such as ATC), it was overlapping in that the final two letters in each triplet formed the first two letters of the next triplet (so, for example, the base sequence ATCCG would contain three triplet words – ATC, TCC and CCG), and, in the language of the time, it was 'degenerate' – an amino acid could be represented by more than one triplet.[35] Because of the overlapping code, any given triplet could be followed by only four different overlapping triplets (so **ATC** could be followed by only **TCA**, **TCC**, **TCG** or **TCT**) and preceded by only four different triplets (**AAT**, **CAT**, **GAT** or **TAT**). Everyone assumed that each triplet corresponded to an amino acid, so that would mean that any amino acid coding for, say, cysteine, could only have four different amino acids on either side of it in the sequence. Brenner's collection of sequence data showed that the actual sequence variability was much greater – cysteine, for example, was preceded by fifteen different amino acids and followed by fourteen different amino acids. Brenner calculated that if the code were overlapping, more than seventy triplets would be required to code the few dozen sequences thus far assembled, whereas only sixty-four combinations were available. Given that the full range of amino acid sequence combinations was probably as high as it could be (in other words, any of the twenty amino acids could precede or follow any other), Brenner's conclusion was simple: 'all overlapping triplet codes are impossible'. 'It seems clear that non-overlapping equivalence between nucleic acids and proteins must exist', he wrote, but it was not at all obvious what this might be.[36] Brenner had shown what the code could not be, not what it was.

Gamow took Brenner's demolition job in good part and ensured that a slightly expanded version of the younger man's article appeared the following year in *Proceedings of the National Academy of Sciences*. Brenner recalled: 'I'm proud of that paper because that was just really spurred by just knowing how to divide by four!'[37] In rewriting the article for a public audience, Brenner framed the problem in the most abstract way – given it was still not certain how protein synthesis took place, and what exactly were the relations between DNA and RNA and protein, he carefully avoided this issue in the article and above all added a new subtitle, taken from some of Gamow's articles, which focused on the role of triplet codes 'in information transfer from nucleic acids to proteins'.[38]

Here was the great advantage and the fatal weakness of the three years that the RNA Tie Club had spent trying to come up with theoretical answers to the coding problem. The work of the Club had elevated coding to a completely abstract level, far away from the molecular jiggery-pokery that was going on every second in every cell all over the planet. That meant that the code was not explicitly about the mechanism of protein synthesis – it was about information transfer, although nobody had spelt out what that information was. But all this abstraction would remain at the level of talk unless it could be turned into biochemistry. In the end, experimental data had the last word, not the theoretical schemes of Gamow and his clever drinking pals in the RNA Tie Club.

*

Twice, Boris Ephrussi asked Francis Crick a question that he could not answer, and both times it was the same question. The first occasion was at Woods Hole in 1954, the second was in a Paris café a year later. Each time, Ephrussi wanted to know why Crick was so certain that the DNA sequence encoded the sequence of amino acids.[39] This pointed question went to the heart of everything that had been done since Watson and Crick had discovered the double helix – they had been avoiding the central issues of what the sequence of bases did, or to put it another way, what information they contained. Everyone assumed that there was what was called a colinearity between the sequence of bases in a DNA molecule and the sequence of amino acids in the corresponding protein, but Ephrussi's question underlined that, in the light of the available evidence, that information could equally be linked to something as mysterious as the overall form of the protein, or it could be something else altogether. As Crick later recalled:

> There was no evidence, you see; there was *absolutely no evidence* that a gene made a difference in the sense that you could actually determine the amino-acid sequence and show that with a mutation it was changed.[40]

Goaded and intrigued by Ephrussi's challenge, in the spring of 1955 Crick began searching for an example of protein variation that he could tie to genetic variation. He looked for a protein that was widely available and was from an organism that could be studied genetically. The protein he chose was an enzyme called lysozyme, which is found in chicken eggs and in human tears. Together with the German-born biochemist Vernon Ingram, Crick tried to crystallise lysozyme from the eggs of various bird species, and even used onions to make himself cry to provide a source of tears. The project was a failure – they could get lysozyme from chickens but not from other species, meaning that a comparison between species would be impossible. So they tried looking for variation in lysozymes from different individual chickens, hoping to eventually relate such variation to genetic differences. Again they found nothing. As Crick wrote to Brenner:

> Attempt by Vernon Ingram and myself to find two hens with different lysozymes so far completely negative … It is all rather discouraging. Even if we find a difference we shall still have to show it's due to amino acid composition, and also do the genetics.[41]

But then Ingram came up with something much more dramatic,

which provided the answer to Ephrussi's question and showed the potential applications of the work that Crick and his colleagues were doing. In 1949, two articles had appeared in *Science* reporting new findings on sickle-cell anaemia, a genetic disease that predominantly affects people of African origin and can lead to debilitating weakness or death. People with sickle-cell disease have strange sickle-shaped red blood cells – hence the name. It had first been described as a genetic disease in 1917, but it was only in July 1949 that James Neel of the University of Michigan showed that the best explanation of the pattern of inheritance of sickle cells was that the trait was caused by a single gene.[42] Four months later, in November 1949, Linus Pauling and his group published a study of the haemoglobin molecule found in the blood of patients with sickle-cell anaemia, with the dramatic title 'Sickle cell anemia, a molecular disease'.[43] They described two forms of haemoglobin, one associated with sickle-cell anaemia (later called the S form) and the other with normal individuals (the A form). These two forms could be distinguished by electrophoresis – when the two haemoglobins were placed in a gel and subjected to an electric field, they moved at different rates, so after a certain time they were found at different points. Pauling's group concluded that the way the two forms moved under electrophoresis suggested that there was a difference in the shape of the two molecules, with the S form having more positive charges than the A form. For the first time, a disease had been shown to have a molecular basis.

Max Perutz's group at Cambridge had been studying haemoglobin for years, and had begun to look at the structure of the S form. Encouraged by Crick and Perutz, in the summer of 1956 Vernon Ingram looked at some samples of the S form of haemoglobin that had been left unused by another researcher, Tony Allison. Allison's field work in Uganda had led him to suggest why sickle-cell anaemia still exists despite having potentially lethal consequences in patients with two copies of the sickling gene: individuals who have one normal and one sickling gene have a lower load of malaria parasites.[44] Having one sickling gene is actually a good thing in malaria-ridden regions.

Ingram's aim was to detect the precise molecular difference between the S and A forms that had been described in general terms by Pauling. The haemoglobin molecule was too long for Fred

Sanger's sequencing method, which remained relatively primitive. First, Ingram had to snip the molecules into smaller bits by using an enzyme; then he divided the components on filter paper, first separating the two forms by applying an electrical current just as Pauling had done and then applying a solvent at 90° to the direction of the current. Once a chemical had been sprayed on the paper to reveal the invisible components of the haemoglobin molecule by turning them purple, the result was a two-dimensional 'finger print' of the molecule.[45]

The differences in electrical charge shown by the two forms of haemoglobin had two potential explanations. Pauling thought it was most likely that the overall shape of the molecule was the source of the differences, and that in turn was produced by the way in which the protein was folded and shaped during synthesis. As he argued in 1954:

> The interesting possibility exists that the gene responsible for the sickle-cell abnormality is one that determines the nature of the folding of polypeptide chains, rather than their composition.[46]

The other possibility was that the difference in charge was simply to do with a difference in the sequence of amino acids in the two forms.

Ingram's 'finger prints' of the two forms of haemoglobin each produced about thirty marks on the filter paper; these were identical in both cases, except for one small blob that was present in the S form and absent in the A form. He therefore analysed this particular component in great detail, and was able to show that the difference between the two molecules was due to a small part of the protein. In an article in *Nature* published in 1956, Ingram concluded:

> there is a difference in the amino-acid sequence in one small part of the polypeptide chains. This is particularly interesting in view of the genetic evidence that the formation of haemoglobin S is due to a mutation in a single gene. It remains to be seen how large a portion of the chains is affected and how the sequences differ.[47]

Shortly before the article appeared, Ingram presented his

findings to the meeting of the British Association in Sheffield, in August 1956. The science correspondent of *The Times* was there and immediately sniffed out the story, describing Ingram's work in some detail and highlighting its significance:

> He described how in the past six weeks he had shown for the first time how a mutation in a single gene, the unit of heredity, can modify chemical structure in a substance in the body for which that gene is responsible.[48]

Within a year, Ingram was in the pages of *The Times* again, following publication of another article in *Nature* that this time revealed the exact cause of the peptide difference between the two forms: it was all due to a single amino acid.[49] In the S form, a valine molecule replaced the glutamic acid found in the normal A form. This minor change, in the most fundamental component of a protein, caused the changes in the behaviour of the molecule that led to the debilitating disease. Although the genetic code was still a mystery, Ingram's work had shown that there was a relation between a mutation in a gene, which was made of DNA, and a change in the amino acid sequence of a protein. Ephrussi's question had been answered. For *The Times*, this was a discovery that was on a par with Mendel's observations that led to the foundation of genetics. The article concluded:

> Dr Ingram's demonstration of a single, identified difference between genetically determined haemoglobins is thus the nearest that has been got to a direct view of one of Mendel's genes in action. This is indeed a landmark.[50]

Ingram's discovery was a brilliant confirmation of the new understanding of gene function. Previously it had been thought that the gene shaped the protein, like a three-dimensional mould. Ingram, inspired by Crick, had now shown that a gene could alter a single amino acid in a one-dimensional protein sequence, and that in turn would alter the function of the protein. A new vision of protein synthesis was appearing, and it had consequences for how the gene and the genetic code were understood.[51] If the gene contained information and a change in that information led to a change in an amino acid sequence, this suggested that genetic information corresponded to nothing more than the amino acid sequence in a protein produced by a gene.

EIGHT

THE CENTRAL DOGMA

In September 1957, Francis Crick gave a lecture at University College London. He had been invited by the Society for Experimental Biology to speak at a symposium entitled 'The Biological Replication of Macromolecules'.[1] For nearly a year, Crick had been musing about what genes actually do, thinking about the mechanisms of protein synthesis, trying to tease out the biochemical steps and their theoretical consequences. The London conference was his opportunity to present his ideas about protein synthesis. It was the most influential lecture he ever gave.

The French molecular geneticist François Jacob was in the audience and recalled his impression of Crick:

> Tall, florid, with long sideburns, Crick looked like the Englishman seen in illustrations to nineteenth-century books about Phileas Fogg or the English opium eater. He talked incessantly. With evident pleasure and volubly, as if he was afraid he would not have enough time to get everything out. Going over his demonstration again to be sure it was understood. Breaking up his sentences with loud laughter. Setting off again with renewed vigour at a speed I often had trouble keeping up with. ... Crick was dazzling.[2]

Crick's recollection was not quite so positive – 'I ran overtime, and didn't get it over very well', he felt.[3]

Crick's lecture led to two articles – one in *Scientific American* that appeared at the same time, and another more detailed piece that was published in the symposium collection in 1958. This second paper has been cited more than 750 times and is still cited frequently.[4] The bold proposals Crick made in his lecture continue to play a fundamental role in modern debates over the nature of the genetic code and the evolutionary process.

Behind its dull title, 'On protein synthesis', Crick's talk explained some of the new ideas about what genes do and how they do it, all wrapped up in an elegant, conversational style that still beguiles the reader. At every step he admitted the limits of his knowledge, and distinguished clearly between established fact and logical conjecture. It was these conjectures that proved so influential. Crick's starting point was his assumption that the role of genes is to control protein synthesis, even though, as he put it with disarming simplicity, 'the actual evidence for this is rather meagre':

> I shall ... argue that the main function of the genetic material is to control (not necessarily directly) the synthesis of proteins. There is a little direct evidence to support this, but to my mind the psychological drive behind this hypothesis is at the moment independent of such evidence. Once the central and unique role of proteins is admitted there seems little point in genes doing anything else.[5]

Crick claimed that the involvement of nucleic acids in protein synthesis was 'widely believed (though not by every one)'.[6] The caveat was significant – even at this date, there were still those who found it hard to accept that all genes were made of DNA. A year earlier, at a symposium on 'The Chemical Basis of Heredity' held in June 1956, Bentley Glass noted the reluctance of some geneticists to abandon the protein part of the old nucleoprotein theory of heredity, but he was nonetheless sure that 'most persons' accepted that DNA (or RNA in some viruses) was the primary genetic material.[7] This obviously left open the possibility that there might be other, secondary and non-nucleic acid forms of heredity. At the same meeting, George Beadle focused on this problem in his opening address, underlining that there was no experimental evidence that DNA was the genetic material in organisms other than viruses and bacteria. In his talk, Steven Zamenhof, who had worked with Chargaff and had been an early supporter of Avery's claim that DNA was the genetic material in bacteria, accepted that although 'extensive evidence' supported the argument that the pneumococcal transforming principle was DNA, and that 'no evidence to the contrary had ever been presented', there was nevertheless 'no absolute proof'.[8] With his fellow-scientists still haunted by doubt, Beadle argued that 'it is assumed as a working hypothesis that the primary genetic material is DNA rather than protein.'[9] What now appears evident was merely a 'working hypothesis' in 1956.

As Crick outlined in both his symposium talk and in his *Scientific American* article, the direct evidence for the involvement of nucleic acids in protein synthesis came from two recent sources – Ingram's discovery of the molecular basis of sickle-cell anaemia, and work on the tobacco mosaic virus (TMV), which was now known to use RNA as its genetic material. In 1956, Heinz Fraenkel-Conrat, a biochemist working at Berkeley, had shown that it was possible to separate the RNA and protein components of the TMV and then reassemble them to produce a functional virus. Fraenkel-Conrat then went on to recombine protein from one TMV strain and RNA from another strain; the recombinant strains produced viruses with proteins that were typical of the RNA donor strain and never like those found in the protein donor strain. In his 1957 lecture, Crick described this finding and concluded: 'the viral RNA appears to carry at least part of the information which determines the composition of the viral protein'.[10] For the first time, Crick spelled out the precise implications of the use of the term information in genetics: 'By information I mean the specification of the amino acid sequence of the protein.'[11]

This was not information as Shannon had described it – Crick was not referring to a mathematical measure of the degree of uncertainty of one particular message. He was talking about something much more tangible, straightforward and immediately understandable: the sequence of amino acids in a protein. A gene somehow carried the code that could produce a particular amino acid sequence, and it was also capable of passing that code to the next generation. The information necessary to include a particular amino acid in a protein was encoded by the sequence of bases in the DNA that made up the gene.

Crick then repeated an assertion that he had recently been touting at various conferences: the sequence of amino acids alone was significant in determining protein function. Although the three-dimensional shapes of proteins were ultimately the explanation of specificity, of the myriad functions of proteins, these complex structures were in fact contained within the one-dimensional message of DNA. The three-dimensional protein structure simply emerged out of the one-dimensional DNA code through the process of protein synthesis, argued Crick. Nothing else mattered except the sequence of amino acids, which was in turn determined by the order of bases in the DNA molecule: the genetic code. As Crick put it: 'It is of course possible that there is a special mechanism for folding up the chain, but the more likely hypothesis is that the *folding is simply a function of the order of the amino acids*'.[12] Crick called this view the sequence hypothesis:

> In its simplest form it assumes that the specificity of a piece of nucleic acid is expressed solely by the sequence of its bases, and that this sequence is a (simple) code for the amino acid sequence of a particular protein. This hypothesis appears to be rather widely held.[13]

Crick may have though it was 'widely held', but plenty of scientists were deeply suspicious of Crick's view.

In June 1956, Erwin Chargaff, in typically contrarian and acerbic mood, complained that too much attention was being paid to nucleic acids, and he still wondered whether there was something else apart from DNA that was giving proteins their final shape:

> It is obvious that sequence cannot be the sole agent of bio-logical information. Even if the arrangement of an entire poly-nucleotide would be written, a third dimension would be

lacking: the operative three-dimensional shape of the molecular aggregate, in which perhaps not only numerous nucleic
acid molecules take part, but also proteins and possibly even
other polymers'.[14]

Another doubter was Macfarlane Burnet, who had just published a short book called *Enzyme, Antigen and Virus: A Study of Macromolecular Pattern in Action*. Burnet's book had made an impression
on Crick because it took an opposing stance to him on many points.
In his book, Burnet declared that it was 'quite impossible at the present time' to envisage how a nucleic acid could specify a linear polypeptide sequence; like Chargaff, instead of relying totally on DNA,
Burnet wanted to 'leave open the possibility of some associated factor, histone possibly, which allows a sufficient complexity to carry
the needed coding.'[15] Even leading scientific figures were clinging to
the idea that proteins must play an essential role in genetics.
    Part of the problem was that no one knew exactly how a cell took
a nucleic acid sequence and turned it into a blob of protein made up
of amino acids. It was known that the main steps of protein synthesis took place in the cell's cytoplasm, which surrounds the nucleus.
On the one hand, DNA was known to stay in the nucleus, so it was
clearly not directly involved; on the other hand, RNA was found
throughout the cell in a variety of forms, apparently associated with
the synthesis of proteins. It looked as though the genetic message
passed from DNA to RNA, but how this happened was unclear. It
was equally uncertain how the amino acid chain was assembled
to make a protein, although small recently discovered RNA-rich
structures called microsomal particles seemed to be involved. Informal discussion at a conference in 1958 led to these particles being
re-baptised 'ribosomes', by which name they are known today.[16] A
few weeks before Crick's talk, Mahlon Hoagland and Paul Zamecnik at Harvard had shown that if amino acids were radioactively
labelled, proteins throughout the cell were eventually found to be
radioactive, indicating that the amino acids had been assembled into
a protein.[17] On shorter time-scales, however, radioactivity was found
only in the ribosomes, strongly suggesting that amino acids had to
pass through the ribosome to be combined into a protein.[18] It seemed

likely that the RNA in the ribosome was the actual site where the protein was made. This raised the question of the nature of the link between DNA and RNA, and how each amino acid found its way to the ribosome.

In his lecture, Crick turned his brilliant mind to both these issues and publicly described the idea he had worked up with Brenner: there must be an unknown class of small molecule, which they called an adaptor, which would gather each of the twenty amino acids and take them to the ribosome, so that the protein could be assembled there. The most likely hypothesis was that there was one adaptor for each type of amino acid, and that it would contain a short stretch of nucleotides – a tiny bit of the genetic code that was able to bind to the RNA template in the ribosome, just like base pairing between the two strands of the DNA double helix. At the same time, on the other side of the Atlantic, Hoagland and Zamecnik were isolating what was later identified as Crick's adaptor – eventually known as transfer RNA or tRNA – without knowing anything about Crick's hypothesis.[19]

To put all this speculation into a theoretical context, Crick explained to his audience that there were three ways of understanding the processes involved in protein synthesis – 'the flow of energy, the flow of matter, and the flow of information.'[20] He focused on the final, most elusive, and most radical aspect – the flow of information. In both his lecture and the *Scientific American* article that appeared at the same time, Crick used a memorable term to describe a fundamental feature of genes: he outlined what he called 'the central dogma' of genetics. Crick explained this dogma as follows:

> once information (meaning here the determination of a sequence of units) has been passed into a protein molecule it cannot get out again, either to form a copy of the molecule or to affect the blueprint of a nucleic acid. The idea is not universally accepted, however. In fact, Sir Macfarlane Burnet, the eminent Australian virologist, persuasively argued another point of view in a very interesting little book which he published recently.[21]

On the basis of the available molecular evidence, Crick was arguing that there were four kinds of information transfer that were likely to take place routinely: DNA → DNA (as in DNA duplication), DNA → RNA (as in the first step of protein synthesis), RNA → protein (as in the second step of protein synthesis) and RNA → RNA (presumed to exist because of the existence of RNA viruses such as TMV, which used RNA both to store information and to synthesise proteins and were able to copy themselves). Crick agreed that there were also two possible information transfers that might conceivably take place, but for which there was no evidence: DNA → protein (this would occur if protein synthesis took place directly on the DNA molecule, which seemed unlikely) and RNA → DNA (there was no evidence for this, nor any biological process that seemed to require it, but it was not seen as structurally impossible).

There were several genetic information transfers that Crick and his colleagues considered to be impossible: protein → protein (this had been disproved by the work of Avery, Hershey and Chase and others), protein → RNA and, above all, protein → DNA. There was no evidence for any of these information flows, nor was there any conceivable mechanism for the sequence of amino acids being back-translated into a DNA or RNA code. As Crick later put it, 'I decided, therefore, to play safe, and to state as the basic assumption of the new molecular biology the non-existence of [these] transfers'.[22] This was the 'central dogma': once information had gone from DNA into the protein, it could not get out of the protein and go back into the genetic code.

*

Crick had first come up with the 'central dogma' phrase and its underlying concept in October 1956, in a set of notes entitled 'Ideas on protein synthesis'.[23] These were not circulated – not even to the RNA Tie Club – but they formed the basis of his discussions with his colleagues and his thinking over the following months. In those original notes – but not in either of the published forms of his talk – Crick included a little diagram to show what he meant.

On the original note, Crick playfully wrote 'The doctrine of the

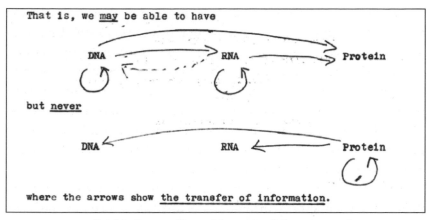

That is, we may be able to have

DNA → RNA → Protein

but never

DNA ← RNA ← Protein

where the arrows show the transfer of information.

5. Crick's first outline of the central dogma, 1956,
http://profiles.nlm.nih.gov/ps/access/SCBBFT.pdf

Triad' (DNA, RNA and protein), but he soon coined the more dramatic term 'central dogma'. As was evident from his presentation of the idea in 1957, it was not, strictly speaking, a dogma (a fundamental belief that cannot be questioned). It was instead a hypothesis, and rather than being based on any a priori position, it was simply based on the available data. Crick later recalled:

> I called this idea the central dogma, for two reasons, I suspect. I had already used the obvious word hypothesis in the sequence hypothesis, and in addition I wanted to suggest that this new assumption was more central and more powerful. I did remark that their speculative nature was emphasized by their names. As it turned out, the use of the word dogma caused almost more trouble than it was worth. Many years later Jacques Monod pointed out to me that I did not appear to understand the correct use of the word dogma, which is a belief that cannot be doubted. I did apprehend this in a vague sort of way but since I thought that all religious beliefs were without any serious foundation, I used the word in the way I myself thought about it, not as most of the rest of the world does, and simply applied it to a grand hypothesis that, however plausible, had little direct experimental support.[24]

As well as predicting the results of future experiments, Crick was unwittingly supporting two of the central tenets of twentieth-century biology. First, there was the assumption, suggested by August Weismann in the 1890s, that in animals the organism contains two entirely separate cell lines, one devoted to the development of the body (the somatic line) and the other to reproduction and the transmission of hereditary characters (the germ line) (no such division exists in plants or, obviously, in single-celled organisms; it is also absent in some animals). According to Weismann, these two cell lines did not interact. As a result, it was impossible for any character that was acquired during an animal's life and which affected the somatic line, to have any effect on the germ line, on heredity. In Crick's language, information could not go in the direction protein → DNA.

By providing a molecular basis for Weismann's position, Crick was also reinforcing the widespread opposition to the suggestion by the nineteenth-century naturalist Lamarck (and also Darwin) that characters acquired by an organism during its lifetime could have an effect on its offspring by altering its hereditary constitution. In Weismann's model, there was simply no route for this to occur. During the 1920s and 1930s, Weismann's division of cell types became a cornerstone of what was known as the neo-Darwinian synthesis as genetics and evolutionary theory fused, changing the way in which scientists looked at evolution by natural selection. In 1957 Crick said essentially the same thing as Weismann, but in the latest language and with a far greater import, because it applied to all organisms, not just animals: once information had got into the protein, it could not get out again. It could not go back into the DNA. Although not a dogma, this was a very strong assertion. As Crick later reflected: 'In looking back I am struck ... by the brashness which allowed us to venture powerful statements of a very general nature.'[25]

Crick did not consciously set out to support the neo-Darwinian position – that would indeed have been dogmatic. Instead he developed his ideas on the basis of the experimental data, which revealed no potential mechanism for information to go from protein to DNA.[26] Crick recalled:

$$\begin{array}{ccc} \textbf{DNA} & \longrightarrow & \textbf{DNA} \\ | & & \\ \textbf{(?PR)} & & \\ \downarrow & & \\ \textbf{RNA} & & \\ \uparrow \quad \downarrow & & \\ \textbf{LIPID} \longleftarrow \textbf{F.PR} \longrightarrow \textbf{POLYSACCHARIDE} \\ \vdots \\ \textbf{STRUCTURAL} \\ \textbf{PROTEIN} \end{array}$$

6. Burnet's view of the pathway from DNA to protein, from Burnet (1956). PR = protein, F.PR = functional proteins, such as enzymes.

Nobody tried to go from protein sequence back to nucleic acid, because that wasn't on. You see. But I don't think it was ever *discussed.*[27]

As Crick explained in both his *Scientific American* article and his lecture, he was struck by Macfarlane Burnet's adoption of a rather different position in *Enzyme, Antigen and Virus*. Burnet later described this as 'a rather bad over-ambitious book', primarily because he very soon changed his views on its main subject – the way in which antibodies are created.[28] Whatever the book's virtues, it contains a diagram outlining his view that may have been the inspiration for Crick's sketch.

Given that Burnet stated that he was trying to apply information theory to cell communication, it is notable that his arrows do not show the flow of information, but instead represent what he called specific pattern (a rather vague concept that seemed to mean something like specificity). In the accompanying text, Burnet described how the system was capable of 'transferring coded information in pattern on one medium to a different coding of pattern on another medium'.[29] Burnet's main difference with Crick, apart from the systematic importance he gave to enzymes (given in the diagram as 'F.PR = functional proteins'), was that the gene product can alter the genetic message, in the shape of the RNA. In Burnet's schema, information can get from the protein into RNA, and hence into the way in which genetic information is represented. Although Crick implied that Burnet argued that proteins could affect DNA, this was not in fact what Burnet stated. Burnet's explanation of his hypothesis was complex and was based on his then-current model of how the body is able to generate vast numbers of different molecules that are used in the immune response as a way of differentiating self from non-self. Burnet argued that foreign proteins or antigens prompted the body to produce a specific antibody by becoming incorporated into the RNA; this was the nature of the arrow going from 'F.PR' to RNA.[30] Within three years, Burnet had abandoned this view in favour of his new clonal theory of antibody generation; a year after that he won the Nobel Prize in Physiology or Medicine for his work on viruses and the immune response.

In the 1960s, Crick's central dogma was rendered truly dogmatic by Jim Watson, who included a simplified version in his textbook *Molecular Biology of the Gene*, converting it to the form first outlined by Boivin in 1949 and Dounce in 1952: DNA → RNA → protein.[31] Crick's initial view, which allowed for the unlikely possibility of information transfer from RNA to DNA, was largely forgotten. Crick's dramatic and mistaken use of words ended up undermining his aim – for some scientists, the central dogma became a dogma, and not merely a hypothesis.[32]

*

Like Burnet, Crick was later critical of his own work. In his autobiography, Crick described 'On protein synthesis' as 'a mixture of good and bad ideas, of insight and nonsense'.[33] As with Burnet, Crick's self-criticism was aimed at those detailed areas of the mechanism that he got wrong. On the big picture, Crick was absolutely right. And in one area, both he and Burnet were positively visionary. Although neither Burnet nor Crick were evolutionary biologists, they each had insights into the way that evolution affects genetic information, insights that are still valid today. In his 1957 talk, Crick pointed to the handful of proteins that had thus far been sequenced in more than one organism and made a leap of the imagination that eventually transformed how we study evolution:

Biologists should realise that before long we shall have a subject which might be called 'protein taxonomy' – the study of the amino acid sequences of the proteins of an organism and the comparison of them between species. It can be argued that these sequences are the most delicate expression possible of the phenotype of an organism and that vast amounts of evolutionary information may be hidden away within them.[34]

Crick was right. Today, protein fragments from the depths of time, such as bits of collagen from *Tyrannosaurus rex*, can be used to study evolution.[35]

Burnet's contribution to evolutionary biology was less dramatic but equally insightful. In the pages of *Enzyme, Antigen and Virus*, Burnet described the Watson–Crick model of the genetic code – that is, of a relation between a sequence of four nucleotide bases in the nucleic acid and the near-infinite structure of proteins – as 'faintly unsatisfactory'.[36] What exactly was 'unsatisfactory' – apart from the fact that the genetic code was still unbroken – Burnet did not explain. Instead he did some quick back-of-the-envelope calculations about the amount of DNA in an average cell and came up with a problem. Assuming that a human cell contained 40,000 genes, each composed of 3,000 bases, that would still account for only 1 per cent of the DNA that was estimated to be present in the nucleus of a human cell; this raised the question of what the other 99 per cent was doing.[37] Although Burnet's guesstimate that an average gene was 3,000 bases long was entirely gratuitous (genes are in fact often much, much bigger), his question was entirely valid – much of the DNA in our cells seems to be doing nothing. By the 1970s, when it became obvious that this really was a problem, Burnet's insight had long been forgotten.

Crick's 1957 speech and the articles that accompanied it were enormously influential, in terms of both the ideas they contained and the words they used. Crick's framework – seeing genes and proteins in terms of information flow – rapidly became the accepted way of understanding the fundamental processes of cells. It might be expected that this shift would have been accompanied by a flourishing application of information theory in the realm of molecular biology – Crick's information could surely be studied using Shannon and Wiener's equations. In fact, at the same time as Crick came up with the idea of the central dogma, it began to become evident that information theory was not going to transform biology.

*

In 1942, 3,000 people who lived and worked on a plateau near Oak Ridge in the Appalachian Mountains were ordered to leave their homes. Within a year, the US government had built a small city on the site, devoted to the production of weapons-grade uranium and plutonium as part of the drive to build the atomic bomb. Thousands of workers were involved in this top-secret work, which cost more than $1bn to set up. After the war, Oak Ridge National Laboratory focused on the development of nuclear reactors and on the production of radioisotopes for use in medicine and in research, as well as occasional crazy projects such as a plan to build a nuclear-powered aircraft. New groups were set up at the laboratory, including a Biology Division to study the effects of radiation and a Mathematical and Computing Section, armed with the latest von Neumann computers, to analyse the data that the site generated. Researchers at Oak Ridge had a broad range of scientific interests, and in 1950 some of them organised the meeting on the chemistry of nucleic acids at which Chargaff showed that DNA was not a boring molecule.

At the end of October 1956, scientists at the laboratory held a symposium on the links between information theory and biology, following in the footsteps of Henry Quastler's 1952 meeting. The conference, hosted by the Manhattan Project physicist Hubert Yockey, was entitled 'A Symposium on Information Theory in Health Physics and Radiobiology', and half of the talks focused on how radiation and ageing affected biological tissues and processes.[38] Some of the other presentations ambitiously tried to find evidence for negative feedback in liver regeneration, or studied the role of protein synthesis and information transfer in the development of the chick embryo, but none came to any real conclusion. The main issues that were discussed in the autumn Appalachian air related to the application of information theory to the genetic code. Gamow and Yčas were there, although they were the only members of the coding community who attended – none of the experimentalists who were involved in trying to crack the code made the trip.

Yockey began uncontroversially by outlining what he called the coding chain: protein specificity was encoded by the order of amino acids, which in turn was encoded by the order of base pairs, which in turn implied the existence of a code that translated from four 'letters' (A, C, G and T) to twenty 'words' (the amino acids). Then, with the confidence of a theoretician unhindered by experimental facts, Yockey suggested that cracking the code was a problem for mathematicians, not biologists:

> Thus by following the logical consequences of purely biological, or perhaps biochemical, problems, one is led directly to a problem purely mathematical in character.[39]

Yockey claimed that 'the central ideas of this paper are independent of much of the detail embodied in Watson and Crick's papers' – the coding problem had no necessary link with biochemical reality, he seemed to be suggesting. Nonetheless, he boldly claimed that it was possible that 'the role information theory will play in biology will parallel that played by thermodynamics in physics and chemistry.'[40]

Gamow and Yčas were less optimistic. Yčas looked at the information content of proteins, treating them as a text and applying cryptographic techniques to work out the genetic code that lay behind it. To Yčas's dismay, there was no apparent consistency in the organisation and length of protein molecules, and an audacious attempt to predict protein diversity from RNA diversity failed dismally.[41] In a joint paper with Yčas, Gamow summarised the molecular model of gene function that flowed from Watson and Crick's work, but ultimately could only express his weary conviction that as more protein sequences became known, 'this problem will be solved in one way or another'. The only way in which Gamow could conceive of cracking the code was by continuing to treat it as a mathematical puzzle that would eventually reveal a solution if probed with sufficient ingenuity.

In contrast, the scientists at the meeting who were most directly involved in information theory were beginning to question whether the informational approach to biology was the right one at all. In so doing, they began to undermine some of the main points of Shannon's original vision. It was widely known that Shannon was not enthusiastic about the application of his ideas to fields other than the strict realm of communication. Henry Quastler, who had organised the 1952 meeting and had been the intellectual driving force behind the biological application of Shannon's ideas, countered this criticism by showing that information theory was an appropriate tool for biology: biological systems involve control, control depends on communication, and communication depends on information. This neat summary tied in with cybernetics, but brought in another problem: meaning.

The key point in cybernetics was that certain signals – for example negative feedback – had greater significance than others in determining how a system functioned. Meaning had been anathema to Shannon, who had insisted that information had to be considered at a completely abstract level for his theory to exercise its full power. But it seemed that in biological systems it was impossible to avoid meaning. In 1953 this point had been the focus of a highly critical paper presented to a meeting on cybernetics by the philosopher Yehoshuua Bar-Hillel.[42] At the 1956 Oak Ridge meeting, the physicist Leroy Augenstine took up the argument and pointed out that not all 'bits' of information were equivalent – their meaning and context had to be taken into account:

> It seems very likely that one bit of potential structural information will not always transmit the same amount of information; rather, the efficiency of transmission will depend upon the context within which the performance is measured. ... This is somewhat analogous to the relative difficulties of determining whether a symbol is 0 or 1, or to determining whether one should get married or not![43]

To understand messages, information theorists needed more than bits; they had to introduce context and meaning into their calculations. Quastler highlighted the problem by using the example of a conversation between two individuals. At first sight, it might appear that during a conversation, information is transmitted by words. This is undoubtedly the case, but many other factors are involved, such as the selection of particular words from various potential synonyms, the tone of voice, the timing of the utterance, speech volume and so on. Identifying all these levels of potential information seemed impossible. What appears to be a simple example of information transmission is in fact extremely complex. Quastler concluded: 'In such a situation we have obviously no hope ever to obtain a precise, unequivocal, and incontestable measure of information content.'[44]

Similar problems were encountered at the molecular level. Leroy Augenstine argued that it would be futile to calculate the amount of information in any chemical communication system, because 'only a small fraction of the potential information on the surface of the molecule is actively utilized in information transfer.'[45] Quastler's summary of the situation undermined the idea that information was a strictly objective measure: 'information applications are relative and not absolute; hence, any information measure associated with a given set of biological objects will depend on the set itself and on the scientist who does the estimating.'[46] Paradoxically, this excess of potential information made it almost impossible to apply the concept in a biological context. As Quastler put it:

> Every kind of structure and every kind of process has its informational aspect and can be associated with information functions. In this sense, the domain of information theory is universal – that is, information analysis *can* be applied to absolutely anything. The question is only what applications are useful.[47]

As Quastler recognised, concrete examples of such useful applications of theory were turning out to be few and far between. After the symposium had finished, a small group of participants met in the evening to discuss their impressions. A slightly depressed Quastler summed up the general feeling:

> Information theory ... has not produced many results so far;
> it has not led to the discovery of new facts, nor has its appli-
> cation to known facts been tested in critical experiments. To
> date, a definitive and valid judgement of the value of informa-
> tion theory in biology is not possible.[48]

If the leading lights of information theory were uncertain as to its
usefulness, the idea of cracking the genetic code by simply applying
it to protein sequences, as Gamow suggested, seemed unlikely to
succeed.

Biologists were also expressing doubts about the usefulness
of information theory. In 1954, the zoologist J. Z. Young, who had
talked extensively about information in his 1950 Reith Lectures, surveyed the field and concluded:

> In spite of much discussion of the application of information
> theory in biology there is still a considerable uncertainty about
> the status of the analogy and about its usefulness.[49]

In 1955, Joshua Lederberg began a correspondence with von Neumann, in which the pair explored the similarities between von Neumann's model of self-replicating automata and theories about the
origin of life. The two men soon realised that each of them did not
understand what the other meant by 'information', and Lederberg
eventually concluded that this was because they were thinking 'at
very different levels'. For biologists, he argued, the 'propagation,
and evolutionary elaboration, of complexity is self-evident' – they
were interested in the detail of how such a system could work. The
logician van Neumann, however, was 'looking for the foundations
of an axiomatic theory of reproduction' – something much more
abstract and not necessarily linked to biology at all. Summarising
von Neumann's views, Lederberg confessed that he 'could not begin
to say whether they would be helpful in genetic analysis.' Although
the two men agreed to meet to discuss the question further, von
Neumann's ideas had no perceptible effect on Lederberg's biology.[50]

Macfarlane Burnet encountered similar difficulties when he was
writing *Enzyme, Antigen and Virus*. As he explained:

> This monograph was originally conceived as an attempt to
> develop something analogous to a communications theory
> that would be applicable to the concepts of general biology.
> However, it has not been found possible to make any serious
> use of the already extensively developed concepts of informa-
> tion theory in the strict sense.[51]

The main reason that Burnet gave for his failure was that 'only the
most generalised sketch of an outline has yet been given of how
information theory at the strict level can be applied to biology.'

J. Z. Young solved the problem by deciding he was not in favour
of considering organisms as examples of Shannon's simple lines of
transmission, but instead as something much more like a computer.[52]
This analogy was also explored by the Bristol psychologist Frank
George at a 1959 Society for Experimental Biology symposium on
'Models in Biology'. George argued that the main contributions of
cybernetics were analogies and metaphors, particularly 'the central
importance of feedback – whereby organisms can be likened to, and
described in the same manner as, inanimate systems'. But instead
of using an approach based on information theory, George thought
that 'the analogy with computers, suitably modified and adapted, to
a good one, and one that is essentially testable.'[53] This kind of vague
metaphor, no matter how supposedly testable, was far removed
from the rigours of Wiener and Shannon's equations.

At the same time, the MIT electrical engineer Pete Elias tried
to make sense of the multiple ways in which information theory
had been transformed in the previous decade, in the realms of coding, communication and cybernetics. Although Elias was convinced
of the relevance of Shannon's ideas to electronic communication,
like Shannon he was less certain when it came to the extension of
information theory to other fields, including biology. In 1958, Elias
made a plea to his fellow-theoreticians not to apply information
theory willy-nilly to other disciplines.[54] A year later, Elias expressed
his doubts about using information theory to understand chemical
specificity, arguing that in this field information was used 'either as a
language for the discussion of purely combinatorial problems or as a
useful statistic, but ... [not] in any coding sense which would imply

that the informational treatment was at all necessary or unique.' Elias was even more dismissive of the application of information theory to the genetic code: 'Although informational ideas may be useful here,' he wrote, 'it seems unlikely that they are essential.'[55] A consensus was emerging, without fanfare and without any public declarations: it was difficult, if not impossible, to apply information theory to biology, even to the genetic code.

There was a lone exception: in 1961, the Japanese theoretical population geneticist Mitoo Kimura published an article entitled 'Natural selection as the process of accumulating genetic information in adaptive evolution'. Writing when the exact nature of the genetic code was still unclear, Kimura showed how new genetic information arises, by focusing on the consequences of one form of a gene ('allele') being replaced by another form (this is the technical definition of evolution), through natural selection. Using some very shaky guesstimates, Kimura calculated that in the animal lineage leading to higher mammals, 10^8 bits of genetic information had accumulated since the Cambrian explosion around 540 million years ago, which saw the appearance of most groups of animals.

Intriguingly, Kimura noted that it was probable that the human chromosomes contained around 10^{10} bits of information; he concluded that either the genetic code was highly redundant or the genetic information we have accumulated is a small fraction of that which we could store in our chromosomes. We now know that the human genome contains approximately 3×10^9 bases, or one-third of Kimura's estimate. Irrespective of the validity of Kimura's calculations, both of his explanations were correct, and he spent much of the rest of his career trying to understand the evolution of the bulk of the genome, which is apparently not immediately shaped by natural selection.

Despite the fundamental role of the information metaphor in molecular genetics, and its vital position in Crick's central dogma, the 1956 Oak Ridge conference marked the swan song of information theory's influence on biology. By this time, what remained of the cybernetics group had drifted off into studies of psychiatry and human behaviour, which were probably the least likely to produce important results because of their complexity.[56] The neurophysiologist Warren McCulloch, who had been present when Wiener gave his paper on 'Behaviour, purpose and teleology' back in 1942, turned down an invitation to the 1956 Oak Ridge meeting, arguing that it was not certain where information actually resided in biological systems, including in genes:

> I doubt whether information theory is yet properly attuned to the complexities of biological problems. To apply the theory in its present state except in a most rudimentary fashion we need to crack the code in genetics as surely as we do in the Central Nervous System.[57]

Biological codes would not give up their secrets merely by being shown Shannon and Wiener's fiendish equations. Experimentation, not theory, would be needed. For more than a decade, the twin theoretical approaches of information theory and cybernetics had beguiled researchers and entranced the general public with their promise of a new way of looking at the world, linking biology with the growing wave of electronic devices that were beginning to permeate all aspects of society, from warfare to work. But in the end, the concrete application of both of these theories seemed to have come to nothing. No new scientific disciplines were founded through the development of information theory or cybernetics – there were no new cross-disciplinary journals, no research institutes, no annual conferences, no new funding sources focused on this area.[58]

That did not mean that cybernetics and information theory had no influence on the development of biology or on the cracking of the genetic code. They both had an impact, but not in the way in which their partisans might have hoped for. In 1961, Martynas Yčas accepted that 'no explicit, and especially no quantitative use of information theory has ... been made in practice.'[59] Instead, as was suggested at the close of the 1956 meeting, Yčas thought that biologists found it 'preferable to use information theory only in a semi-quantitative fashion', as a metaphor.[60] In this metaphorical form – infuriatingly vague and imprecise for mathematicians and philosophers but extremely powerful for biologists – information came to dominate discussions of the genetic code and its meaning, down to the present day. Although the cybernetics group had fallen apart, its fundamental ideas played an important role in the development of radical new views that changed the way in which scientists understood gene function and the meaning of the genetic code. This came about through the work of Parisian researchers influenced by the cybernetic approach.

NINE

ENZYME CYBERNETICS

In the summer of 1947, the French bacteriologist Jacques Monod visited Cold Spring Harbor Laboratory, where Max Delbrück's training course in phage genetics was in full swing. Monod was good-looking and dynamic, with thick hair combed back from a high forehead; he had fought in the French Resistance and was a charismatic figure.[1] Monod later recalled that as he sat in on one of the lectures, there was a 'short fat man' in the front row of the audience, who seemed to be asleep. Monod thought that the man, 'with his round face and fat belly, looked like a petty Italian fruit-merchant, dozing in front of his shop'.[2]

The 'fruit-merchant' was the physicist Leo Szilárd, who in 1929 had made the link between information and entropy. In 1933, a Jewish refugee from Hitler, he came up with the idea of the nuclear chain reaction, which led him six years later to write a letter to President Roosevelt, co-signed by Einstein, calling for the US to develop atomic weapons. This initiative was at the origin of the Manhattan Project, in which Szilárd was heavily involved. But after the surrender of Germany in May 1945, Szilárd, like so many other Manhattan Project physicists, had a profound change of heart and argued vociferously against using the bomb to attack Japan. After the destruction of Hiroshima and Nagasaki, Szilárd turned away from physics. In middle age, he decided to study biology and soon found that his interests coincided with those of Monod.[3]

Monod wanted to understand how bacteria responded to being placed on different kinds of food: since 1900 it had been known that bacteria were able to start producing an enzyme necessary to digest a particular sugar, even if they had never previously encountered that type of food. This phenomenon was known at the time as enzymatic adaptation, and no one understood how it worked; the bacteria seemed to be able to sense what was needed to survive in a particular environment. Apart from its intrinsic interest, adaptation was an extremely attractive system for anyone interested in working out exactly what genes did when proteins were synthesised, because it provided a tool for initiating the process under controlled conditions.

When Szilárd and Monod were introduced at Cold Spring Harbor in 1947, Monod was surprised to learn that Szilárd had read all his papers. Szilárd fired off several 'unusual, startling, almost incongruous' questions, leaving the Frenchman bewildered but delighted at his new friend's knowledge and interest.[4] In 1954, Szilárd met Monod's future collaborator, François Jacob, with exactly the same result. Jacob later recalled:

> At our first encounter, at a colloquium in the United States, he led me over to a corner to ask me about my work. At each response he cut in to reshape my answers to suit his style, to force me to speak his language, to use his words, his expressions. He carefully noted each answer in a notebook. At the end, he said, 'Sign there!' Two years later, during another encounter, he asked 'Is what you told me a few months ago still true?' And he noted: 'Still true!'[5]

Monod's study of adaptation focused on how bacteria grown on the sugar lactose were able to synthesise an enzyme called β-galactosidase (the 'β' is pronounced 'beta'), which could break down lactose. Understanding this response proved difficult: Monod and his team discovered that other sugars, which could not be broken down by β-galactosidase, nevertheless acted as β-galactosidase inducers. This meant that bacteria produced β-galactosidase even when there was no lactose present, and the enzyme therefore could have no function.[6] Because of this troubling result – bacteria could be induced to produce apparently useless enzymes – the term adaptation was abandoned and replaced by the more neutral word induction.* From 1953 these proteins were therefore called inducible enzymes.[7] The situation soon became more confusing: within a few years, Monod had created mutant strains of bacteria in which, just as different inducers could lead to the production of the same enzyme, so, too, different enzymes could be induced by the same inducer – lactose. New data were making induction more complex, not less.

*

Crick's 1957 'central dogma' lecture emphasised that the race to decipher the genetic code was closely intertwined with the attempt to understand protein synthesis. Although neither Monod nor Szilárd was involved in the coding problem, they were both convinced that studies of bacteria would provide insights into how genes function, by understanding the main thing they do, which is to enable the cell to produce proteins. At a conference in 1952 Szilárd and his Chicago colleague, Aaron Novick, described their hypothesis for how protein synthesis was controlled, focusing on how the cell knew when to stop synthesising a particular amino acid: 'somehow the increased concentration of each amino acid depresses the rate of the individual steps of synthesis leading to the formation of that amino acid.'[8] As more of a particular amino acid is produced, the rate at which it is synthesised slows down. Novick and Szilárd thought that protein synthesis involved a negative feedback loop, just like those seen in the cybernetic devices studied by Wiener a few years earlier.

In 1954, Szilárd explained his idea to Monod. The Frenchman later admitted he found it 'a rather startling assumption' and did

---

*This is the explanation given by Cohn *et al.* (1953b), who proposed the change in terminology. Historians have suggested that this change flowed from Monod's highly-public opposition to Lysenko's claim that organisms could adapt to new environments by directly altering their hereditary composition. Although this may be true, there is no direct evidence to support it.

not agree. This was surprising, because a year earlier Monod had shown that the biosynthesis of some enzymes was suppressed by their respective end-products – a negative feedback loop – but had been unable to explain his finding.[9] It took several years for Monod to realise the significance of what he had discovered. By the end of the decade, Szilárd's idea loomed large in Monod's thinking, eventually influencing how we understand gene function.

Other scientists were also looking at protein synthesis in bacteria and were groping towards the same kinds of interpretation as Novick and Szilárd. In 1954, Richard Yates and Arthur Pardee, from the University of California at Berkeley, described an investigation into the biosynthesis of the pyrimidines uracil and cytosine in *Escherichia coli* bacteria. They found that once pyrimidines began to appear in the cell, their presence led to a decrease in the levels of the enzymes involved in the biosynthetic chain. Their 1956 report of their experiment concluded:

> Inhibition by an end-product of its own synthesis appears to be a common control mechanism in the cell.[10]

Although Yates and Pardee had clearly described a negative feedback mechanism, they used the term 'feedback' only in the title of their article, and beyond asserting that such systems were widespread in the cell, they did not explore the matter any further. At about the same time, Edwin Umbarger of Harvard Medical School used negative feedback to interpret a study of *E. coli* in which the presence of the end-product of a biosynthetic pathway inhibited its own biosynthesis. In 1956, Umbarger published an article in *Science* that began with a statement that revealed the influence of Wiener's cybernetics on the new generation of scientists known as molecular biologists:

> Recent developments in automation have led to the use in industry of machines capable of performing operations that have been compared with certain types of human activity. In the internally regulated machine, as in the living organism, processes are controlled by one or more feedback loops that

prevent any one phase of the process from being carried to a catastrophic extreme. The consequence of such feedback control can be observed at all levels of organization of a living animal'.[11]

For Umbarger, the relatively simple system of bacterial biosynthesis provided an opportunity to explore the molecular mechanisms involved in negative feedback, and he was sure that the example he described – the biosynthesis of isoleucine – was just one case among many.

Although all these researchers contributed to the shift in thinking about how protein synthesis worked, the people who linked the ideas of feedback and genetic information, changing our view of life and of the genetic code, were Jacob and Monod, the new Paris team.

Monod had realised that for a full understanding of protein synthesis he needed the help of a geneticist, so he contacted a colleague at the Institut Pasteur in Paris, François Jacob. Jacob, a physician who had signed up with de Gaulle's Free French in 1940 and had been severely wounded after D-Day, joined André Lwoff's laboratory at the Institut Pasteur in 1950 to study the interactions between bacteriophages and their single-cell hosts.[12] Jacob's skills in bacterial genetics and his wide-ranging philosophical interests formed a perfect complement to Monod's more biochemically centred approach and his interest in existentialism. The result was an intellectual partnership that rivalled that of Watson and Crick and which in some ways surpassed it in terms of providing a model to young researchers around the world – Crick himself called it 'the great collaboration'.[13]

Jacob and Monod's joint research, which began in 1957, took place in the heart of Paris, in an Institut Pasteur attic laboratory that was nicknamed the *grenier* (loft). Jacob and Élie Wollman had been studying how bacterial mating ('conjugation') affected the growth of bacteriophage viruses; with Monod, Jacob now used conjugation to explore the genetic basis of induction in bacteria. They carried out these experiments in late 1957 and early 1958 with one of many US visitors to the Institut, Arthur Pardee – the studies became widely known as the PaJaMo (Pardee, Jacob and Monod), or, more colloquially, Pajama (or even Pyjama), experiments.[14]

Pardee had arrived in Paris in September 1957, and began studying how one of Jacob and Monod's bacterial mutants responded to induction.[15] Normal bacteria could digest lactose by producing induced β-galactosidase. These cells were called *lac+* – *lac* (short for lactose) referred to a region of the bacterial chromosome containing several genes involved in this complex phenomenon, and the '+' indicated that this was the normal, or wild, type. Mutant bacteria – known as *lac−* – could not grow on lactose unless they acquired the relevant genes by mating with a *lac+* individual. Pardee showed that when the *z+* gene, which produced the β-galactosidase enzyme, was transferred into a *lac−* individual, it became active within minutes. This implied that there was an immediate chemical signal that passed directly from the introduced gene to the host cell's protein synthesis system. Over the next year or so, the Paris group became focused on the nature of this mysterious messenger molecule, which they called X.[16]

The PaJaMo experiments also investigated bacteria that continually produced β-galactosidase in the absence of an external inducer molecule. These were known as constitutive strains because they produced the enzyme as part of their constitution. A single gene called *i* seemed to be involved: *i+* bacteria were inducible, whereas *i−* individuals were constitutive (the system also involved another gene, *y*, which controlled the action of an enzyme, permease, that allowed lactose into the cell). Things got interesting when the group explored the interaction between the *i* gene and the *z* gene that allowed bacteria to produce β-galactosidase. When Pardee introduced both *i+* and *z+* genes into bacteria carrying the *i−* and *z−* forms, the bacteria initially produced high, constitutional, levels of β-galactosidase, showing the action of the *z+* gene. But then something odd happened: β-galactosidase production declined rapidly. The *i+* gene seemed to start repressing the activity of the *z+* gene.[17] To the untutored eye, this complex set of results seems either bewildering or boring, or both. But what Jacob and Monod did with these data – the way in which they interpreted them – altered our understanding of what genes do.

The first step forward in this new view of life came at the beginning of 1958, when Leo Szilárd was visiting Paris as part of a tour of laboratories during which he presented his negative feedback model of protein synthesis.[18] Five years earlier, Szilárd had proposed to Monod that a form of negative feedback might explain induction; now he made a mind-twisting suggestion that expressed the idea at an even more complex level. Perhaps, he said, 'induction could be effected by an anti-repressor rather than repression by an anti-inducer?'[19] Szilárd was proposing that induction might work by stopping the action of a molecule (a 'repressor') inside the bacterial cell that normally repressed enzyme formation. The effect would be a bit like releasing a brake. This could be understood as an example of two negatives producing a positive or, as Monod put it in his 1965 Nobel Prize lecture, a 'double bluff'.[20]

Szilárd had not come up with this idea himself. In April 1957, the bacteriologist Werner Maas had given a lecture in Chicago that Szilárd had attended; Maas, who was studying inducible enzymes, had hypothesised that:

> inducers which enhance the formation of an enzyme when added to a growing bacterial culture may perhaps be capable of doing so only because there is a repressor present in the cell, and that the inducer might perhaps do no more than inhibit some enzymes.[21]

Maas later stated that on hearing this, Szilárd 'jumped on the hypothesis ... and became quite excited'.[22] Szilárd wanted to publish the idea straight away, but Maas refused, because he had no evidence to back it up. At the same time, Henry Vogel of Yale University submitted an article in which he suggested that a common theoretical framework could understand both induction and negative feedback inhibition of biosynthesis – they involved what Vogel called regulator molecules.[23] Although Maas's idea of a repressor and Vogel's regulatory framework were both based on interactions between proteins, not genes, the ideas of regulation and repression were in the air.

By the time that Szilárd visited Paris less than a year later, he was clearly convinced that induction might not be a positive effect, but rather a 'de-repression'. The Paris group were intrigued by this suggestion and they briefly outlined the concept alongside the first publication of the PaJaMo results (in a French journal in May 1958, where it was described as an 'initially surprising hypothesis') and again in a lecture by Jacob in June 1958. Even though the team were prepared to go public with the idea, they had made no experimental test of the hypothesis and it was still not certain that there were any general implications beyond the narrow world of bacterial genetics.[24]

Jacob later recalled that the decisive moment came on a Sunday afternoon late in July 1958. All his colleagues were on holiday, while he remained in Paris with his wife, Lise, preparing for a lecture he had to give in New York on how phage viruses hijack the genetic apparatuses of their host. Unable to work, Jacob went to the cinema with his wife. He could not concentrate on the film, so he closed his eyes and suddenly in 'a flash', he realised that the two experiments he had been thinking about – the PaJaMo experiment and his own work with Élie Wollman on phage reproduction – were in fact fundamentally identical. He now understood that they both involved the modulation of gene activity by directly affecting the DNA. Jacob later described his almost mystical experience as he realised the connection between his two problematic experiments:

> Same situation. Same result. Same conclusion. In both cases, a gene governs the formation of a cytoplasmic product, of a repressor blocking the expression of other genes and so preventing either the synthesis of the galactosidase or the multiplication of the virus. ... Where can the repressor act to stop everything at once? The only simple answer, the only one that does not involve a cascade of complicated hypotheses is: on the DNA itself! ... These hypotheses, still rough, still vaguely outlined, poorly formulated, stir within me. Barely have they emerged than I feel invaded by an intense joy, a savage pleasure. A sense of strength as well, of power. As if I had climbed a mountain, attained a summit from which I saw in the distance a vast panorama. I no longer feel mediocre or even mortal. I need air. I need to walk.[25]

It was over a month before Jacob could share his insight with Monod. When the two men finally met up, in September 1958,

Monod was initially unconvinced. The link between the two experiments was tenuous, and the idea that the repressor acted directly on DNA seemed outlandish. Genes had previously been seen as pure and abstract entities, solidly placed in the background of the cell, passive repositories of information and nothing more – as Monod later put it, they were thought to be as inaccessible 'as the material of the galaxies'.[26] According to Jacob's new view, genes were intimately involved in the messy reality of cellular processes. Over weeks and months of increasingly intense argument and repeated cycles of experimentation, Jacob and Monod explored their ideas, testing hypotheses by creating new mutants and predicting how they would behave if the model were correct. Jacob later recalled their discussions 'moved at top speed, in bursts of brief retorts, like a ping-pong match'. As they scribbled feverishly on the blackboard, the model and its predictions became clearer.[27] The language they were using changed, too.

Monod began to describe gene function in terms of information transfer, and he explicitly embraced cybernetic thinking by describing protein synthesis in terms of patterns of control. This shift took its clearest shape in a plan for a book entitled *Essays in Enzyme Cybernetics*, which he began writing with Melvin Cohn of Washington University towards the end of 1958.[28] Even though the project was never completed, its very existence was extremely revealing. Ephrussi and Watson's weak satirical joke that had fooled the editors of *Nature* six years earlier had become a reality: cybernetics was being used to understand biology. This was not a reflection of some fashionable trend; it was a powerful interpretative approach that had very real advantages for understanding biological processes. However, as with information theory, it was the general framework, rather than the precise mathematical detail, that was being employed. For biologists, cybernetics was becoming an analogy, a metaphor, a way of thinking about life in terms of flows of control and information, a way of thinking about how genes worked.

Jacob used a military analogy to explain the genetic 'circuit' he had discovered with Monod, unwittingly returning to Wiener's original recognition of the role of feedback through his work on anti-aircraft fire control:

We saw this circuit as made up of two genes: transmitter and receiver of a cytoplasmic signal, the repressor. In the absence of the inducer, this circuit blocked the synthesis of galactosidase. Every mutation inactivating one of the genes thus had to result in a constitutive synthesis, much as a transmitter on the ground sends signals to a bomber: 'Do not drop the bombs. Do not drop the bombs.' If the transmitter or the receiver is broken, the plane drops its bombs. But let there be two transmitters with two bombers, and the situation changes. The destruction of a single transmitter has no effect, for the other one will continue to emit. The destruction of a receiver, however, will result in dropping the bombs, but only by the bomber whose receiver is broken.[29]

A few months later, Monod gave a lecture in Germany and described the model in more precise terms, adopting the vocabulary of information and control: 'We could imagine that the z locus contains genetic information relating to the structure of the galactosidase protein, while the i locus determines the conditions under which this information is potentially transferred to the cytoplasm.' By implication, Monod was arguing that that the 'one gene, one enzyme' idea – less than 20 years old – could not explain the complex reality of protein synthesis. Instead, he suggested there were two kinds of genes: structural genes, which contained the information necessary to make proteins, and regulator genes, which determined when that information was employed, by synthesising a specific repressor that inhibited the expression of the structural gene.[30]

In March 1959, the final version of the PaJaMo experiment was submitted to a new academic publication, which had a title that was a manifesto for the new science: the *Journal of Molecular Biology*. In the paper, which was more developed than the original French publication, the trio acknowledged that they were 'much indebted to Professor Leo Szilárd for illuminating discussions'. The paper presented the dense details of their experiment and outlined the repressor hypothesis, suggesting that it was a general model for protein synthesis and showing the parallels with the phage work of François Jacob and Élie Wollman. Above all, they highlighted two points that they could not answer: the nature of the repressor and how it worked.[31] As they explored these issues, Jacob and Monod changed the vocabulary that was used for talking and thinking about genes and what they – and the code they contain – might do.

*

For much of the 1950s, scientists had felt uncomfortable about the word 'gene'. In 1952, the Glasgow-based Italian geneticist Guido Pontecorvo highlighted the existence of four different definitions of the word that were regularly employed by scientists and which were sometimes mutually contradictory. A gene could refer to a self-replicating part of a chromosome, the smallest part of a chromosome that can show a mutation, the unit of physiological activity or, finally, the earliest definition of a gene – the unit of hereditary transmission.[32] Pontecorvo questioned whether the gene could any longer be seen as a delimited part of a chromosome, and suggested instead that it was better seen as a process and that the word gene should therefore be used solely to describe the unit of physiological action.

Although Pontecorvo's suggestion was not taken up, scientists recognised the problem. The debate over words and concepts continued at the Johns Hopkins University symposium on 'The Chemical Basis of Heredity', which was held in June 1956. By this time it was generally accepted as a working hypothesis that all genes in all organisms were made of DNA and that the Watson–Crick double helix structure was also correct. Joshua Lederberg, a stickler for terminology, declared audaciously that '"gene" is no longer a useful term in exact discourse'.[33] He would no doubt be surprised to learn that it is still being used, more than half a century later. At the same meeting, Seymour Benzer came up with a solution:

> The classical 'gene', which served at once as the unit of genetic recombination, of mutation, and of function, is no longer adequate. These units require separate definition.[34]

Since 1954, Benzer had been studying the structure of genes, focusing on the *r*II genetic region of the T4 phage virus.[35] From today's perspective, it is noteworthy that at the 1956 meeting Benzer was still hedging his bets over the nature of the genetic material in T4: all he was prepared to say at that point was that in this virus DNA 'appears to carry the hereditary information'.[36] By screening hundreds of mutants in the rII region of the virus's DNA and then carrying out thousands of crosses to see whether they were part of the same functional unit, Benzer was able to construct an extremely detailed genetic map, down to a single pair of nucleotides. Like Rutherford showing that the atom had an internal structure through his work in Manchester at the beginning of the twentieth century, Benzer was able to show that the gene was not a single, indivisible unit.[37] Rather than the classic image of the gene as a bead on a string, Benzer saw the gene as a one-dimensional stretch of DNA, which would reveal its secrets by dissecting it down to the molecular level. He found that not all parts of the gene were equal – some areas were much more prone to spontaneous mutations, and mutations in different areas produced different effects. Long before the development of DNA sequencing, Benzer's pioneering and painstaking study showed that genes have an internal structure.

On the basis of this work, Benzer came up with new words, focusing on what genes actually did: the unit of genetic recombination was a 'recon', the smallest unit of mutation was a 'muton', and the unit of function was a 'cistron'.* Although only 'cistron' survived into common scientific usage for a while (it is now very much on the wane), Benzer's attempt to reconceive the gene in molecular terms was highly influential. Widely praised at the time – at the Johns Hopkins symposium, George Beadle described it as 'very beautiful work' – Benzer's approach helped fuse the structural insights of Watson and Crick with the traditional approaches of genetics, creating the new subject of molecular genetics.[38]

This new field was reinforced in 1957 by two PhD students,

---

*Benzer later recalled that at the 1956 meeting he was criticised by a French geneticist for his choice of terms: 'At this meeting at Johns Hopkins, I was attacked by Elie Wollman, who said these are very unfortunate names because they don't translate well into French – cistron sounds like a lemon, and muton sounds like a sheep, recon is a dirty word. [Laughter] But I had already announced these names, so I stuck with them. [Laughter] Wollman was kind of upset about that.' (Benzer, 1991, p. 51).

Matthew Meselson and Frank Stahl, who carried out what has been described as 'the most beautiful experiment in biology'.[39] One of the problems raised by the double helix structure was how the DNA molecule copied itself. The complementarity of the base pairs on the two strands suggested that the cell used each strand as a template to create two identical double helices, but how this worked was unclear. At the same 1956 Johns Hopkins meeting at which Benzer spoke, Max Delbrück outlined three models for DNA replication – 'conservative', in which the original DNA double helix remained intact and was entirely copied into a completely new molecule; 'semi-conservative', in which one strand of each molecule was copied, producing two daughter molecules, each of which had one old and one new strand (this was the model suggested by Watson and Crick); and Delbrück's preferred view, 'dispersive' replication, whereby bits of each DNA strand were copied and then reassembled, producing two double helices, each strand of which was made up of a mixture of old and new.[40] Meselson and Stahl's experiment was designed to distinguish between these three hypotheses.

They began planning their experiment in 1954; they decided to use a heavy isotope of nitrogen (^{15}N; normal nitrogen has an atomic weight of 14) to distinguish between the original strand and its copy. Nitrogen is an important atomic component of DNA, so DNA that was synthesised with ^{15}N was slightly heavier than its ^{14}N counterpart. After three years of preliminary experiments, they grew *E. coli* bacteria for several generations on a medium rich in ^{15}N, thereby creating a stock of bacteria with ^{15}N-based DNA. The bacteria were then transferred to medium containing normal ^{14}N and were allowed to reproduce for several cycles before their DNA was extracted. Meselson and Stahl spun the samples in an ultracentrifuge machine at 44,700 r.p.m. for 20 hours – any double helix containing ^{14}N ended up at a different position in the tube from its ^{15}N equivalent, simply because it was lighter.

The results were very clear. After one cycle of reproduction, in which the bacteria had made only one copy of their DNA, two kinds of DNA were detected: a heavy band, composed of the original ^{15}N-based DNA, and a slightly lighter band composed of a mixture of ^{15}N and ^{14}N-based DNA. This showed that 'conservative'

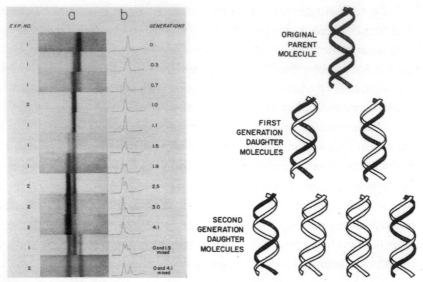

7. Figures from Meselson and Stahl's 1958 paper. Left: DNA produced
by two identical experiments (1 and 2), shown as a band (1) and a
graph (r). At the beginning there was only ¹⁵N DNA. As the bacteria
reproduced on ¹⁴N-rich culture, two lighter forms appeared, one after
one generation, the other after two generations. By generation 4,
most of the DNA was of the lightest type. Right: Meselson and Stahl's
interpretation. In each round of reproduction the bacterial DNA was
copied, using nitrogen from the ¹⁴N medium. After one generation,
each DNA double helix was composed of a new ¹⁴N strand and an
old ¹⁵N DNA strand. At the second generation, some molecules were
composed entirely of ¹⁴N-rich DNA, and were therefore lighter again.

replication was wrong – that model predicted that only ¹⁴N-based
DNA would be present in the newly synthesised molecules. The
next step was decisive: if the bacteria were allowed to reproduce for
another cycle, then a new, lighter, band composed only of ¹⁴N-based
DNA appeared. These molecules must have been created by copying
from a whole strand of ¹⁴N DNA that had been created after the first
round of replication. This showed that Watson and Crick were right
and Delbrück was wrong: DNA replication was 'semi-conservative'
with, most probably, the whole of each strand being copied simultaneously to make new daughter molecules.

By resolving the thorny problem of DNA replication, Meselson and Stahl's elegant and precise experiment represented the final confirmation of the significance of the double helix structure of DNA. As it closed one phase of the history of molecular biology, it opened another, showing that DNA molecules – and hence the genes they contained – could be investigated by using the latest analytical techniques.

Three studies from the late 1950s therefore pointed the way to the future: Meselson and Stahl showed that DNA could be labelled and tracked down the generations, Benzer's fastidious work revealed that it was possible to investigate the molecular structure of the gene by exploring its tiniest components, and Ingram's discovery that the sickle-cell mutation in the haemoglobin gene produced a single amino acid change hinted that the nature of the genetic code itself might be within reach.

*

Even before Jacob and Monod finally described their view of gene function in 1961, their work had led to an important discovery. On Good Friday in 1960, a small group of researchers, including Crick and Jacob, gathered in Sydney Brenner's rooms in King's College, Cambridge, as a kind of 'after' meeting following a conference in London on the previous day. Although the Cambridge and Paris groups were on very friendly terms, they were not on precisely the same wavelength. As Brenner later recalled, 'You see, the Paris people were interested in regulation. We essentially were interested in the code. So we had a slightly different approach.'[41] But that day, the two approaches suddenly fused as Jacob explained the latest results from Paris, focusing on the puzzle of how the z^+ gene that enabled bacteria to produce β-galactosidase was able to synthesise such high levels of the enzyme so soon after it was introduced into a z^- cell. One of the possibilities that the Paris group had briefly considered was that the gene synthesised a handful of very efficient ribosomes, which then churned out the enzyme at a high rate. But, as Jacob explained, Pardee had recently done an experiment that suggested that the z^+ gene did not produce anything stable, but only a rather transitory messenger molecule.

'At this point,' recalled Crick, 'Brenner let out a loud yelp – he had seen the answer.'[42] Jacob vividly described the following minutes:

> Francis and Sydney leaped to their feet. Began to gesticulate. To argue at top speed in great agitation. A red-faced Francis. A Sydney with bristling eyebrows. The two talked at once, all but shouting. Each trying to anticipate the other. To explain to the other what had suddenly come to mind. All this at a clip that left my English far behind.[43]

In that moment, Crick and Brenner had realised that the PaJaMo messenger could explain some recent results from various groups that suggested that at certain points in the reproduction of phage, a short-lived form of RNA was produced. This RNA had the same A:G base composition as phage DNA, indicating that it had been copied from the phage, and differed from the ribosomal RNA that was found in the host cell. The two Cambridge men immediately seized on the possibility that this short-lived RNA was the mysterious messenger that the Paris group had hypothesised. This would make the ribosome an inert structure in the cell – Crick described it as a reading head, like in a tape recorder.[44] Messenger RNA, as Jacob and Monod called it that autumn (this was soon abbreviated to mRNA), was a tape that copied information from the DNA and then carried that information to the ribosome, which read it off and followed the instructions to make the appropriate protein.

Jacob and Brenner immediately began planning how to test the hypothesis. That evening, Crick and his wife held one of their many parties. Jacob recalled the scene clearly:

> A very British evening with the cream of Cambridge, an abundance of pretty girls, various kinds of drink, and pop music. Sydney and I, however, were much too busy and excited to take an active part in the festivities. … It was difficult to isolate ourselves at such a brilliant, lively gathering, with all the people crowding around us, talking, shouting, laughing, singing, dancing. Nevertheless, squeezed up next to a little table as though on a desert island, we went on, in the rhythm

of our own excitement, discussing our new model and the preparations for experiment ... A euphoric Sydney covered entire pages with calculations and diagrams. Sometimes Francis would stick his head in for a moment to explain what we had to do. From time to time, one of us would go off for drinks and sandwiches. Then our duet took off again.[45]

Jacob and Brenner's proposed experiment needed the help of Meselson and his ultracentrifuges at Caltech to determine whether phage infection led to the creation of new ribosomes or, as they predicted, to a new transient form of RNA that simply employed the old host ribosomes to turn its message into protein. After a tense month in California, endlessly fiddling with the experimental conditions, Jacob, Brenner and Meselson got the experiment to work. As they had hoped, no new ribosomes appeared; instead, RNA that had been copied from the phage DNA was associated with old ribosomes that were already present in the bacterial host. Other researchers were on the same track – that autumn, when Jim Watson heard of the first results from California, he informed the trio that his Harvard group was working on something similar; meanwhile, Martynas Yčas announced that yeast also produced an unstable RNA, with the same A:G ratios as yeast DNA, suggesting that it had been copied from the yeast chromosomes.[46]

In May 1961, after some delays caused by Watson's group, which had used a different technique but came up with similar findings and wanted to publish simultaneously, the Brenner, Jacob and Meselson paper appeared in *Nature*, accompanied by the paper from the Watson laboratory.[47] As the title of the Brenner paper put it, they had found 'An unstable intermediate for carrying information from genes to ribosomes for protein synthesis'. That intermediate was messenger RNA. The story of how genetic information got out of DNA was complete.*

---

*Despite the importance of this discovery, there was no Nobel Prize for those involved. There may simply have been too many people with an equal claim – no more than three people can share a Nobel Prize.

*

In 1961 Jacob and Monod stepped into the history books with three powerful works of synthesis, summarising and developing their ideas about the nature of gene regulation with a verve, elegance and rigour that still put most scientific articles to shame. First, they published a long review in the *Journal of Molecular Biology* (this was submitted in December 1960, and appeared in May the following year). Then, at the June 1961 Cold Spring Harbor symposium, which was entitled 'Cellular regulatory mechanisms' and was primarily focused on examples of negative feedback and repression, they presented a more data-rich version of their review before Monod closed the conference with a summary of the significance of their approach.[48] Together, these three papers have been cited more than 5,500 times – more than Watson and Crick on the double helix structure of DNA.

In their 1961 review article, Jacob and Monod brought together the ideas that they had been developing in a series of unnoticed publications in French and German, in which they had separated genes into two kinds according to their function. They began with the classic structural gene that coded for a protein and then described the kind of regulatory gene that they had discovered in bacteria. Their vision of what genes actually do, of the meaning of the genetic code, was ultimately framed in terms of information and control:

> let us assume that the DNA message contained within a gene is both necessary and sufficient to describe the structure of a protein. The elective effects of agents other than the structural gene itself in promoting or suppressing the synthesis of a protein must then be described as operations which control the rate of transfer of structural information from gene to protein.[49]

The article described gene action as involving control mechanisms and suggested that an inducer 'somehow accelerates the rate of information transfer from gene to protein'. The role of repression was to inhibit enzyme synthesis, but in a very different manner from the negative feedback loops that had been described by Yates and Pardee or Novick and Szilárd. Classic negative feedback involved the end-product of a reaction curtailing that reaction through some kind of protein–protein interaction. Repression involved the direct action of the repressor on the DNA of the structural gene itself. According to Jacob and Monod, the agent that acted as the repressor was the product (RNA or protein) of another gene, which they called a regulator gene:

> A regulator gene does not contribute structural information to the proteins which it controls. The specific product of a regula-tor gene is a cytoplasmic substance, which inhibits informa-tion transfer from a structural gene (or genes) to protein. In contrast to the classical structural gene, a regulator gene may control the synthesis of several different proteins: the one-gene one-protein rule does not apply to it.[50]

Both repression and induction were controlled by regulator genes.

Jacob and Monod also showed that the response to viral infec-tion was controlled by regulator genes, linking bacterial genetics with the world of the phage, and implying that there was a funda-mental process at work that could be applied to other organisms. They finally provided a theoretical framework to explain how genes work together, with structural and regulator genes interacting as a physiological unit. They gave their discovery a name: 'This *genetic unit of co-ordinate expression* we shall call the "operon"'.[51] Jacob and Monod were arguing that the operon was composed of genes that had been selected to work together as a Darwinian adaptation.

With the discovery of messenger RNA, the mechanism of pro-tein synthesis had become clearer; above all, genes were now seen not simply as producing proteins, but also as controlling the activ-ity of other genes in a coordinated unit – the operon. It was still not known whether the repressor was made of RNA or protein (they ini-tially leaned towards the RNA option; this was eventually revealed to be wrong in this particular case) and whether it acted directly on the DNA of the operator gene or on its RNA product.

Jacob and Monod were quite aware of the implications of their discovery of genetic regulation. As they pointed out, one of the cen-tral mysteries of life is why cells in a body do not express all their genetic information all the time, but instead differentiate and turn

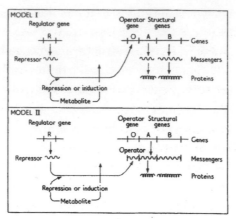

FIG. 6. Models of the regulation of protein synthesis.

8. Jacob and Monod's models of the Operon. In the top model,
gene regulation occurs via RNA; in the lower version, it occurs
via a protein. Taken from Jacob and Monod (1961a).

into different structures with different functions. Their idea of gene
regulation provided a conceptual key that we are still using and
exploring today. They also realised that as well as explaining normal development, gene regulation could also provide an insight into
cancer; they wrote: 'Malignancy can adequately be described as a
breakdown of one or several growth control systems, and the genetic
origin of this breakdown can hardly be doubted.'[52]

The conclusion of their article transformed our view of what
genes do, placing their complex research findings in an innovative
framework that fused molecular genetics, cybernetics and computing, and which unwittingly echoed Schrödinger's view that genes
'are architect's plan and builder's craft – in one'. Jacob and Monod
wrote:

According to the strictly structural concept, the genome is
considered as a mosaic of independent molecular blue-prints
for the building of individual cellular constituents. In the
execution of these plans, however, coordination is evidently
of absolute survival value. The discovery of regulator and
operator genes, and of repressing regulation of the activity of

structural genes, reveals that the genome contains not only a series of blue-prints, but a coordinated program of protein synthesis and the means of controlling its execution.[53]

Jacob and Monod had shown that genomes are not simply blue-prints. Instead, they contain programs that determine the physical and temporal patterns of gene expression, and can also interact with the environment. This demonstration, which was the proof of Schrödinger's theoretical insight two decades earlier, has shaped biology ever since.

As Monod indicated at the Cold Spring Harbor meeting in June 1961, the discovery of regulator genes also held out the possibility of a radically new form of genetics:

> Adequate techniques of nuclear transfer, combined with systematic studies of possible inducing or repressing agents, and with the isolation of regulatory mutants, may conceivably open the way to the experimental analysis of differentiation at the genetic-biochemical level.[54]

It was decades before this vision of genetic manipulation was realised, but it eventually transformed biology and medicine by making it possible to genetically manipulate organisms, and to understand diseases such as cancer.

The discovery of the operon and of the idea of gene regulation was the complex outcome of chances and brilliant insights, as well as a lot of hard work. To mark the fiftieth anniversary of the operon, Jacob gave a graphic description of how he and Monod made their discovery, giving a glimpse of what it was like to be part of such a process:

> Our breakthrough was the result of 'night science': a stumbling, wandering exploration of the natural world that relies on intuition as much as it does on the cold, orderly logic of 'day science'.[55]

In a similar vein, in his 2013 obituary of Jacob, Mark Ptashne captured the intensity of the three years' work that led up to Jacob and

Monod's insight, describing an experience that few scientists have been lucky enough to share:

> The repressor was not so much discovered by Jacob and col-
> leagues as imagined – an entity that would explain disparate
> phenomena as analogous, connected by a similar underlying
> reality.[56]

*

Although the main focus of the 1961 Cold Spring Harbor sympo-sium was gene regulation, both Monod and Brenner made passing references to the coding problem. The recent identification of mes-senger RNA led Brenner to be optimistic that a breakthrough would soon occur, while despite the absence of any real progress Monod confidently suggested that a demonstration of the link between a DNA sequence and a protein sequence (colinearity) would soon be at hand.[57] During these brief discussions, Gordon Tomkins of the National Institutes of Health in Maryland sat still and said nothing, although he must have been bursting. Unlike everyone else in the room, indeed virtually everyone else on the planet, Tomkins knew that the genetic code had been cracked three weeks earlier. He had been the first person in the world to hear the news and had been sworn to secrecy. No one else in the sweltering auditorium knew that the code had been cracked; in fact, no one else had even heard of the two men who had made the discovery.

TEN

ENTER THE OUTSIDERS

At 3.00 a.m. on Saturday 27 May 1961, Heinrich Matthaei began one of the most significant experiments in the history of biology. Matthaei was a 32-year-old German researcher in Marshall Nirenberg's laboratory at the National Institutes of Health in Bethesda, Maryland, and he was about to crack the genetic code. Together with Nirenberg, a biochemist only two years his senior, Matthaei was studying protein synthesis in a test tube. In the middle of the night, he took his protein-synthesising mix, added two radioactive amino acids, phenylalanine and tyrosine, to different tubes and then introduced a long string of man-made RNA that was composed of just one kind of base – uracil (U – this replaces the T found in DNA sequences). The RNA molecule therefore read 'UUUUUUUU …' and was known as poly(U). By seeing which radioactive amino acid was turned into a protein chain by poly(U), Matthaei hoped he would be able to read the first word in the genetic code. It did not matter whether the code used sets of one, two, three, four or more U bases: the 'cell-free' protein synthesis system in the test tube would be able to read the message.

When Gordon Tomkins, the laboratory head, came in at around 9 a.m., Matthaei had the answer. Radioactive protein had been produced in the test tube containing the amino acid phenylalanine.

This must mean that a combination of Us coded for phenylalanine. On 27 May 1961, Heinrich Matthaei had read the first word of the book of life.

*

Nirenberg and Matthaei's discovery transformed the study of the genetic code, both because it was successful and because it used a radically different approach. Until this breakthrough, the campaign to crack the code had begun to look listless. In 1959, at a meeting held at the Brookhaven Laboratory, Crick had summarised what he called 'the present position of the coding problem'. He divided coding research into three phases: the vague phase (up to 1954), the optimistic phase (opened up by Gamow) and the 'present position', which Crick called 'the confused phase'.[1]

The situation was confused because none of the theoretical models matched the increasingly complex experimental findings. For example, a study of nineteen different species of bacteria showed that they had very different ratios of bases in their DNA, but their RNA and amino acid compositions were essentially similar.[2] Crick outlined a number of 'unattractive' explanations of this finding, including the possibility that the genetic code was not universal, or that only part of the DNA in an organism codes for protein, with the rest being 'nonsense'. Crick was optimistic, however, and with Brenner was trying to create viral mutants that would give them an insight into the code. This approach had been given a boost in the summer of 1960, when Heinz Fraenkel-Conrat at Berkeley described the amino acid sequence of the tobacco mosaic virus, and began making mutations with the hope of observing amino acid changes. Although this approach would take a long time and had yet to provide any concrete insight into the code, the press had got interested and was hyping up the idea that the code could soon be cracked. In May 1960 *Time* magazine published two articles on the subject: the first was entitled 'Close to the mystery', and the second proclaimed that Fraenkel-Conrat's work represented the 'genetic Rosetta stone'.

Meanwhile, there were still mathematicians who were convinced that it was possible to crack the code simply by thinking.

Just six weeks before Matthaei completed the decisive experiment, there was a symposium in New York on 'Mathematical problems in the biological sciences', which included molecular biologists such as Max Delbrück and Alex Rich. One of the speakers was Solomon Golomb, a mathematician from the Jet Propulsion Lab in Pasadena, who had previously worked with Delbrück. Golomb described various theoretical schemes that might correspond to the actual genetic code, before concluding, 'It will be interesting to see how much of the final solution will be proposed by mathematicians before the experimentalists find it.'[3] The answer, worked out over the following seven years or so, was simple: not one single part of it.

Nirenberg and Matthaei's radical experimental approach to the coding problem was all the more notable because they were complete outsiders, unconnected with any of the groups that had been struggling with the coding problem over the previous eight years. They were not part of the golden trio of Cambridge, Harvard and Paris, which had produced all of the main discoveries thus far.[4] Nirenberg was so unknown that his application to attend the June 1961 Cold Spring Harbor meeting had been rejected. Ironically, while the great and the good of molecular biology were talking about the genetic code, Nirenberg and Matthaei were cracking it

*

Nothing in Nirenberg's early career suggested that he would be the man who would crack the genetic code. In 1951 he obtained his MSc for a study of caddis-fly biology, then he changed subject and did a PhD in biochemistry. Next he held a two-year postdoctoral fellowship at the National Institute of Arthritis and Metabolic Diseases (part of the National Institutes of Health) in Bethesda, a few miles northwest of Washington DC. After François Jacob and Joshua Lederberg both rejected his applications to work with them, Nirenberg became an NIH research biochemist in the Section of Metabolic Enzymes at Bethesda, led by the charismatic jazz fanatic Gordon Tomkins, who at 35 years old was barely older than Nirenberg.

During those initial years at Bethesda, Nirenberg tried to induce the synthesis of an enzyme called penicillinase in *Bacillus cereus* bacteria. Nirenberg kept detailed lab diaries in which he noted his thoughts and ideas, giving an insight into how he approached his experiments.[5] These diaries reveal that he had been thinking about his revolutionary approach to cracking the genetic code for more than two years before he was finally able to see the result. At the end of November 1958, he described his idea of getting protein synthesis to work in a test tube and he gave an outline of the ideal experiment he wanted to perform:

> would not have to get polynucleotide synthesis very far to break the coding problem. Probably 30 nucleotides & equal number of AA would do it. *Could crack life's code!*[6]

Nirenberg's approach was predicated on recent results from the laboratories of Paul Zamecnik and Severo Ochoa. In the early 1950s, Zamecnik had achieved a technical tour-de-force by making protein synthesis take place in a test tube. Zamecnik's 'cell-free' system, based on the contents of rat liver cells, used radioactive amino acids to show that a new protein was synthesised.[7] The second element of Nirenberg's approach came from the research of Spanish-born Severo Ochoa, who worked in New York and won the 1959 Nobel Prize in Physiology or Medicine for the discovery of poly-nucleotide phosphorylase – an enzyme involved in the metabolism of RNA.[8] The isolation of polynucleotide phosphorylase meant that Ochoa was able to create artificial RNA molecules by incubating the enzyme with the four RNA bases (A, C, G and U). It was not possible to determine the order in which the nucleotides were strung together, but it was relatively straightforward to create a monotonous RNA molecule composed only of one base – known as poly(A), poly(U) and so on.

For most people, it was not immediately obvious what could be done with such molecular freaks of nature – nothing like them had ever been observed in any cell – but Nirenberg glimpsed the opportunity. Nirenberg was clever and he was lucky: Ochoa was synthesising poly(A), poly(U) and so on together with Leon Heppel – the head of biochemistry at Bethesda. Heppel's lab began to produce synthetic RNA molecules in collaboration with another Bethesda researcher, Maxine Singer.[9] Nirenberg found himself in one of the two places in the world that was making these unearthly strings of RNA.

Despite his understandable desire to attack the genetic code immediately, Nirenberg tried to keep his focus on the research he was supposed to be doing. As he reminded himself in an entry in his lab diary from spring 1959: 'My main aim is not to crack protein synthesis but to have everything ready to study enzyme induction.'[10] At the spring 1960 meeting of the Federation of American Societies for Experimental Biology (FASEB), Nirenberg gave a brief talk on his work on induction.[11] His aim was to see whether the same gene was involved in the synthesis of two very similar inducible enzymes, or, as he put it, 'whether a portion of one gene contains information for the synthesis of a protein subunit which might be an integral part of two or more enzymes'. To Nirenberg's disappointment, there was no proof of what he termed 'shared genetic information'.

Although this finding was not particularly interesting (the talk has never been cited), the way in which Nirenberg approached the problem is significant because it highlighted the two ways of looking at life that were now happily coexisting in laboratories all over the world. Everyone thought that genes contained information, but that abstract quality also had a concrete form: it was a nucleotide sequence that made something happen. In this case it was what Nirenberg called 'information for the synthesis of a protein subunit'. Whatever the advantages of the new way of thinking about genes as information, in the end those ideas would have to be translated into detailed, dirty biochemistry.

Nirenberg's outlook changed completely in August 1960, when Zamecnik showed that it was possible to get protein synthesis occurring in a test tube containing the contents of Nirenberg's favoured organism, *Escherichia coli*.[12] Nirenberg immediately began to try the experiment at Bethesda. He wrote in his diary: 'Hurry up exps. Shouldn't take 1 week to know whether system will work. *Work-Work-Work*.'[13] But it would not work. Then he had two strokes of luck: first, Alfred Tissières and François Gros at Harvard published a refinement of Zamecnik's system that was much easier to use, and then a lanky, prematurely balding German called Heinrich

Matthaei joined his lab.[14] Matthaei had obtained a NATO fellowship to work on cell-free protein synthesis, using radioactively labelled amino acids.[15] He had initially been working on carrots but various things had gone wrong, and he eventually found himself assigned to Nirenberg. Matthaei had exactly the technical skills that Nirenberg needed.

Shortly after Matthaei's arrival, Nirenberg abandoned his ideas of looking at induction in bacterial cells and threw himself into an exploration of protein synthesis in cell-free *E. coli*-based systems. Within weeks, the pair had made a technical breakthrough as they were able to make bulk enzyme extracts and then store them, rather than having to make fresh extracts for every experiment. This soon led to a steep increase in the number of experiments they could do.[16]

By end of November 1960, Nirenberg's diaries were full of discussions about cell-free systems, the importance of messenger RNA, and the use of synthetic RNA as a key: 'Can you swamp system with *messenger RNA*?' he wrote.[17] This is striking, because Nirenberg was writing several months before the May 1961 publication of the *Nature* papers by Brenner, Jacob and Crick and by Gros and Watson, which first publicly used the term messenger RNA. 'Cytoplasmic messenger' had been used in the 1959 PaJaMo paper, and by the end of 1960 Jacob and Monod had arrived at the concept of messenger RNA, at which point the term was being bandied about in conferences.[18] But the exact phrase had not yet been used in print. Even though Nirenberg was not part of the inner circle of molecular biology, he had picked up on the term before there was conclusive evidence that the substance existed.

In one respect, this relative distance from the intellectual centres of the work on the genetic code proved an advantage. Nirenberg was blissfully unaware of the debates among those working on the coding problem over the structural restrictions imposed by what was called a commaless code. In 1957, Crick, Leslie Orgel and J. S. Griffith had theorised that if the code was composed of words made of three bases, as many people thought, and there were no bases between the triplets that acted as commas indicating the separate words, then triplets composed of the same base (AAA or UUU, for example) were forbidden because the cellular machinery would not know where to start reading. With various other ad hoc restrictions, partly based on chemistry, Crick's theoretical scheme allowed merely twenty combinations out of the sixty-four possible combinations of bases. As there were twenty naturally occurring amino acids, this was aesthetically very pleasing, but it was entirely speculative. As Crick noted at the time, this 'gave the magic number – twenty – in a neat manner' but the 'arguments and assumptions' behind the theory were 'too precarious for us to feel much confidence in it'.[19] A few months later he admitted:

> I find it impossible to form any considered judgement of this idea. It may be complete nonsense, or it may be the heart of the matter. Only time will show.[20]

Time showed it to be complete nonsense. More importantly, the idea may have restricted what researchers thought was possible, in particular because it ruled out triplets composed of the same base. For most people trying to crack the code, investigating the effect of a polynucleotide containing only one kind of base would have been pointless.

*

By the end of 1960, Nirenberg and Matthaei were working through the night on protein synthesis in *E. coli* extracts. In mid-January, one of Nirenberg's diary entries was headed '*Idea. Approach to code*' and outlined the use of poly(A), poly(U), poly(C) or poly(G) and of poly(AG), etc.; poly(AG) would be composed of equal amounts of A and G, in an unknown sequence. Nirenberg's aim was to put polynucleotides into his cell-free protein synthesis system and use the output to understand the nature of the genetic code, starting with establishing the number of bases involved in coding for an amino acid:

> Might be able to get enough info to establish limits of a code …
> If you need all 4 bases, could not be triplet code.[21]

At the FASEB Meeting in February 1961, Matthaei and Nirenberg gave a brief talk describing the way their system incorporated a ^{14}C-labelled amino acid (valine) into a protein.[22] A few weeks later, on 22 March, they submitted an article on the topic to *Biochemical and Biophysical Research Communications*.[23] This journal had been set up in the previous year to respond to the increasingly fierce competition in the area by offering rapid publication of short articles – it used camera-ready typed copy provided by the authors rather than traditional typesetting, thereby speeding up the whole publishing process.

The article described how the presence of ribosomal RNA in their cell-free system was essential for amino acid incorporation, which 'had many characteristics expected of protein synthesis', and that 'all of the activity appeared to be associated with RNA'.[24] Exactly what kind of RNA Nirenberg and Matthaei were referring to was not entirely clear. They concluded, somewhat confusingly 'It is possible that part or all of the ribosomal RNA used in our study corresponds to template or messenger RNA.'[25] By ribosomal RNA they did not mean the RNA that makes up the ribosome itself, but rather an RNA molecule that was attached to the ribosome. The ambiguity contained in the term 'template or messenger RNA' is not simply a matter of uncertainty over which word to choose. As Lily Kay has pointed out, Nirenberg and Matthaei's use of 'template or messenger RNA' shows that their language was poised at a cusp between the old, physical, way of thinking of specificity – as a structural template – and the new, abstract idea of information being transferred by a messenger.[26] Understandably, these semiotic niceties were not noticed at the time, and the paper made no impression – it was not cited until 1963, by which time all the dust had settled.

In early May 1961, Nirenberg and Matthaei decided to add RNA from the tobacco mosaic virus (TMV) to see if they could get the cell-free system to synthesise TMV protein. It worked like a dream. As Nirenberg recalled in the 1970s, the results were *'superb ... beautiful ... It was superbly active'*.[27] He realised that they would need to collaborate with the Berkeley TMV expert Fraenkel-Conrat if they were to fully exploit this novel approach. In the meantime, they continued to crank through the effects of the various synthetic RNA molecules that they were able to borrow from Leon Heppel's laboratory next door.

9. Matthaei's notebook showing the results of the crucial experiment – reproduced from Kay (2000)

In the middle of May, Nirenberg left to spend a month in Fraenkel-Conrat's lab in Berkeley, getting himself up to speed with TMV. Back in Bethesda, Matthaei started a set of experiments to study the response of the cell-free system when it was seeded with artificial RNAs. On 15 May (his 32nd birthday), Matthaei began an experiment testing the effect of poly(A) (AAAAA...), poly(U) (UUUUU...), poly(2A)U (a ratio of two A nucleotides to one U, randomly distributed through the RNA molecule), and poly(4A)U (a randomly distributed ratio of four A nucleotides to one U). When all twenty amino acids added to the test tube were radioactive, Matthaei obtained a twelvefold increase in radioactivity in the protein product after incubation with poly(U), a small increase with poly(AU), and barely any change with poly(A). Something was going on in the poly(U) tube, which could explain how genetic information leads to a particular protein being created. To detect which radioactive amino acid had been incorporated into the protein that was produced by the cell-free set-up, Matthaei had to test all twenty amino acids systematically. He did this by putting ten radioactive amino acids into a test tube – the other ten amino acids added were the usual 'cold' versions. He then did the poly(U) experiment again. If there was an increase in radioactivity, then clearly one of the 'hot' amino acids was involved. By repeating this process, Matthaei was finally able to narrow the effect down to one of two amino acids: phenylalanine or tyrosine.

On Saturday 27 May at 3.00 a.m., Matthaei began the final experiment. This involved ten test tubes and was labelled '27-Q' in his lab book. In tube number 3 he had nineteen unlabelled amino acids together with radioactive phenylalanine, and in tube number 8 he had nineteen unlabelled amino acids together with radioactive tyrosine. The remaining eight tubes contained various controls to prove that the effect was due to the combination of the poly(U) and one of the two radioactive amino acids. Matthaei allowed the mixture to incubate for one hour at 36°C; then he began the tedious task of isolating the protein produced by the reactions and measuring the radioactivity that had been incorporated. Tube 3, which contained radioactive phenylalanine, produced a protein with a radioactivity level that was more than twenty times higher than the control tubes; the protein from the tyrosine tube showed no increased radioactivity. When Gordon Tomkins came into the lab a few hours later, Matthaei told him the news: poly(U) coded for phenylalanine. The first word in the genetic code had been read.

*

Nirenberg, who was at Berkeley, heard of the breakthrough over the telephone, and by 11 June he was back at Bethesda doing experiments. Matthaei recalled the feelings in the laboratory at the time: 'of course we were excited, because we knew *exactly* what we had. And we knew what we'd wanted to get.'[28] Everyone was sworn to secrecy – nobody was to hear about the finding until the results had been published. This caused some difficulties – at the beginning of June, Sydney Brenner gave a talk at Bethesda, during which he said it was not possible to study messenger RNA in a cell-free system. When Matthaei asked him how he could know this, Brenner astutely fired the question back at Matthaei, asking whether he had any insight into the question. Matthaei said nothing.[29] Above all, the normally garrulous Tomkins had to bite his tongue throughout the Cold Spring Harbor meeting, which took place a week after Brenner's visit. After the meeting was over, Tomkins finally cracked, and at the end of July he told Alex Rich in Boston. The news went no further, because Rich was too busy with his own work to gossip and, anyway, everyone else was either on holiday or was heading off to the International Biochemical Congress in Moscow.

Matthaei had also found it hard to keep quiet. At the end of June he took the phage course at Cold Spring Harbor, during which each student described his or her research. Matthaei initially refused, but eventually outlined his discovery. Delbrück, who was teaching on the course, was amazed and immediately told Jerry Hurwitz at New York University; Hurwitz in turn phoned Tomkins and got confirmation.[30] The secret was out, and by the beginning of August, researchers in Severo Ochoa's laboratory in New York had heard the garbled news that 'someone from MIT' had broken the code.[31]

Meanwhile, Nirenberg was heading for the Biochemical Congress in Moscow, where he was planning to unveil his discovery. Before he left he had two things to tie up. He got married to Perola Zaltzman, a Brazilian biochemist, and he staked his claim to priority by submitting two articles to *Proceedings of the National Academy of Sciences*. At the time, articles in *PNAS* had to be sponsored by a member of the Academy. Hearing that Academy member Leo Szilárd was staying just down the road in Washington, Nirenberg spent a whole afternoon discussing the results with him in the lobby of the Dupont Hotel. Szilárd was reluctant to help out: 'It's too much out of my field,' he said, 'I'm sorry, I can't sponsor it.'[32] It is hard to imagine Szilárd responding in the same way to Jacob or Monod. Nirenberg was clearly an outsider.

The two papers were submitted to *PNAS* on 3 August 1961, with the support of Joseph Smadel, the Associate Director of NIH. Straight afterwards, Nirenberg flew to Moscow. The articles appeared, back to back, in the October issue of the journal, by which time everyone who was anyone already knew all about their stunning content. Both papers contained meticulous descriptions of the protocols involved and above all were characterised by the use of tightly conceived control experiments that enabled the authors to exclude alternative explanations, rendering their conclusions incontestable.

The first, more technical, paper described the characteristics of protein synthesis in cell-free *E. coli* extracts, repeating and expanding the results that had been published earlier in the year. Significantly, Matthaei was the first author on this article, which was destined to be read by fewer people (it has been cited fewer than 300 times). This paper showed that protein synthesis could be disrupted by RNase, which attacks RNA, and – eventually and to a lesser extent – by the DNA-destroying enzyme DNase. Matthaei and Nirenberg correctly suggested that the presence of intact RNA was essential for protein synthesis to occur, and that inhibition by DNase was due to 'the destruction of DNA and its resultant inability to serve as templates for the synthesis of template RNA.'[33]

The second paper was obscurely entitled 'The dependence of cell-free protein synthesis in *E. coli* upon naturally occurring or synthetic polyribonucleotides'. Despite this unappetising opening, it contained the experiment that showed an increase in the amount of radioactive protein when poly(U) was incubated with radioactive phenylalanine – after refining the protocol, they were able to get a roughly 1,000-fold increase over control levels. The article, which has been cited more than 1,400 times, was entirely couched in the language of biochemistry and protein synthesis, referring in a rather old-fashioned way to 'specificity for phenylalanine incorporation'. Only in the final paragraph did Nirenberg and Matthaei frame their discovery in the new language of life, emphasising that the full detail of the code was not yet known:

> One or more uridylic acid residues therefore appear to be the code for phenylalanine. Whether the code is of the singlet, triplet, etc., type has not yet been determined. Polyuridylic acid seemingly functions as a synthetic template or messenger RNA, and this stable, cell-free *E. coli* system may well synthesize any protein corresponding to meaningful information contained in added RNA.[34]

Most people assumed that the code was based on triplets, simply because this gave sixty-four possible combinations for twenty naturally occurring amino acids, but this had not been shown to be true. Nirenberg and Matthaei were rightly hedging their bets – strictly speaking, their data were compatible with the unlikely possibility that just a single base of U coded for phenylalanine. Although they again referred to messenger RNA, they did not cite any of the three recently published papers that had first used this term (the two *Nature* papers and the Jacob and Monod review in the *Journal of Molecular Biology*, all of which had appeared in May). Indeed, for reasons that remain obscure, Nirenberg never cited any of these articles.[35] In a note added shortly before the article was printed, they included a recent result obtained by Matthaei while Nirenberg was in Moscow – poly(C) coded for proline. Two words of the genetic code had now been read, but it was still not clear how many letters each contained.*

*

The Fifth International Congress of Biochemistry took place in Moscow, from 10 to 16 August 1961. It was the largest conference ever held in the USSR – there were more than 5,000 participants, including 3,500 foreigners from fifty-eight countries – and was commemorated by a special Soviet postage stamp. With nearly 2,000 talks and up to eighteen parallel sessions, many of the presentations were poorly attended.[36] Eight large symposia, including one organised by Max Perutz on 'Biological Structure and Function at the Molecular Level', were held in various Moscow University buildings.

The congress opened in the Sports Palace of the Central Lenin Stadium on the outskirts of Moscow, where the slides were poorly projected and it was hard to see anything. One symposium had to be cut short due to a press conference that was held for the second man to orbit the Earth, 25-year-old Gherman Titov, who had returned to Earth on 7 August, after spending more than a day in space. Later, delegates gathered in a sunlit Red Square to see a parade to celebrate Titov's return.[37] This was at the height of the Cold War, and Russian

---

*Lily Kay, the principal historian of this subject and someone not at all sympathetic to the 'great man' view of history, nevertheless wrote: 'the breaking of the code by Nirenberg and Matthaei was one of the most stunning events in the history of modern science. It represented a victory of material ingenuity over the Pythagorean ideals and is a David versus Goliath tale of an obscure young scientist defeating the eminent gray matter of physicists, mathematicians, biochemists, and geneticists, some of them Nobel laureates.' (Kay, 2000, pp. 254–55).

superiority in space was extremely significant. Furthermore, while the congress was taking place, the Cold War got a bit hotter as the Berlin Wall began to be constructed on 13 August.

Like every other non-plenary speaker at the massive meeting, Nirenberg was given a brief ten-minute slot to present his findings, which concentrated on the material from the second *PNAS* paper and concluded, after a last-minute edit, with the phrase he and Matthaei had used in their article: 'One or more uridylic acid residues therefore appear to be the code for phenylalanine'.[38] The tiny lecture theatre was partly filled with a big old-fashioned slide projector, and there were only a couple of dozen people in the audience.[39] Watson later said that he 'heard rumours that there might be an unexpected bombshell talk by Marshall Nirenberg' – this may have been chatter from Delbrück or others, but according to Nirenberg he introduced himself to Watson shortly before the talk and outlined his findings.[40] Whatever the case, Watson was clearly not intrigued enough to go and listen. Instead he sent along his postdoctoral researcher, Alfred Tissières; Matthew Meselson was also there. Meselson, who was younger than Nirenberg, later recalled:

> I heard the talk. And I was bowled over by it. … I went and chased down Francis [Crick], and told him that he must have a private talk with this man.[41]

The next morning, Watson told Jacob what Nirenberg and Matthaei had discovered. Jacob assumed that this was one of Watson's tiresome practical jokes, and refused to believe him.[42] Crick was sharper – on hearing the news from Meselson he immediately decided to invite Nirenberg to present his talk again the next day, in the symposium on molecular structure and function that Perutz had organised and which Crick was due to chair. In typically generous fashion, Crick was offering Nirenberg the opportunity to step into history as the man who had cracked the genetic code.

According to Watson, Nirenberg's plenary talk was 'an extraordinary moment'. At the time, Crick reported that the audience was 'startled' by Nirenberg's announcement – he later described it as

'electrified'.* Even the modest Nirenberg recalled that the audience was 'extraordinarily enthusiastic'.[43] Meselson recalled that
after Nirenberg's second presentation 'I ran up to Nirenberg and I
embraced him. And congratulated him ... It was all very dramatic.'[44]
Nirenberg was deeply touched by this gesture:

> The second time I gave the paper it was to a very large
> audience. The reception was really remarkable, fantastic. I
> remember Matt Meselson, who was sitting right up front. I
> didn't know him at the time, but he was so overjoyed about
> hearing this stuff that he impulsively jumped up, grabbed my
> hand, and actually hugged me and congratulated me for doing
> that. I could have been part of a rock band or something! That
> meant an awful lot to me. It really meant more to me than
> all kinds of awards and what-not because it was genuine and
> spontaneous.[45]

Meselson also recalled the effect that the talk had on the audience: 'it gave some people who were in this field the immediate itch
to get out of Moscow, to get back to the lab.'[46] What they would do
back in the lab was simple – they had to adopt Nirenberg's technique. Jerry Hurwitz, who heard both of Nirenberg's Moscow talks,
recently told me how the new approach changed everything:

> I remember thinking about the ramifications of the Niren-
> berg–Matthaei findings. In early June 1961, at the Cold Spring
> Harbor meeting, it was evident that a number of laborato-
> ries were using specific proteins ... to get at the code. I recall
> thinking that these efforts were now obsolete.[47]

Harold Varmus, who was not even a scientist at the time, found
his life changed because of Nirenberg's talk. Varmus, a student of
English, was accompanying his biochemist friend Art Landy, who

---

*In response to this claim, the joker Seymour Benzer sent Crick a photo of the congress
in which the audience looked utterly bored (Crick, 1988, p. 131). I have been unable to
locate any photo of the symposium – Benzer may well have deliberately sent a picture of
a different meeting.

was attending the congress. Varmus understandably spent the day that Nirenberg spoke 'riding Moscow's fabled ornate subways and roaming Russian art galleries'. But that evening he heard something that made him doubt his career choice:

> listening to Art Landy's excited report at the end of the day in our rooms at Moscow State University, I began to understand that something of fundamental significance had occurred, and I felt that a seed of professional envy had been planted. Scientists seemed likely to discover new, deep, and useful things about the world, and other scientists would be excited about these discoveries and eager to build on them.[48]

That feeling grew, and Varmus soon switched from English to medicine. In 1989, he won the Nobel Prize in Physiology or Medicine for his work on genes and cancer.

Not everyone was convinced, however. When Jerry Hurwitz returned to New York University in the middle of August, he told colleagues about Nirenberg's talk but added:

> several people didn't believe the validity of the data. It seems that, although Nirenberg performed the first basic experiment, there is still a great deal to gain from the application of this new, extremely sensitive method.[49]

Whoever those 'several people' were, within weeks their doubts were blown away.[50]

Matthew Meselson later explained the widespread surprise that was felt about Nirenberg's success, in terms of the social dynamics of science:

> there is a terrible snobbery that either a person who's speaking is someone who's in the club and you know him, or else his results are unlikely to be correct. And here was some guy named Marshall Nirenberg; his results were unlikely to be correct, because he wasn't in the club. And nobody bothered to be there to hear him.[51]

This explanation is reinforced by a private letter to Crick, written in November 1961 by the Nobel laureate Fritz Lipmann, which celebrated the impact of Nirenberg's discovery but nevertheless referred to him as 'this fellow Nirenberg'.[52] In October 1961, Alex Rich wrote to Crick praising Nirenberg's contribution but wondering, quite legitimately, 'why it took the last year or two for anyone to try the experiment, since it was reasonably obvious'.[53] Jacob later claimed that the Paris group had thought about it but only as a joke – 'we were absolutely convinced that nothing would have come from that', he said – presumably because Crick's theory of a commaless code showed that a monotonous polynucleotide signal was meaningless.[54] Brenner was frank: 'It didn't occur to us to use synthetic polymers.'[55] Nirenberg and Matthaei had seen something that the main participants in the race to crack the genetic code had been unable to imagine. Some later responses were less generous: Gunther Stent of the phage group implied to generations of students who read his textbook that the whole thing had happened more or less by accident, while others confounded the various phases of Matthaei and Nirenberg's work and suggested that the poly(U) had been added as a negative control, which was not expected to work.[56]

In fact, Nirenberg and Matthaei were not the only ones to have the idea of using synthetic polynucleotides to crack the code.* It had occurred, separately, to at least two people in Severo Ochoa's laboratory. In 1957, the Yugoslavian scientist Mirko Beljanski was on sabbatical in Ochoa's lab. For nearly a year, Beljanski tried to get synthetic poly(A) to direct protein synthesis in a cell-free system, but

---

*In 1914, the veteran German chemist Emil Fischer had effectively suggested exactly the experiment that Nirenberg and Matthaei carried out. Writing at a time when nucleic acids were known to be an important component of the nucleus, but when their role was still unclear, Fischer discussed recent developments in the synthetic creation of nucleic acids: 'we are now capable of obtaining numerous compounds that resemble, more or less, natural nucleic acids. How will they affect various living organisms? Will they be rejected or metabolized or will they participate in the construction of the cell nucleus? Only the experiment will give us the answer. I am bold enough to hope that, given the right conditions, the latter may happen and that artificial nucleic acids may be assimilated without degradation of the molecule. Such incorporation should lead to profound changes of the organism, resembling perhaps permanent changes or mutations as they have been observed before in nature' (cited in McCarty, 1996). Nothing came of Fischer's insight.

without success.[57] Paul Zamecnik later recalled that Ochoa had sent him some poly(A) to be studied in the cell-free system, which was waiting in the freezer when the news of Nirenberg and Matthaei's discovery came through.[58] It is not obvious what would have happened had Zamecnik tried the poly(A); in 1961, Nirenberg and Matthaei, like Beljanski, failed to get poly(A) to produce any protein. The eventual explanation was quite simple – the protein produced by poly(A), a lysine polymer, interacted with some of the chemicals that were initially used in the cell-free system and gave no apparent result.[59] A biochemical tweak was needed to make poly(A) work, but that would not have been apparent without the previous success of the poly(U) experiment.

At exactly the same time as Nirenberg and Matthaei were tying up their experiments in the summer of 1961, Peter Lengyel, a young researcher in Ochoa's laboratory, came up with the idea of using synthetic polynucleotides while listening to Sydney Brenner's talk about messenger RNA at the Cold Spring Harbor meeting. While Ochoa was on holiday in Europe, Lengyel and three young colleagues planned out the experiments, but these were radically changed at the beginning of August when they heard through the grapevine that Nirenberg and Matthaei had used poly(U) to produce polyphenylalanine. The New York group immediately replicated the result. So at the same time as Nirenberg was stunning the audience in Moscow, Lengyel was showing that, in his hands too, poly(U) led to the incorporation of radioactive phenylalanine into a protein. By the time Ochoa returned to New York at the beginning of September, his laboratory was ready to make up the ground they had lost to the upstarts from Bethesda.

*

In late September, Nirenberg presented his work at a meeting in New York. He had barely progressed since Moscow – in an uncharacteristically unfocused moment, he had spent his time in the laboratory tying up a minor loose end in the protein synthesis pathway that had been activated in the poly(U) tube. He had even allowed himself a 'leisurely, two week vacation' in Copenhagen with his new wife.[60] Nirenberg was therefore devastated when Ochoa stood up at the New York meeting and presented data that showed that the Ochoa group was hard on his heels. They had made huge progress – in the space of about six weeks, Ochoa's lab had managed to get the cell-free system working, had replicated Nirenberg and Matthaei's results and above all had shown that two other artificial polynucleotides – poly(UA) and poly(UC) – were also active. As Ochoa later explained: 'When we heard the news from Moscow we immediately tried it. And other polymers, copolymers we had in the icebox. We got immediate results with four or five.'[61] New words in the genetic code were being read, but not in Nirenberg's lab. Ochoa reported that poly(UC) led to the incorporation of radioactive phenylalanine, serine and leucine into a protein, and poly(UA) led to the incorporation of phenylalanine and tyrosine. Ochoa later described the excitement of discovery:

> Lengyel, Speyer, and I were watching the counter and were thrilled. This result, obtained for the first time anywhere, showed that the incubation of *E. coli* extracts with copolynucleotides containing C or A besides U residues promoted the synthesis of polypeptides containing serine, leucine and tyrosine, along with phenylalanine. I remember this as one of the most exciting moments of my life.[62]

Understandably, Nirenberg's response was less enthusiastic: 'It floored me, that Ochoa had made such advances. Things developed much faster than I would ever have dreamed that they'd develop.'[63] As he later recalled:

> I flew back to Washington feeling very depressed because, although I had taken only two weeks to show that aminoacyl-tRNA is an intermediate in protein synthesis, I should have spent the time focusing on the more important problem of the genetic code. Clearly, I had to either compete with the Ochoa laboratory or stop working on the problem'.[64]

A colleague remembered the scene:

It was a Saturday afternoon in the Indian summer of 1961, and almost no one was around. Marshall was sitting alone at a table with his head bowed and his eyes glassy, obviously upset and depressed. ... How could Marshall and Heinrich keep up with a lab that had nearly 20 scientists?[65]

After a cordial discussion with Ochoa in which it became obvious that collaboration was not an option, Nirenberg decided to fight back (he later said, 'to my horror, I found that I enjoyed competing'), and he enlisted the help of Bob Martin and Bill Jones at NIH in synthesising polynucleotides for cracking the rest of the code.[66] The race was on.

*

Francis Crick was immensely impressed by Nirenberg's breakthrough – four months later, he described it on the BBC as 'spectacular' – but he did not shift his focus at all.[67] Refreshed from a month-long holiday in Morocco just before the Moscow congress, excited by the new phase of discovery that had just been opened, Crick returned to Cambridge determined to settle the vexed question of whether the genetic code was a triplet code or not. Together with Brenner, he dreamt up the idea of studying mutants in the rII region of the T4 phage, which had been induced by a chemical that deleted single bases. According to Crick and Brenner's thinking, a single base deletion would alter how the information was read, potentially rendering the message after that point nonsensical, because what they called the reading frame of the message would now be out of sync. For example, if there were a triplet code sequence such as ATG CAT CCC TGA ... and the first C were deleted, then the sequence would become ATG ATC CCT GA ... The first codon would be the same but the remaining codons would be altered. As Crick put it:

The simplest postulate to make is that the shift of the reading frame produces some triplets the reading of which is 'unacceptable'; for example, they may be 'nonsense', or stand for 'end the chain', or be unacceptable in some other way to the complications of protein structure.[68]

They found that some point deletions in the *r*II region did indeed stop the phage from functioning. The trick was to combine a number of these deletions and thereby put the reading frame back in line. If three deletions restored viral function to the manipulated phage and it was able to infect bacteria (this would be detected by the plaques that infected bacteria formed in the Petri dish), that would provide very strong evidence that the code was based on triplets.

In the autumn of 1961, Brenner went to Paris, leaving Crick to do the experiment with Leslie Barnett, a 42-year-old microbiologist. Crick recalled the moment:

> And all we had to do was look at one plate. And see if it had any plaques on it. So we came in late at night, ten o'clock at night, or something, and there were plaques on the plate! So I said to Leslie, 'Let me check; we may have got the plates mixed up,' and she checked it, and then I told her, 'We're the only two to know it's a triplet code!'[69]

These findings were published in *Nature* at the end of December 1961, with a title that was full of Crick's flair: 'General nature of the genetic code for proteins'. The paper combined the experimental detail of the study of *r*II mutants and a summary of research on the coding problem (including Nirenberg and Matthaei's work), and was infused with Crick and Brenner's theoretical insight. The article contained four fundamental conclusions that are now taught in schoolrooms and university lecture theatres all around the world:

(a)   A group of three bases ... codes one amino acid.
(b)   The code is not of the overlapping type.
(c)   The sequence of the bases is read from a fixed starting point. ...
(d)   The code is probably 'degenerate'; that is, in general, one particular amino-acid can be coded by one of several triplets of bases.[70]

Strictly speaking, not all of these conclusions had been proven. As the paper explained, it was technically possible that the number of bases in each group was six, or some other multiple of three, although this was highly unlikely. Second, although they did not yet have evidence that the code was 'degenerate', this 'could also account for the major dilemma of the coding problem, namely, that while the base composition of the DNA can be very different in different micro-organisms, the amino-acid composition of their proteins only changes by a moderate amount.'[71]

Crick was later dismissive of the significance of the article, pointing out that 'it was pretty obvious it was *likely* to be a triplet code … the fact is, if we'd shown that the code was a *quadruplet* code, *that* would have been a discovery.' He even suggested, 'I think you could have deleted the whole work and the issue of the genetic code would not have been very different.' Nirenberg did not agree. In a letter to Crick written in January 1962, he described the article as 'beautiful'. More than 700 papers have subsequently cited it, and in 2004 the original manuscript sold at Christie's for £13,145. Crick's conclusion was audacious, conveying the optimism that swept through the scientific community after Nirenberg and Matthaei's transformation of the field:

> If the coding ratio is indeed 3, as our results suggest, and if the code is the same throughout Nature, then the genetic code may well be solved within a year.[72]

This view was shared by Nirenberg, who shortly afterwards claimed that 'within another six months or so most of the genetic code will be cracked.'[73] Both men severely underestimated the difficulties ahead.

*

In January 1962, Crick gave a talk on the BBC that outlined the importance of Nirenberg's discovery. He concluded by putting it into context, and posing some questions that we now know the answers to, and others that are still unanswered today:

We still don't know whether the code is universal. The same 20 amino acids are used in proteins throughout nature, from virus to man, but it is not yet certain that the same triplets code them in all organisms, although preliminary evidence suggests this is probable. If so, we shall have the key to the molecular organisation of all living things on Earth.

But on Mars, I wonder? Will there be life, or the remains of life, on Mars? And will it be DNA and RNA and protein all over again? The same languages perhaps, with the same code connecting them? Who knows?[74]

In the coming years, Crick continued to be generous towards Nirenberg and Matthaei, recognising that their discovery had altered the course of history and paying glowing tribute to the importance of the work of the two outsiders. As he put it in 1962:

We are coming to the end of an era in molecular biology. If the DNA structure was the end of the beginning, the discovery of Nirenberg and Matthaei is the beginning of the end.[75]

ELEVEN

THE RACE

The weeks after Nirenberg's dramatic revelation saw the beginning of a frenetic scientific race to crack the rest of the genetic code. Nirenberg and Matthaei's papers appeared in *Proceedings of the National Academy of Sciences* in October, quickly followed by a paper from the Ochoa lab, also in *PNAS*. This was the first of nine articles by Ochoa's group that were collectively entitled 'Synthetic polynucleotides and the amino acid code'.[1] The pace with which the Ochoa lab produced their material was extraordinary. The first five papers in the series were submitted in the space of about eighteen weeks – today it is hard to imagine this level of productivity when a group begins work in a new field. Irrespective of the fact that the articles would not have gone through any form of peer review – at the time, Academy members like Ochoa could treat the *PNAS* more or less as a private publishing service – this represented a remarkable avalanche of activity, and it nearly crushed Nirenberg, who was also distraught at the death of his parents around this time.[2]

Ochoa's dominance was expressed both in the data and in the tone and detail of his group's publications. Nirenberg and Matthaei's contribution was downplayed – they were not even cited in several of the papers – and the opening salvo in the nine-paper bombardment closed with a statement that seemed to claim the breakthrough as Ochoa's own:

These and other results reported in this paper would appear to open up an experimental approach to the study of the coding problem in protein biosynthesis'.[3]

Faced with the aggressive output of the Ochoa group, Nirenberg became more savvy about the presentation of his work – his last paper in 1961 was entitled 'Ribonucleotide composition of the genetic code', a far clearer indication of what his work implied than the technical titles he initially employed.[4]

This wave of discovery, coupled with Crick's *Nature* paper on the 'General nature of the genetic code for proteins', which appeared in December 1961, finally attracted the attention of the world's press. For nearly six months, the scientific world had been buzzing with excitement – in a reference to the shooting down in 1960 of an American U-2 spy plane over the Soviet Union, Rollin Hotchkiss quipped that 'The U-2 incident started the cold war, the U3 incident started the code war.'[5] Now the general public got to hear about it. On Xmas Eve 1961, the *New York Herald Tribune* announced 'The code of life finally cracked', explaining the four-month press silence since Nirenberg's Moscow announcement by claiming that 'the news did not leak into the newspapers until last week'. A few days later, the *Sunday Times* took a similar line – 'Scientists have cracked the code of life' – but chose to emphasise the role of British scientists and the importance of Crick's recent *Nature* paper.[6] Crick was embarrassed by the coverage, writing to Nirenberg to explain that he had done his best to set the record straight.[7] Nirenberg's response was typically relaxed. It also shows that the way in which the media treat scientific breakthroughs has not changed that much:

I haven't seen the English newspapers but the American press has been saying that this type of work may result in (1) the cure of cancer and allied diseases (2) the cause of cancer and the end of mankind, and (3) a better knowledge of the molecular structure of God. Well, it's all in a day's work.[8]

*

The widespread excitement was stoked further by what seemed to be rapid progress in the race to crack the code. With the new focus on manipulating RNA rather than DNA, there was an unstated shift in the way that the code was now thought of. It was no longer located solely in the double helix, but equally in the molecular transcription of the gene, in the shape of messenger RNA. As a result, the letters of the code increasingly became those of the RNA molecule – A, C, G and U. Today, as researchers study genomes and their DNA content, the code is tending to be presented in terms of DNA bases.

Nirenberg and Ochoa's groups both used the same method: they synthesised RNA polynucleotides with known ratios of bases but unknown sequence, put them into the cell-free protein synthesis system and interpreted what came out in terms of the ratios of bases that had been put in. For example, if the 'coding unit' was a triplet, a randomly assembled polynucleotide composed of five parts U to one part C – 'poly(5U1C)' – would contain the coding sequences UUU UUC UCU, UCC, CUU, CUC and CCC in varying proportions. The CCC combination would be present in much smaller amounts than UUC, because there was far less C present in the mixture than there was U. However, the three '2U1C' triplets – UUC, CUU and UCU – would be expected to be present in the same proportions, so their effects could not be distinguished.

When Ochoa's laboratory reported that with poly(5U1C) they got high levels of phenylalanine and lower levels of proline and of serine, this was interpreted in terms of the relative proportions of the possible coding triplets. Starting with the known fact that UUU coded for phenylalanine, they claimed that proline was coded by 1U and 2Cs (either CCU, UCC or CUC) and serine was coded by 2Us and 1C (either UCU, UUC or CUU).[9] Although this was not too far from the truth – proline is indeed coded by CCU, and serine by UCU – in both cases the method used by the Ochoa group could not distinguish between the three alternative triplets. Furthermore, if one of those alternatives coded for phenylalanine, this would not be detectable, because it would be assumed that the phenylalanine was encoded by UUU, about the only part of the code that everyone agreed on. In fact, this was exactly what was happening – UUC, like UUU, codes for phenylalanine. Comparing the theoretical frequencies of different triplets and the proportions of amino acids they seemed to code for could only get you so far in cracking the code.[10]

By the beginning of 1962, the two competing laboratories had identified the potential base compositions of the RNA code for nearly all twenty naturally occurring amino acids.[11] They had no idea of the sequential order of the nucleotides, nor of how many nucleotides there were in each coding unit; although most people assumed that the code was based on triplets, there was still no absolute proof that this was true. In February 1962, Ochoa told Crick that he expected that fewer than '30 triplets would stand for amino acids' – it was widely thought that the code generally involved a one-to-one correspondence between a single RNA coding unit and a single amino acid, with the other forty-odd possible triplets coding for 'punctuation' or for nothing.[12] But within weeks it was evident that many amino acids were coded by more than one RNA coding unit, suggesting that the code was degenerate or, as we now put it, redundant.[13]

And something else was odd. As Ochoa's group noted in March 1962: 'A striking feature of the code triplets is that they all contain U.'[14] A month later, Nirenberg's lab made the same observation: 'a surprisingly high proportion of U has been found in coding units thus far'.[15] Attempts to get amino acid incorporation with poly-nucleotides that did not contain U failed repeatedly, leading some researchers to conclude that the twenty-seven possible triplets that did not contain U were 'nonsense' triplets, with no meaning. But when the composition of virus RNA was studied, U was not present in particularly high levels, implying that coding units not containing U must exist somewhere in nature, or that something strange was going on in the test tube. Confusion was increased when both Ochoa and Nirenberg's labs reported that poly(U), which everyone agreed led to the incorporation of phenylalanine, also seemed to code for leucine and valine.[16] Everyone assumed that this must be an experimental error (it was), but until that could be explained, there was a real danger: if a triplet coded for more than one amino acid, then all existing ideas about coding would have to be scrapped. The whole code edifice could come crashing down.

*

The cascade of new data, and the lack of clarity about what it all meant, encouraged the theoreticians to return to the coding problem. As Crick put it in 1966, during this period there was 'a flurry of theoretical papers, most of which are best forgotten'.[17] Forgettable they may have been, but they reveal the thinking of scientists at the time and highlight how they were groping their way towards the right answer. In September 1961, Richard Eck re-raised the possibility that the code might be overlapping, with the bases in one coding unit also forming part of the subsequent unit (so, for example, the sequence CACGU would contain three triplets – CAG, ACG, and CGU).[18*] Brenner had disproved this idea to most people's satisfaction in 1957; his criticism had been reinforced by the fact that mutational studies of viruses had shown that changes to a single base only ever altered a single amino acid – if the code were overlapping, then two or more amino acids should be altered.[19] Nevertheless, as Crick later put it, it was possible with ingenuity to come up with a complex overlapping code.[20] Another theoretician took as his starting point the complementary coding groups that would be found on the two strands of the DNA molecule, for example TAC and ATG, and argued they must code for the same amino acid, to avoid the problem of the cell having to know which strand of DNA to use.[21] This bold and mistaken vision underestimated the skill of the cell, which can indeed distinguish between the two strands.

The most sophisticated attempt to crack the code by theoretical means was made by Carl Woese (pronounced 'Wose'). His starting point was that any code had to be compatible with the known nucleotide composition of RNA in a variety of organisms, which was known not to be rich in U. After a tortuous set of calculations, Woese produced a code that used only twenty-four potential triplets out of sixty-four, with most amino acids being coded by both a triplet that was low in G + C and one that was high in these two bases.

---

* Eck began his paper by accepting the slight possibility that the hereditary material was made of nucleoproteins, not DNA. This was the latest doubt about the genetic role of DNA that I have found.

Like all the other theoretical schemes, this one was ingenious, but wrong.[22]

Other less complicated theoretical codes were dreamt up. Richard Roberts found an easy solution to the problem of the U-rich data coming from Ochoa and Nirenberg's labs – the Us were simply not relevant, he argued, because the code was really composed of only two bases. Although this fitted with the known RNA base composition of viruses, which was not U-rich, it had the disadvantage of providing only sixteen possible combinations, when at least twenty were needed to code for the naturally occurring amino acids. To get out of this dilemma, Roberts suggested that the code was composed of both doublets and triplets, with some doublets, such as AA or GG, indicating the start of a triplet. There was no evidence to support this, but as Roberts cunningly pointed out, there was no evidence against it, either.[23] At the beginning of 1963, Thomas Jukes put forward a variant of this idea, suggesting that each triplet had what he called a 'pivotal' base, which could change without altering the amino acid that was coded for. He announced that U was the 'pivotal base' in all triplets containing U (it is not).

In a return to the numerology that had dominated the coding problem in the 1950s, Eck published a four-page article in *Science* in which he claimed to detect a symmetrical pattern in the attribution of triplets to amino acids – four amino acids were coded by four triplets, with the remaining sixteen each being coded by two.[24] Eck said all he had to do was to tabulate the known distribution of triplets 'and the puzzle practically solved itself'. But the solution was based entirely on conveniently fitting as-yet unallocated triplet/amino acid combinations into the schema. The pattern was in Eck's head, not in the data.

Finally, the pioneer of the biological application of information theory, Henry Quastler, came up with a schema based on data from amino acid changes induced by mutations. He was unimpressed by the cell-free studies, arguing that they did not necessarily measure protein synthesis, and above all he emphasised that in most cases the precise nature of the polynucleotides was unknown.[25] Crick was scornful of Quastler's paper, claiming that it consisted of 'a rather poor fit to some very doubtful data' and was based on 'an unspecified technique'. All of the triplets predicted by Quastler were wrong, with the exception of UUU = phenylalanine, which was hardly a prediction.

The real answer to the conundrum of the predominance of U nucleotides in the cell-free data was inadvertently provided by Solomon Golomb.[26] He performed various calculations and concluded that it was not possible to deduce anything about the role of non-U sequences without doing an experiment. Which is what the biochemists did, and by the middle of 1962, RNA with no U bases had been shown to encode amino acids.[27] The bewildering preponderance of U-rich polynucleotides was an artefact due to the solvents that were initially employed in the cell-free system.[28]

*

With the smell of competition in everyone's nostrils, two meetings took place in the summer of 1962 at which progress on cracking the code was discussed. In July, a 'Colloquium on Information in Contemporary Science' was held in the glorious surroundings of the thirteenth-century Royaumont Abbey to the north of Paris. As one of the participants recalled, 'the gardens, the musical evenings, and supper by candlelight' were almost as significant as the discussions.[29] This meeting involved philosophers, mathematicians, sociologists and biologists and was one of the last attempts to explore the usefulness of information theory across scientific disciplines.

In his brief introduction on 'the concept of information in molecular biology', André Lwoff of the Pasteur Institute set out his position quite trenchantly, unwittingly repeating the critique of information theory as applied to biology that had been made at similar meetings in the US a few years earlier. Lwoff argued that it was not useful to calculate the information contained in a DNA sequence, using either Shannon's equations or Wiener's negative entropy (following Brillouin, Lwoff called it negentropy), because such calculations did not deal with the meaning or function of that information in the organism. As he put it: 'the calculation of negentropy using Shannon's formulas cannot in any way be applied to an organism.' It would be like trying to calculate the information content of a tragedy by Racine, he said. For a biologist, argued Lwoff, the only meaning of information was 'a sequence of small molecules and the set of functions they carry out'.[30] Wiener and the philosophers who were present could not see what the problem was, thereby inadvertently illustrating the gulf between the information theoreticians and the biologists.

Similar mutual incomprehension was revealed in the other sessions, which were often fractious. The mathematician Benoît Mandelbrot suggested that such information-focused cross-disciplinary meetings were pointless:

> The implications of the strict meaning of information have sufficiently explored for its consequences to be quite clear. What remains is so difficult that it can usefully be discussed only in private ... we must consider that its scientific usefulness has ceased, at least for the time being'.[31]

Alongside these rather sterile plenary discussions there were workshops in which experts in the various fields explored their topic in more detail. The workshop on information theory in biology included a session on the genetic code, chaired by Delbrück, with contributions from Crick, Nirenberg, Woese, Jacob and Ochoa's student Peter Lengyel. During the discussion, Crick introduced the term 'codon' to describe the group of bases that codes for an amino acid – the word had been invented by Brenner, apparently partly as a spoof on the other '-on' words that had been coined by Benzer and by Jacob and Monod.[32] It stuck, and is still in use today.

Much of the discussion at the workshop focused on the uncertainty of the results from the cell-free system: some participants questioned whether the polynucleotides truly contained the proportions of bases that Ochoa and Nirenberg's groups assumed they did. Woese outlined his proposed code, framed in terms of the informational content of the different bases, and Jacob described protein synthesis in terms of a 'theory of informational transfer and regulation'. Whatever the insights these presentations may have had, they were not published and left no trace on subsequent research. As most people agreed, the influence of information theory on molecular biology had passed its peak. Information had now become a vague but essential metaphor, rather than a precise theoretical construct.

This was reflected in the 'Symposium on Informational Macro-molecules', which took place at Rutgers University in New Jersey, at the beginning of September 1962. Despite the title, there was very little direct exploration of the informational content of the macromol-ecules that were the subject of the meeting – DNA, RNA and proteins. When Ochoa opened the conference, he nodded in the direction of the new vocabulary, referring to 'information coded into the DNA mol-ecule' that was 'transferred to an RNA tape', but his real position was made abundantly clear in his very first sentence: 'This symposium deals essentially with the molecular mechanisms concerned with the genetic control and regulation of protein synthesis.'[33] The focus was biochemistry, not information.

The first of two sessions on the genetic code was chaired by the veteran geneticist Ed Tatum, who referred back to his discovery of the 'one gene, one enzyme' principle with George Beadle, twenty years earlier:

> I think back to the time when we started our work, so many years ago. I think we would not have been able to anticipate that we would, in this relatively short time, be present at a sym-posium on informational macromolecules. This is something that most of you take for granted, but I can assure you – and I think I speak for Dr Beadle too – that this is really an extraordi-nary phenomenon in the development of molecular biology.[34]

Around 250 people attended the meeting, but only 13 were from out-side the US; neither Crick nor Brenner, nor anyone from Watson's group, attended, and only François Gros was there from the Pas-teur Institute. The stars of the show were the new kings of the code: Nirenberg and Ochoa.

Dozens of speakers summarised data from a range of species (including bacteria, mice and algae) that indicated that the genetic code was universal; they outlined the growing conviction that only one of the two strands in the DNA double helix was used to make protein via RNA; and they described the recent discovery that non-U-containing polynucleotides could code for amino acids. But on the question of questions – the nature of the genetic code – there was no agreement. During the coffee breaks and at mealtimes, attendees argued about whether the genetic code had been cracked or not.[35] Nothing was certain, beyond the fact that UUU coded for phenylalanine, AAA for lysine and CCC for proline.

During the meeting, both Ochoa and Nirenberg toyed with Roberts's combined doublet–triplet code. Nirenberg assumed that the code was based on triplets, but warned, 'it is not possible at this time to distinguish between triplet and doublet codes'.[36] As he put it with disarming clarity: 'Almost all amino acids tested can be coded by polynucleotides containing only two bases.' Ochoa was even clearer – during the discussion of Nirenberg's paper he said:

> I must say I have been very impressed by Dick Roberts' ingenious doublet code idea. … It almost looks as if that third base does not matter and, in this regard, I cannot help but think of the possible significance of Roberts' proposal.[37]

In a summary of the meeting that appeared in a book collecting the talks, the organisers of the symposium suggested that the status of the genetic code at the time was something like that of the periodic table first published by Mendeleev – it was fragmentary and not all of its predictions were correct, but 'nevertheless, a fundamental system had been discovered!'[38]

*

Francis Crick was frustrated by the mixture of unclear experimentation, loosely argued theory and guesswork that had begun to infest studies of the genetic code. In the summer of 1962, as scientists involved in the coding race were either recovering from Royaumont, preparing to go to Rutgers, or both, Crick wrote a long, highly critical, review article on the topic. In typical patrician style it was entitled 'The recent excitement in the coding problem'. Crick summarised the work of the Ochoa and Nirenberg labs, praising their results as being 'of very considerable interest' before changing gear and pulling no punches:

There are so many criticisms which can be brought against this
type of experiment that one hardly knows where to begin.[39]

Crick's critique was rock solid: the composition of the polynucleotides apart from poly(U), poly(A), etc. was completely unknown (it
was not even certain that the incorporation of the different bases into
synthetic RNA molecules was truly random), the levels of amino
acid incorporation in the 'cell-free' experiments were often worry-
ingly low, and even the strongest effect – poly(U) coding for phenylalanine – was weakened by the fact that poly(U) sometimes seemed
to code for leucine. Having grudgingly accepted that two codons
could be reliably identified (only one – the inevitable UUU = phenylalanine – was in fact correct), Crick concluded that the methodological problems he had outlined 'make the allocation of further triplets
very precarious'.[40] He continued:

> although not one single codon can be said to be known with
> certainty we do know something: one codon for phenylala-
> nine contains Us, one for proline contains Cs, and so on. The
> coding problem has moved out of the realm of rather abstract
> speculation into the rough and tumble of experimentation.[41]

Crick was not even convinced by the evidence that the code possessed redundancy – the evidence, he said, 'is of two types, direct
and indirect, and with one exception, none of it is satisfactory'.[42]
Having surveyed the various options, including Roberts's code,
Crick put his finger on what was the true situation: 'if the code is
really a pure triplet code its degeneracy makes it look at times more
like a doublet one'.[43]

Crick was not scornful of theory – after all, it was theory that
had underpinned most studies of the coding problem over the previous nine years, including many of his own contributions – but ultimately, more precise experimentation was needed. No matter how
elegant a theoretical solution might be, the data would determine
whether it was correct. Crick recognised that his own pursuit of
theoretical models, such as the commaless code, had not led to any
breakthroughs (he subsequently described the commaless code as

'one of those nice ideas which is, nevertheless, completely wrong'[44]). He gave his readers a clear outline of how he thought research on the topic should proceed:

> In the long run we do not want to guess the genetic code, we want to know what it is. ... The time is rapidly approaching when the serious problem will not be whether, say, UUC is likely to stand for serine, but what evidence can we accept which establishes this beyond reasonable doubt. What, in short, constitutes proof of a codon? Whether theory can help by suggesting the general structure of the code remains to be seen. If the code does have a logical structure there is little doubt that its discovery would greatly help the experimental work. Failing that, the main use of theory may be to suggest novel forms of evidence and to sharpen critical judgement. In the final analysis it is the quality of the experimental work which will be decisive.[45]

About a week after sending off his article, Crick heard that he, Jim Watson and Maurice Wilkins had won the 1962 Nobel Prize in Physiology or Medicine, for their work on the structure of nucleic acids and its significance for information transfer in living material.[46] The debates on the Nobel Prize committee are closed, but it seems probable that the renewed interest in the significance of the sequential structure of DNA produced by the cracking of the code convinced the committee that Watson, Crick and Wilkins's time had come.

*

In early June 1963, just two years after Nirenberg and Matthaei's discovery, Cold Spring Harbor Laboratory held its annual meeting under the title 'Synthesis and structure of macromolecules'. This time, not only was Nirenberg allowed to attend, he had pride of place on the programme. In the ten years since gangly Jim Watson had presented the double helix structure of DNA in a stifling Cold Spring Harbor lecture theatre, the field of molecular biology had been utterly transformed – this was the largest ever meeting held at Cold Spring Harbor, with more than 300 scientists attending, about one-fifth of them from outside the US.

The seventy-four presentations at the meeting were focused on DNA, on various forms of RNA and on protein synthesis, and the framework was resolutely biochemical. Nirenberg's talk was entitled 'On the coding of genetic information', but after the introductory paragraph he immersed himself in the biochemical details, even reverting back to the old vague language of specificity rather than giving any content to the idea of information. Nirenberg's talk revealed that, as Crick had pointed out, the race to crack the code had hit an experimental bottleneck. The techniques that were employed – a combination of synthetic RNA of unknown sequence and data from the effects of mutations on viruses – could not crack the code. Worse, they could not even settle the question of whether the code was composed of groups of two, three or more bases. Furthermore, Nirenberg sounded a new note of caution about the technique that had made his name: it was possible, he argued, that natural messenger RNA found in cells might not use all sixty-four potential triplet codons; as a result, the randomly ordered synthetic molecules might 'test the cell's potential to recognise code words', he said.[47]

Progress was certainly being made – increasingly detailed experiments and more accurately assembled synthetic RNA allowed Nirenberg's group to suggest that a number of amino acids, including proline and phenylalanine, were coded by more than one codon, but there was still no absolute proof. The actual code remained out of reach, because the sequence of bases on the RNA molecules used in the cell-free system remained unknown. At the end of his talk, Nirenberg described how his group and that of Indian-born biochemist Gorind Khorana, who was based at the University of Wisconsin, were separately using two techniques for synthesising short bits of DNA, or oligodeoxynucleotides (oligo- is a Greek prefix meaning few). When these pieces of DNA were transcribed into RNA, it was shown that molecules composed of only four bases could still produce detectable levels of amino acids in the cell-free system. Nirenberg's understated conclusion pointed the way forward:

It is possible that defined oligodeoxynucleotides may be useful in the determination of nucleotide sequence and polarity of RNA code words, and also in the study of control mechanisms related to DNA-directed protein synthesis.[48]

The next talk was by Joe Speyer from Ochoa's laboratory, who summarised their two-year-long attempts to correlate the theoretical frequency of different triplets in a synthetic RNA molecule with the levels of different amino acids.[49] There was little new there, beyond a summary of research from a variety of species indicating that the code was universal. The Ochoa group had made a substantial impact in the field, recovering the initiative from Nirenberg, and showing what focused, large-scale molecular research could achieve, but the limits of their techniques were now apparent. Two weeks later, Ochoa gave a talk in Switzerland in which he inadvertently outlined the impasse his group was in; unlike Nirenberg, he had no solution to the problem.[50]

The summer of 1963 represented a double shift in the race to crack the code. New techniques had been developed for creating small RNA molecules of known sequence, while the competing laboratories had changed. Ochoa's group effectively bowed out of trying to determine which triplets coded for which amino acids; Gorind Khorana, the expert in RNA synthesis, took their place.

Over the next two years, Khorana's group refined its technique for creating small RNA molecules of a known sequence, and in 1964 Nirenberg's laboratory solved the problem from the other direction – they worked out how to identify the nucleotide sequence on a piece of RNA that had just led to the incorporation of a particular amino acid into a protein chain. This bit of heavy-duty biochemistry involved trapping a complex of molecules – radioactive transfer RNA (tRNA; this was Crick and Brenner's adaptor molecule), nucleotides and ribosomes – on a Millipore filter. Using this technique, Nirenberg and his colleague Phil Leder were able to show that a UUU triplet led to the binding of phenylalanine tRNA, whereas a UU doublet did not.[51] There was no evidence for any of the fancy doublet-based codes that had been suggested in the previous couple of years – a codon was composed of three bases.

With the help of NIH colleagues and highly skilled visitors such as Marianne Grunberg-Manago, Nirenberg's lab was soon able to use a variety of techniques to synthesise triplets with known composition, and then put them through the Millipore 'plater' device to demonstrate which amino acid they coded for. One of Nirenberg's trusted technicians, Norma Heaton, recalled that during this period the atmosphere in the laboratory was 'intense … busy … crowded … competitive'. She described how they would suck up radioactive reagents with their mouths ('that would never be allowed today', she said) and that members of the lab would crowd round the data coming out of the radioactivity counter, eager to know what new codon had been discovered, 'Then you would hear this shout, like "Oh, we discovered a new one."'[52]

Heaton also gave some insight into the prosaic work that takes place in a laboratory – exactly what has to be done to obtain the data that is interpreted to make scientific breakthroughs. The routine she described resembles the precise, repetitive gestures of a worker on a production line, which, in a way, is what she was:

> Initially we used single platers. It was a little round, stainless steel tube, just big enough to hold the Millipore filter, and about so high, and it screwed onto a base. You had a glass Erlenmeyer flask connected to a vacuum, and then you had a rubber gasket at the top, and you plunked this thing down.
>
> Then you had the vacuum on, and you took one of the test tubes that had your experiment in it, and you would precipitate the complex with TCA. Then you would pour it through the plater and the precipitated complex would be collected on the filter.
>
> Then you would unscrew it, take out the Millipore filter with forceps, and put it in order onto a piece of aluminium foil. Initially, we used what was called a Nuclear-Chicago planchet counter. You placed the dried filter onto little copper or aluminium planchets, about so big around, they had a little, tiny lip, and you would put the filter on that, and then you would stack them up and you would put them into the Nuclear-Chicago, and they would drop down and as they went across, the level of radioactivity would be counted. …

1. Erwin Schrödinger (left) with the Irish President Hyde (in wheelchair) and Prime Minister De Valera (far right), at the official opening of the Institute for Advanced Studies, Dublin 1943.

2. Norbert Wiener in the 1950s, accompanied by some of his fiendish equations.

3. Oswald Avery at a laboratory party, Christmas 1940.

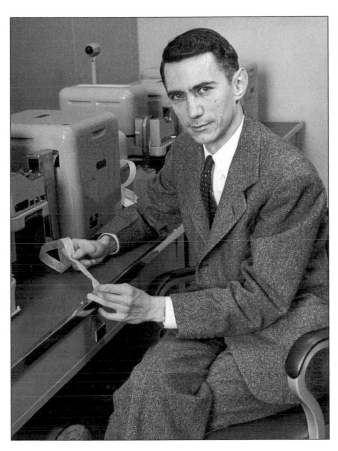

4. Claude Shannon with paper tape used for programming computers.

5. Harriett Ephrussi-Taylor, Boris Ephrussi and Leo Szilárd at the 1951 Cold Spring Harbor Symposium.

6. Alfred Mirsky (left) and Masson Gulland in conversation at the 1947 Cold Spring Harbor Symposium. Four months later, Gulland was killed in a train crash south of Berwick.

7. André Boivin (left) discusses with Joshua Lederberg at the 1947 Cold Spring Harbor Symposium.

8. Salvador Luria (standing) and Max Delbrück at Cold Spring Harbor, 1953.

9. Leo Szilárd (left) and Al Hershey in the rain at the Cold Spring Harbor Symposium, 1951. They look like characters in a film by the Coen Brothers.

10. Team photo of Al Hershey's laboratory at Cold Spring Harbor, 1952. Martha Chase is second from the left, Hershey is standing next to her.

11. Rosalind Franklin on holiday in Tuscany, 1950.

12. Maurice Wilkins with
an X-ray crystallography
apparatus in the 1950s.

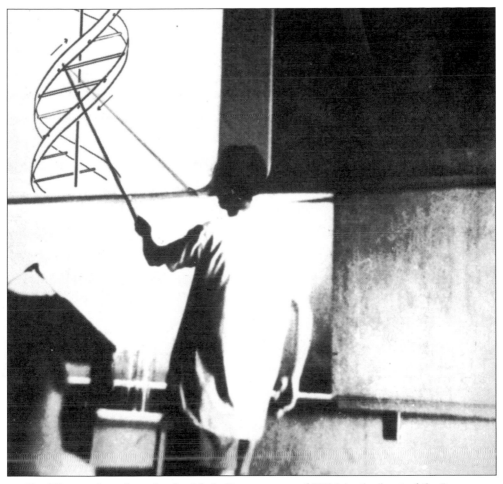

13. Jim Watson describes the double helix structure of DNA in the heat of the June 1953 Cold Spring Harbor Symposium.

14. Watson and Crick with the double helix, photographed by Antony Barrington Brown, in May 1953 in their office at the Cavendish Laboratory in Cambridge. The photograph was not published at the time.

GEORGE GAMOW
19 THOREAU DRIVE
BETHESDA, MARYLAND

It just accured to me today that another way of geting 20 different _loops_ would be the ~~usse~~ use of _triangular_ combinations with three arbitrary bases at vertices:

$\triangle_1^1 , \triangle_2^2 , \triangle_4^3$ ect. $\frac{4 \times 5 \times 6}{1 \times 2 \times 3} = 20$.

And, indeed, the W&C model permits such shapes because of helical nature of DNA. On the surface of a cylinder the bases will form a system of rombs with three independent and one dependent vertices. A, B, and C can be any of the four bases, while D is determined completely by B (D=2 if B=1, ect)

[Two sugar-phos. chains.]

15. One of many letters sent by George Gamow. In this letter to Linus Pauling, written in November 1953, Gamow explains his 'diamond' model of the genetic code.

16/17. The RNA Tie Club in the 1950s. Right: George Gamow, wearing his Club tie and tie-pin. Below, left to right: Francis Crick, Alexander Rich, Leslie Orgel and Jim Watson in 1955. Orgel was apparently not playing the game – he was not wearing his RNA Club tie.

18. Heinrich Matthaei (left) and Marshall Nirenberg, 1961, after they had cracked the genetic code.

19. Moscow Biochemistry Congress 1961: Commemorative stamp and photo showing Crick (left) and Benzer (to his right), with Jacob (second from right). All five men are wearing Congress badges.

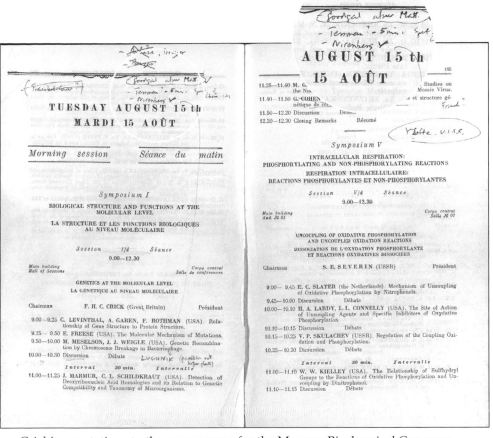

20. Crick's annotations to the programme for the Moscow Biochemical Congress session he chaired on 15 August 1961, showing his handwritten inclusion of Nirenberg during the 10:00 Discussion period. This gem was discovered by Bob Goldstein.

21. François Jacob and his wife, Lise, on the beach at La Tranche-sur-Mer, on the west coast of France, August 1962.

22. Jacques Monod (left) and Sydney Brenner in the 1960s.

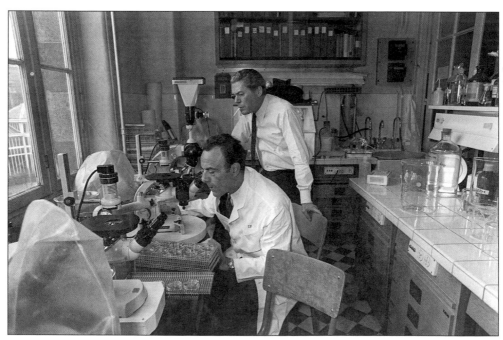

23. François Jacob (left) and Jacques Monod in the laboratory in the early 1970s. Monod died of leukemia in 1976.

24. Francis Crick speaking at the 1963 Cold Spring Harbor Symposium. On the blackboard is a diagram of the central dogma.

25. Francis Crick (lying down with back to camera) hosts a 'flower party' at his house in Cambridge, in the mid-1960s. The house was named The Golden Helix. According to Henry Selby-Lowndes, the son of the photographer, Guy Selby-Lowndes, the music included hits by Herman's Hermits and Herb Alpert.

26. Jacques Monod (left) and Leo Szilárd discuss the operon, Cold Spring Harbor, 1961. Monod described Szilárd as looking 'like a petty Italian fruit-merchant'.

27. Francis Crick (left) and Seymour Benzer at a Symposium on Nucleic Acids held in Hyderabad, India, in January 1964.

<del>Dr. A. Lwoff</del>
✓ Dr. J. Monod
<del>Dr. F. Jacob</del>

Mon cher collègue,

    Please accept my sincere condolences on the occasion of
your being forced to share the Nobel Prize with those other
two jerks when, in my opinion, you alone fully deserve it.

                With love,

                Seymour

                Seymour Benzer

SB:bl

28. Exchange of letters between Seymour Benzer and François Jacob, André Lwoff and Jacques Monod, on the occasion of the French trio being awarded the Nobel Prize, in 1965. Benzer was renowned for his sense of humour.

                31.12   65

Dr. Seymour BENZER
Division of Biology
California Institute of Technology
PASADENA, Calif. 91109

Dear Dr. Benzer,

    You alone understand just as I
alone deserve the Prize.

F. Jacob   A. Lwoff   J. Monod

29. Banner put up in Marshall Nirenberg's laboratory at the National Institutes of Health in Bethesda, Maryland, when news came through of his 1969 Nobel Prize.

30. Asilomar conference on recombinant DNA, 1975. Left to right: Maxine Singer, Norton Zinder, Sydney Brenner and Paul Berg. The possibility of using CRISPR to change the human germ line has recently led to calls for a 'new Asilomar' to debate the ethical and technical questions involved.

It all had to be timed. When you got good at this, you knew how many seconds it took you to unscrew this single plater, take it out, put it down, set it up with a new Millipore. I think I got so I could do it every thirty seconds, or maybe every twenty-five seconds.[53]

As the data came tumbling out of the Nirenberg laboratory, Crick, like so many before him, tried to find the reason why some amino acids were coded by more than one codon – the logic behind the degenerate nature of the code – simply by thinking about it. He wondered how the codon on the messenger RNA molecule bound with a complementary set of bases – what he called an anticodon – on the small transfer RNA (tRNA) molecule.* It was not clear whether there was one tRNA molecule per RNA codon (so sixty-four different versions), or one molecule per amino acid (in which case there would there would be twenty), or some intermediate situation. There was some experimental evidence that the tRNA that attaches to phenylalanine could recognise both the UUU and the UUC codons; to explain this curious phenomenon, Crick resorted to the precise molecular modelling that had preoccupied him during the race to discover the structure of the DNA molecule at the beginning of 1953. He came up with the idea that there was a degree of what he called molecular wobble in the binding of the third base in the RNA codon with its equivalent in the tRNA anticodon. Crick provided a masterly survey of the situation at the time, and then concluded with a smile: 'In conclusion it seems to me that the preliminary evidence seems rather favourable to the theory. I shall not be surprised if it proves correct.'[54]

It was correct – we now know that most organisms have more than twenty but less than sixty-four tRNAs (for example, there are forty-eight tRNA 'anticodons' in humans, but only thirty-one in bacteria), which is explained by the wobble in the anticodon's ability to recognise more than one base in the third position of the codon.

By the middle of 1965, Nirenberg's group had identified the

---

*For example, an ACU codon in mRNA would bind with a UGA anticodon on a tRNA molecule.

function of fifty-four out of the sixty-four RNA codons; at around the same time, Khorana confirmed these data by using synthetic codons of known sequence.[55] None of the theoretical schemes that had been so carefully developed over the previous decade proved to be correct. The genetic code is highly redundant, so that in many cases a base in a codon can alter without changing the amino acid that is being coded for. Most of these silent changes in DNA occur in the third base – this was the reason why theoreticians had wondered whether in fact the code was basically a doublet code. In some cases the third base in the codon provides no additional information because all four alternatives code for the same amino acid, as a result of the wobble in codon–anticodon binding.

Three interlinked issues remained: understanding which way the genetic message is read, and finding out how the cell knows where the genetic message begins and ends. A sequence of bases can be read in either direction, with completely different meanings: a DNA codon reading AGG codes for serine, whereas GGA codes for proline. Furthermore, because of the complementary nature of the two DNA strands a given stretch of DNA contains four possible alternatives – in this example, AGG and GGA on one strand, and TCC and CCT on the complementary strand. The genetic code seemed to be becoming even more complicated, but this mystery was soon solved.

By 1963 a series of experiments using radioactively labelled mRNA showed that for each gene only one strand was read by the cell's machinery, while in 1965, Ochoa's group confirmed that the message was read by the ribosome in the same direction as mRNA was synthesised from the DNA strand, in what is known as the $5' \rightarrow 3'$ direction (pronounced 'five prime' and 'three prime').[56] These numbers refer to the way that ribose molecules, which are composed of five carbon molecules in a ring, are chained together in RNA and DNA by phosphate molecules attached to the fifth carbon of one ribose molecule, the third carbon of the next, the fifth carbon of the following, and so on (bases – A, C, G and T/U – are attached at the first carbon of each ribose).

The direction in which the genetic message was read was now known, and it seemed logical to assume that there had to be some way in which the message set its reading frame, enabling it to be read correctly. To everyone's surprise, in 1966 it was found that the sole RNA codon for methionine – AUG – also acts as a start codon for protein synthesis if it is at the beginning of a sequence.[57]

A final enigma was the presence of three RNA codons that did not appear to code for any amino acids – nonsense codons. The first of these (UAG) was a mutation that had been identified in 1962 in phage and given the enigmatic name 'amber', apparently after the German translation of the name of one of the American scientists involved, Bernstein. The other two nonsense codons were identified by the Brenner laboratory, and given colour names to go with the original – ochre (UAA) and opal (UGA).[58] In 1964, Brenner's group showed that the amber codon – UAG – coded for stop. This also provided evidence to support one of the widespread assumptions of molecular genetics, namely that the gene and the protein it creates are colinear. By creating amber mutants at different points in the gene coding for the phage head protein, they showed that the lengths of the corresponding protein fragments in each mutant were correlated with the position of the amber mutation. At almost exactly the same time, a group at Stanford led by Charles Yanofsky also provided evidence of colinearity between the DNA sequence and protein structure, in *Escherichia coli*.[59]

All of the assumptions that had underpinned the work on the genetic code over the previous years had proved correct: the code was universal, it possessed redundancy, its fundamental unit – the codon – was composed of three bases, the message was read in a particular direction, there was a reading frame, gene and protein were colinear, and the sequence contained simple instructions to start and stop reading. The detail had often been wrong, but as researchers groped towards the truth, using a mixture of theoretical insight and experimental ingenuity, the central principles they had clung to were all shown to be true.

It was not until 1967 that the last of the sixty-four words in the genetic code was read, appropriately in an article co-signed by Francis Crick.[60] This was the opal codon – UGA. Like the amber and ochre codons, it read 'stop'.

*

The June 1966 Cold Spring Harbor symposium was entirely devoted to the genetic code. The opening talks were by Nirenberg, Matthaei and Khorana. By choosing these speakers, the organisers – who included Watson and Crick – recognised the way in which the field had been transformed in the previous few years. The sessions looked at the direction in which the code was read, the role of 'punctuation' (initiation and stop, rather than commas), control of gene expression, a detailed exploration of the structure and function of tRNA, and a discussion of mutation and errors (under the title 'Infidelity of information transfer'). None of the talk titles at the meeting even mentioned 'information', and all the presentations were framed in terms of biochemistry, not information theory. It was biology, and not mathematics, that dominated the celebration of the cracking of the code. The final two talks outlined two of the future developments that would preoccupy the field, down to the present day: the use of DNA sequences to establish patterns of evolution, and the origin and evolution of the genetic code.[61]

The published proceedings of the meeting opened with an article by Crick that was a magisterial overview of the previous fifteen years' work, entitled 'The genetic code – yesterday, today, and tomorrow'. Crick looked back to the earliest ideas of Caldwell and Hinshelwood in 1950, Dounce in 1952 and above all Gamow in 1953, when the 'coding problem' suddenly came into focus with the discovery of the double helix structure of DNA. Crick did not refer to Schrödinger's code-script idea – he clearly did not feel that this insight had any direct effect on subsequent events. Nor did he cite Boivin's insight into the relation between DNA, RNA and protein; he may not have known of it.

Crick's opening words accurately summed up the situation: 'This is an historic occasion.'[62] To all extents and purposes, the genetic code was known. Cracking the code underlined the power of experimentation: none of the attempts to work out the code theoretically had got it right. Crick argued that it had been inevitable that they would fail:

We can see now, from the known code, that it would have been
almost impossible to have deduced it correctly at the time.

None of the hypothetical codes dreamt up by the theoreticians were
correct, because they made assumptions that were logical, rigorous
and hopelessly wrong. The physicists' appetite for elegance and the
biochemists' naive assumptions about natural selection led them to
assume that the code had to be extremely economical, that it would
look as though it had been designed along logical principles. But that is
not how biology works. The genetic code is a product of biology and is
messy, illogical and inelegant. It is highly redundant, but to bewilder-
ingly varied degrees: one amino acid (leucine) has six codons, whereas
another (tryptophan) has only one. Explaining this pattern on the basis
of chemical, physical or mathematical principles has so far proved difficult. Whatever logic there may have been has been overlain by billions of years of evolution and chance events. As Jacob put it in 1977,
natural selection does not design, it tinkers with what is available.[63]

Much of the work that contributed to the code brought world-
wide acclaim. The Nobel Prize committee repeatedly rewarded
those who had made the essential breakthroughs. Joshua Lederberg,
George Beadle and Ed Tatum won Nobel prizes in 1958 for their
work on microbial genetics, which had transformed the way that
genes were understood and could be studied, and Watson, Crick and
Wilkins won in 1962 for the double helix structure of DNA. In 1965,
Jacob, Monod and Lwoff were awarded the prize for their work on
the repressor and the genetic regulation of protein synthesis, and in
1968 Nirenberg, Khorana and Robert Holley won for their work on
the genetic code, nucleotide synthesis and tRNA structure respectively. To celebrate, Nirenberg's lab hung a banner across a corri-
dor that read 'UUU are great Marshall'.[64] In the following year, the
founders of the phage group – Delbrück, Hershey and Luria – were
awarded the prize for their work on virus genetics, through which
they had set the stage for many of the fundamental discoveries of
molecular genetics. Other participants in the race either did not win
the prize (Benzer, Matthaei) or were awarded one for other work
(Brenner, Ochoa). By the end of the 1960s, a period had closed. Many
of the leading figures, such as Crick, Benzer and Nirenberg, turned

their attention to neurobiology, while Brenner became interested in developmental biology. In 1968, phage group member Gunther Stent gave an overview of the period under the elegiac title 'That was the molecular biology that was'.[65]

*

The discoveries made in this period of science have transformed the whole of biology, and enabled us to make massive advances in the development of new medical treatments. As well as producing a revolution in our knowledge, the twenty-two years that separated Avery and Schrödinger from Crick, Nirenberg, Jacob and Monod also produced a revolution in our thinking. Everyone now knows that genes contain information, and that they function as part of complex networks, controlling the production of proteins and the activity of other genes. The informational and cybernetic theories that flourished in the late 1940s and 1950s have left little direct trace in the thinking of today's scientists, but their influence remains, in terms of the metaphors and concepts that we all use to think about some of the most fundamental features of life. The way in which we now think about genes and how they work would have been incomprehensible to a scientist from the 1930s. Cracking the genetic code was a leap forward in humanity's understanding of the natural world and our place within it, akin to the discoveries of Galileo and Einstein in the realm of physics, or the publication of Darwin's *On the Origin of Species*. These comparisons are not the fruit of hindsight, they were made at the time.[66]

The discovery of the code did not occur through the genius of a single person or even a handful of thinkers. Instead, it required brilliant insights, audacious experiments and above all a lot of hard work by a great many people. The range of techniques that was used was enormous, with particular emphasis on the interface between physics and biology. Above all, cracking the code represented the triumph of experimentation over theory. None of the purely theoretical approaches was able to uncover the secret, which finally revealed itself through the probing of the experimenters, not the pencil-chewing of the theoreticians.

Unlike previous collective scientific breakthroughs, such as the Manhattan Project, or subsequent ones, such as the landing of men on the Moon, the Large Hadron Collider or the Human Genome Project, there was no concerted organisation of the research. Institutions such as the Medical Research Council in the UK, the NIH in the US, and the Institut Pasteur and the Centre national de la recherche scientifique (CNRS) in France all supported key researchers, but the work had not been coordinated; there had been no committee or council that oversaw the project. There had been funding for the wartime work that led to cybernetics and information theory, but at the time the potential implications of that research were unknown, beyond their immediate practical use in the war effort. Whereas Cold War governments poured substantial amounts of money and resource into nuclear research and rocket science, and the private sector spent billions developing computers and new medicines, the amount of money that flowed into molecular biology and cracking the code was tiny, and several researchers – Avery, Watson and Crick, Nirenberg – initially received minimal support for their work.

Apart from the meetings at Cold Spring Harbor, there had not even been a regular meeting-place for researchers to discuss progress, and the clubby nature of those meetings had excluded the two unknowns who ultimately made the breakthrough. Equally, there was no undisputed leader driving the project forward with a clear vision. Gamow had inspired the RNA Tie Club, but his broader influence was less strong. Crick was omnipresent, but he did not have the power to promote or reorient areas of research or particular researchers and he had no control over funding.

Above all, unlike many other great examples of collective research, the solution to the problem was not known in advance – this was not like the Manhattan Project, for which the end point was clear from the outset. The research into the genetic code was not an engineering problem: it was pure research, which could not simply be cracked by hard work, and the researchers involved could not be given a small part of the problem to solve, knowing that it would contribute to the eventual outcome. Crick was not the Oppenheimer of the genetic code. His intelligence, criticism and encouragement did much to create the essential insights that shaped how the science developed, but although he came up with the idea that 'the precise sequence of the bases is the code which carries the genetical information', the thirteen-year-long campaign to crack the code had not been led by him, nor did he foresee even the outline of the eventual solution.

The path taken by the scores of researchers who had been involved in the coding problem was as unpredictable and on occasions as illogical as the code itself seems to be. Cracking the code was an example of what Jacob called 'night science', in which intuition and audacious guesswork accompany strict logic.[67] In 1966, John Cairns, the director of Cold Spring Harbor Laboratory, accurately summarised the significance of what had just taken place:

> The effort that has gone into this decipherment, the strange sense of urgency, and the remarkable variety of approaches that have together led to the solution, must be without parallel in the history of biology.[68]

UPDATE

Nearly half a century has passed since the final word of the genetic code was read. In the intervening years, science has made substantial advances, some of which seem to challenge the fundamental discoveries that were made in the heady years of 1944–67. The closing chapters bring the history of the genetic code up to date, showing what happened in the intervening decades.

TWELVE

SURPRISES AND SEQUENCES

All of the researchers involved in cracking the genetic code agreed on two basic principles. First, they assumed that there was a one-to-one correspondence between the DNA sequence of a gene and the corresponding amino acid sequence – what Crick called colinearity. Second, they considered that the genetic code and the way in which genes functioned were universal, so 'anything found to be true of E. coli must also be true of elephants', as Jacques Monod put it at Cold Spring Harbor in 1961.[1] Neither of these principles was required for genetics to work or for the genetic code to be cracked, but they made sense and they gave a universal significance to the models and interpretations that were being developed. They also ensured that the new science of molecular genetics fitted into the Darwinian framework according to which all life had a single origin and could therefore be assumed to share fundamental processes. Crick later said that those who were studying the genetic code had 'a boundless optimism that the basic concepts involved were rather simple and probably much the same in all living things.'[2]

Within ten years of the final word in the genetic code being read, it became obvious that such boundless optimism was unfounded, as the assumptions of colinearity and the universality of the genetic code were proved to be wrong.

*

In autumn 1977, several linked papers appeared in *Proceedings of the National Academy of Sciences* and in the new journal *Cell*, which had been set up three years earlier with the ambitious aim of being 'a journal of exciting biology'.[3] For once, the reality lived up to the hype, as the articles announced that, in viruses, genes were not necessarily continuous stretches of DNA but instead could be spread out along a sequence, split into several pieces.[4]

What Watson called the bombshell discovery of split genes had first been announced at the Cold Spring Harbor meeting in the summer of 1977, and the scientific community was abuzz with the implications. It was soon found that mammalian genes shared this property, which contrasted sharply with the strictly continuous organisation of genes in bacteria. The surprise and excitement felt by researchers is shown by the unprecedented language used in the title of the first paper in that issue of *Cell*, by Louise Chow, Richard Roberts and their colleagues, which described 'An amazing sequence arrangement' of viral nucleic acids. Scientists rarely use words like 'amazing' in their professional publications.

The reason for the excitement was simple: nearly twenty-five years of assumptions about gene structure had been overthrown by a completely unexpected discovery. Within a few months scientists were revelling in what was widely described as a revolution (Crick called it a mini-revolution).[5]* In eukaryotic cells (that is, in cells with a nucleus, so in all multicellular organisms and in single-celled organisms such as yeast) it turns out that genes often contain many bases that are not used to make a protein. As a result there is often no colinearity between the DNA sequence and the amino acid sequence of the protein. Between the start and the stop codons of a gene, there may be huge chunks of non-coding DNA that have no relation to the final protein. In an article in *Nature*, Wally Gilbert named these

---

*In 1988 the historian Jan Witkowski pointed out that there had been no historical study of this period. Although that might not have been unusual in 1988, it is surprising that nearly 40 years after the discovery there has still been no detailed historical analysis of the revolution and its implications.

apparently irrelevant non-coding sequences introns (from 'intra-genic regions'); the DNA sequences that are expressed in protein were called exons.[6] Most eukaryotic DNA is a patchwork of exons and introns. Introns are generally around forty bases in length, but they can be very large – for example, one of the introns in the human dystrophin gene is more than 300,000 bases long.[7] In some rare cases, the intron of one gene can even contain a completely separate, pro-tein-encoding gene.[8]

The existence of introns means that the cell has to process the genetic message before it can be turned into protein. The first tran-scription of DNA into RNA was named pre-mRNA – it initially con-tains all the irrelevant introns, but these are immediately snipped away and the two new ends of the mRNA molecule joined together ('spliced') to form a messenger RNA sequence that corresponds to the final amino acid product of the gene, along with untranslated regions at the beginning and end of the mRNA sequence, which tell the cell how the gene is to be expressed and processed. This splicing is done by tiny cellular structures made of RNA and protein, known clumsily as spliceosomes (some RNA molecules can splice them-selves, without the aid of a spliceosome). The beginning and end of an intron are marked by specific sequences that are recognised by the spliceosome and which indicate which bits of the pre-mRNA molecule need to be snipped out.[9] It is this spliced version of mRNA, called mature mRNA, that contains an RNA sequence that is colinear with the amino acid sequence of the protein and is used by the cell in protein synthesis.

Despite the initial amazement of the scientific community, the existence of all this non-coding DNA in eukaryotic organisms was soon welcomed by researchers, as it seemed to provide an answer to the nagging suspicion, first voiced by Burnet in 1956, that not all of the DNA in a genome actually contributes to producing proteins.[10] If important chunks of the genome were composed of what Wally Gilbert called 'a matrix of silent DNA', this would explain the situa-tion, even if in 1959 Crick had described this possibility as unattract-ive.[11] Since Gilbert's first description, there has been a long-running debate about where introns come from – some scientists have argued that even the earliest genomes had introns, but most now think that introns appeared with the evolution of the eukaryotes, because there is no evidence that any prokaryotic organism – single-celled organisms with no nucleus – ever had introns, or possessed the complex cellular machinery required for splicing them out.[12] Why introns evolved is still unclear.

Splicing is not just a matter of snipping out a few irrelevant bases. It allows the production of different proteins from a single gene, because under different conditions different exons can be spliced together – this is called alternative splicing. A single DNA sequence can give rise to several mRNA sequences, depending on a variety of external factors, including the type of cell that the gene is expressed in. Currently, the largest known number of mRNAs that can be produced by a single gene is 38,016. These mRNAs are encoded by the *Drosophila* gene *Dscam*, which has four clusters of exons, each of which has twelve, forty-eight, thirty-three or two alternative splices.[13] Many of the 38,016 potential Dscam proteins differ only slightly, but this variability is of major functional significance because they mean that the fly's neurons differ. The consequence is that Dscam proteins help determine the intricate way that those neurons interconnect, shaping the brain.[14] The DNA sequence can contain an astonishing degree of complexity.

Until the discovery of introns, it had been assumed that gene mutation primarily involved point mutations – changes in a single base that would either lead to a different amino acid being inserted into the protein, or, if the base were deleted, would produce a frameshift mutation in which the remaining bases of the genetic sequence would be read in a novel series of triplet codons, which would often be nonsense, as Crick had suggested in 1961. With the discovery of introns, it was realised that a mutation at the beginning or end of an intron could radically alter the structure of the translated protein by allowing new DNA sequences from the intron to be included in the coding region of the gene, thereby providing an additional source of genetic novelty. The two principal researchers involved in the discovery of what were initially called 'split genes' or 'genes in pieces' were Richard Roberts of Cold Spring Harbor and Phillip Sharp of MIT, and in 1993 they won the Nobel Prize in Physiology or Medicine for their work.

*

Two years after the discovery of introns, the scientific world was shaken yet again. The last word of the genetic code to be deciphered was the stop codon UGA (nicknamed opal), in 1967. In November 1979, a group at Cambridge discovered that in human mitochondria – small energy-producing structures found in all eukaryotic cells, which contain their own DNA and ribosomes – UGA does not encode stop but instead produces an amino acid, tryptophan.[15] The genetic code is not strictly universal; even more surprisingly, the same organism – you – contains two different genetic codes, one in your genomic DNA, the other in your mitochondrial DNA.

This fact tells us something fundamental about the history of life on our planet. In 1967, the US biologist Lynn Margulis began arguing that mitochondria were not merely micro-structures within eukaryotic cells but were remnants of a single-celled organism that had fused with the ancestor of all eukaryotic organisms, billions of years ago, probably as part of a symbiotic relationship. She was not the first to come up with this idea – in the early years of the twentieth century, both Paul Portier and Ivan Wallin suggested that mitochondria might be symbionts.[16] Margulis argued that these symbiotic bacteria subsequently found themselves trapped in every one of our cells and lost all their independence, but not their own, separate genome – a tiny ring of DNA about 16,500 base pairs long (in comparison, the human nuclear genome contains about 3 billion base pairs). (Genes and genomes are measured in 'base pairs' because of the two strands of the DNA double helix – for each base there is a complementary base on the other strand, forming a base pair.)

It appears that all mitochondria, in all the eukaryotes on the planet, have a common ancestor that was alive more than 1.5 billion years ago. The ancestors of plants subsequently incorporated another microbe in the same way, thus gaining their power-generating chloroplast organelles and the ability to gain energy from sunlight.[17] In the cases of both mitochondria and chloroplasts, there are arguments over exactly what kind of microbe fused with what, and above all the speed with which the fusion took place, but most scientists now think that in each case there was a single event that enabled what was effectively a hybrid organism to grow larger and to acquire the energy required by more complex organisms.[18] The extremely small nature of the mitochondrial genome, and its peculiar use of codons, can be explained in terms of the history of this symbiotic relationship. The mitochondrial genome codes for very few proteins – most of the other genes were lost before or shortly after fusion with our ancestors or were incorporated into the genomic DNA of the host – so the appearance of a new function for a codon in mitochondrial DNA through mutation would not have had an important effect on the symbiont, most of whose needs were provided by the host cell.

Mitochondria are not alone in having a non-standard genetic code. In 1985, it was discovered that single-cell ciliates – tiny organisms such as *Paramecium* – show variants of the nuclear genetic code that have appeared several times during evolution. In some species of ciliate, UAA and UAG code for glutamate rather than stop, with only UGA encoding stop; in others, UGA codes for tryptophan.[19] Sometimes UGA and UAG have been recoded by natural selection to code for extra amino acids, not generally found in life – selenocysteine and pyrrolysine, respectively.[20] This can occur by altering the genetic code only in particular genes. For example, the human genome contains a handful of genes in which UGA has been recoded to encode selenocysteine.[21] In these cases part of the mRNA for these genes instructs the cell to insert selenocysteine when it reads UGA; in all our other genes, UGA retains its normal stop function.[22]

A recent study of 5.6 trillion base pairs of DNA from more than 1,700 samples of bacteria and bacteriophages isolated from natural environments, including on the human body, revealed that in an important proportion of the sequences, stop codons had been reassigned to code for amino acids, and an investigation of hitherto unstudied microbes revealed that in one group UAG had been reassigned from stop to code for glycine.[23] There are even cases of novel codons being used to start translation, for example, in 2012, it was discovered that in some unusual circumstances in the mammalian immune system, the genetic message does not begin with the normal AUG codon but can be initiated from a CUG codon, which normally codes for leucine.[24]

More than fifteen alternative or non-canonical genetic codes are known to exist, and it can be assumed that more remain to be discovered.[25] The non-canonical codes generally involve the reassignment of stop codons; this may indicate that there is something about the machinery involved in stop codons that makes them particularly susceptible to change, or it may simply be that as long as the organism can still code stop using another codon, reassigning one stop codon to an amino acid does not cause those organisms any major physiological or evolutionary difficulties.[26]

The exact process by which codon change takes place has been the focus of a great deal of theoretical and experimental research, and several hypotheses have been put forward to explain how variant codes might arise. The current front-runner is called the codon capture model, and was first put forward in 1987 by Jukes and Osawa. According to this model, random effects such as genetic drift can lead to the disappearance of a particular codon in a given genome; similar effects then lead to that codon being 'captured' by a tRNA that codes for another amino acid.[27] A recent experimental study of genetically engineered bacteria in which some codons had been artificially replaced supported this model, and even suggested that reassignment of codons could be advantageous in some circumstances, providing the organism with expanded functions.[28]

The non-universality of the genetic code and the existence of introns were both completely unexpected, and went against all the assumptions of all the researchers who had been studying the genetic code. These discoveries showed that, strictly speaking, Monod was wrong – what is true for *Escherichia coli* is not necessarily true for an elephant in all respects. Nevertheless, the basic positions established during the cracking of the genetic code remain intact. The strict universality of the code and the linear organisation of genes were not laws, or even requirements. The only requirement is that any divergence from these assumptions can be explained within the framework of evolution, and through testable hypotheses about the history of organisms. This has been amply met for both the non-universality of the code and the existence of introns.

Although the genetic code is not strictly universal, this has not altered our view of the fundamental processes of evolution at all. There is no dispute that life as we know it evolved only once, and that we all descend from a population of cells that lived more than 3.5 billion years ago, known as the Last Universal Common Ancestor, or LUCA.[29] Because all organisms use amino acids with a left-handed orientation and RNA is universally used as a way of stringing amino acids together to make a protein, scientists are convinced that this hypothesis is true. In 2010, Douglas Theobald calculated that the hypothesis that all life is related 'is $10^{2,860}$ times more probable than the closest competing hypothesis.'[30]

The variations in the code that have been discovered are in fact quite minor and can be explained either in terms of the deep evolutionary history of eukaryotes – thereby revealing the thrilling fact that our evolution has hinged on the chance fusion of two cells to create the eukaryotes – or in something recent and local in the life-history of a particular group of organisms such as the ciliates. Similarly, although eukaryotic genes are profoundly different from those of prokaryotes, because they are 'split', they still work according to the same principles. All that has happened is that the cellular machinery for taking the information in genomic DNA and turning it into protein has been revealed to be very complicated in a group of organisms that we are particularly interested in, because it includes ourselves. Our basic understanding of how the information in a DNA sequence becomes an amino acid sequence has not been altered; although things are far more complex than the code pioneers could have imagined, the basic framework they developed still stands. The simple models developed in the 1950s and 1960s were not universally correct, but they were a necessary step for the development of our current understanding. And they remain true for the oldest and most numerous organisms on our planet, the prokaryotes.

This final point highlights the power of the reductionist approach adopted by Crick, Delbrück, Monod and the others. They chose to use the simplest possible systems – bacteria and viruses – to understand fundamental processes. In so doing, they gambled that their findings would be applicable to all life. The models that they came up with were simple, elegant and susceptible to experimental testing. Had they been studying mammals and the tangled web of molecules and processes that lead from DNA to protein in these species, it is unlikely that much progress would have been made.

*

Over recent decades, the study of the genetic code has been transformed by one of the most significant technological changes that
have taken place in biology – our ability to determine the sequence
of DNA and RNA molecules. The breakthrough came with the work
of Fred Sanger, who won the Nobel Prize in Chemistry twice, first
in 1958 for determining the structure of insulin and other proteins,
then in 1980 for sequencing nucleic acids (he shared the second prize
with Wally Gilbert, who came up with a less widely used technique
for sequencing DNA).

Sanger was not the first to sequence a nucleic acid – a small
transfer RNA was sequenced in 1965, using techniques similar to
those that had previously been used to sequence proteins.[31] But
Sanger's method made it possible to sequence up to 300 bases of a
piece of DNA (in reality 200 bases was more often the limit), mark-
ing DNA chains of varying lengths with radioactive phosphorus-
containing bases (A, C, G or T), and then visualising these fragments
on an electrophoresis gel. Sanger obtained these DNA chains by
carrying out four separate reactions to copy a DNA molecule. Each
test-tube included four normal nucleotide bases (A, C, G and T),
enzymes used to copy the DNA molecule, together with a radioactively labelled variant of one of the bases (hence the need for four
reactions). As well as being radioactive, these special bases had been
chemically modified so as to stop the chemical reaction when they
were incorporated randomly into a new DNA chain. Because a typical extract contains so many identical copies of the DNA molecule
and the radioactive base was incorporated at a random point in each
new chain, the result was a large number of DNA molecules that were
of different lengths and which were radioactive, and could therefore
be detected on the gel. Each reaction (A, C, G and T) was then loaded
side by side onto a gel and the electric current was turned on. Different lengths of DNA migrated at different speeds and so ended up at
distinct points on the gel, enabling the sequence to be read by eye.

Sanger later described this technique, known variously as the
chain termination method, dideoxy sequencing or, more simply,
Sanger sequencing, as 'the best idea I have ever had'.[32] The rest of the

scientific community seems to agree – his 1977 paper describing the method has been cited more than 65,000 times, a staggering number that makes it the fourth most cited article in the history of science.[33]

Using this technique, in 1978 Sanger and his colleagues sequenced the first complete genome, that of a bacteriophage. It was 5,386 base pairs long and represented months and months of work.[34] The technique soon became well established even though it was tedious and repetitive. It was also dangerous: as well as the omnipresence of radioactivity, the electrophoresis gel was made of toxic material and various steps in the procedure involved nasty chemicals that unravelled the DNA in the sample and, potentially, in the experimenter's body. Despite these hazards, by 1984 researchers had sequenced the full genome of three viruses – two bacteriophages, and the Epstein–Barr virus, which causes glandular fever in humans. The Epstein–Barr virus sequence, which was described in Cambridge, was 172,282 bases long or thirty-two times the length of the first genomic sequence. This was a major feat, representing years of work, and involving what was then a large team of twelve researchers.

Sanger's method became widely used in the late 1980s with the development of the polymerase chain reaction (PCR), which allows tiny samples of DNA to be amplified in a test tube. This method was invented by Kary Mullis, who was working at the biotech company Cetus Corporation in California, in a flash of insight during a night-time drive with his girlfriend.[35]* PCR involves heating a sample to very high temperatures (up to 95° C); this separates the complementary DNA strands. The sample is then cooled slightly, DNA polymerase enzymes begin to copy the DNA molecules and the complementary strands then pair up. A single cycle doubles the amount of DNA in the sample. By repeating this cycle of heating and cooling dozens of times, even minute amounts of DNA can be amplified millions of times over in a couple of hours.

Mullis had a problem though – he needed a polymerase enzyme

---

*In his Nobel Prize address, Mullis said that when he realised what he had dreamt up, he stopped his car at mile marker 46.7 on Highway 128 and scribbled down the essential elements of the technique.

that could resist the relatively high temperatures his experiment required. As luck would have it, such an enzyme had recently been described in *Thermus aquaticus* (generally known as *Taq*), a bacterium that lives in ocean thermal vents.[36] The final addition to this procedure is that by adding to the test tube short pieces of DNA, fifteen to twenty bases long, which mark the beginning and end of a DNA sequence of interest, it is possible to target the PCR and thereby amplify only the section of DNA that you are interested in.

PCR rapidly overtook the previous technique of inserting a DNA fragment into a phage genome, then infecting bacteria and allowing the bacteria to reproduce, thereby amplifying the DNA. PCR is much simpler, and even a complete novice can soon amplify minute quantities of DNA. In 1993, less than a decade after his invention, Mullis was awarded the Nobel Prize in Chemistry. The initial application of the technique was diagnosis, and it is now routinely used in medicine as a tool for identifying diseases, both infectious and genetic. Coupled with sequencing, PCR has transformed the way in which biology and medicine work.

The practical application of DNA technology really took off in 1984, when Alec Jeffreys of the University of Leicester discovered the existence of small stretches of DNA that can be easily identified and which represent a unique genetic 'fingerprint' of each individual. The significance of these bits of DNA, known as minisatellites, was instantly obvious to Jeffreys, and he immediately wrote down a series of potential applications that included forensics, conservation biology and paternity testing. In less than a year, the technique was used to determine the outcome of an immigration case by showing that a young Ghanaian boy was indeed the son of the woman who claimed to be his mother; as a result the child was allowed back into the UK.[37]

Jeffrey's technique soon proved more flexible and simple than the previous method for identifying genetic variants, which involved snipping bits of DNA at defined locations, using special proteins called restriction enzymes. If the population being studied contained variability for the length of DNA between the two sites where the restriction enzymes acted, then those variants could be detected on an electrophoresis gel. This was first demonstrated in

1980 by David Botstein and his colleagues, who were working on the human genetic disorder Huntington's disease.[38] Outside of medicine, the use of restriction enzymes proved invaluable for mapping genes and for the development of recombinant DNA biotechnology.

All around the world, DNA fingerprinting is now routinely employed by the judicial system to convict criminals and to prove the innocence of the wrongly accused. The routine collection of DNA samples by the police, and the existence of databases permitting the identification of individuals, has led to a continuing ethical debate of the conflict between liberty and justice, with state forces arguing that only the guilty have something to hide, whereas more libertarian arguments underline the potential dangers.

By the late 1980s, machines were able to read DNA sequences, using a system based on fluorescence rather than radioactivity, but still using Sanger's sequencing method.[39] Sequences were now detected in tiny capillary tubes rather than on huge heavy gels, opening the possibility of simultaneously carrying out many parallel reads, and the sequence could be read in real time, as the reaction took place, rather than waiting for the gel to run and then detecting the radioactive products using a photographic plate. At the beginning of the 1990s, these technical developments led to the creation of a series of projects for sequencing the genomes of multicellular organisms, with the ultimate objective being the sequencing of the human genome. The first animal genome to be completed, in 1998, was that of the nematode worm, *Caenorhabditis elegans*, closely followed by that of the tiny vinegar fly, *Drosophila melanogaster*, in 2000. These projects provided vital information about two widely used laboratory organisms and were testing-grounds for different technical and commercial approaches to genome sequencing. The *C. elegans* genome project, led by John Sulston, was entirely funded by public money, whereas the *Drosophila* genome was a joint effort between publicly funded researchers and a company called Celera Genomics, led by Craig Venter, a molecular biologist turned entrepreneur.

Despite the very different motivations of the public and private researchers, the *Drosophila* genome project was a success. In contrast, the Human Genome Project, which took place in parallel, was the focus of clashes of scientific and commercial outlook as well as of personality.[40] The human genome contains around 3 billion base pairs, far more than that of *C. elegans* (100 million base pairs) or *Drosophila* (140 million base pairs). The size of the human genome and the large stretches of repetitive sequences it contains posed new difficulties that were exacerbated by the very different approaches taken by the public and private researchers.

The publicly funded International Human Genome Sequencing Consortium, led first by Jim Watson and then by Francis Collins, had been working since 1990 to produce a full sequence of every base in the genome, and its members were resolutely hostile to the idea of patenting genes. In contrast, Craig Venter and Celera initially focused on sequencing only genes that were known to be expressed in certain tissues or under certain conditions, with the hope of finding patentable products. They did this by collecting mature mRNA that was present in the cell or tissue of interest, transcribing that back into what is known as complementary DNA (cDNA) and then sequencing this cDNA molecule.

This approach had the great advantage of focusing on genes that were apparently important in a given tissue and meant that researchers did not waste time sequencing the millions of bases in the huge non-coding regions that can be found between genes, or even sequencing the introns of the gene of interest, which had been stripped out by the cellular machinery during the synthesis of RNA from the genomic DNA. Using this method, Venter showed that it was possible to identify genes involved in vital processes with the tantalising possibility of gaining insight into novel medical treatments. While this was extremely productive and held the promise of financial gain, it was at odds with the aim of the publicly funded project, which was to sequence every base in the human genome.

Venter's group then used a different approach, which was initially opposed by the publicly funded researchers but ended up dominating the field and has since been used in sequencing the genomes of many organisms. Known as shotgun sequencing, the technique involves identifying the bases on many short pieces of DNA and then assembling them into huge long sequences. Sequencing short stretches of DNA is easier, but it leads to a substantial difficulty: which of the resultant hundreds of thousands of short sequences follows on from which – how to reassemble the puzzle? This was particularly problematic when it came to dealing with the immense stretches of apparently functionless DNA to be found between genes, which could consist of featureless repetitions of two bases, such as ACACACAC....

To resolve this problem, Venter and his Celera colleagues enlisted computer scientists to develop algorithms for assembling the sequence, and they were able to prove the validity of their approach with the *Drosophila* genome. Despite hostility from many scientists around the world, Venter was probably right to argue that this method would make it possible to complete the project. Nevertheless, problems remained – even with the cleverest algorithms in the world, it is not possible to join up all of the bits of sequences. To get over this problem, recalcitrant parts of the genome were amplified in bacteria to try and bridge the gap. This does not always work – some sections of the human and the *Drosophila* genomes have still not been joined up, fifteen years after the sequences were published.

Despite the continuing clashes, the completion of the draft human genome was announced by President Bill Clinton in 2000, even though it was in fact nowhere near finished. Collins and Venter stood on either side of Clinton in the White House, while the ambassadors of the UK, Japan, Germany and France were in the audience. Meanwhile Tony Blair, along with Fred Sanger, Max Perutz and other British scientists, appeared at the end of a video feed from Downing Street. In a jarring counterpoint to the celebration of human ingenuity and the power of evolution that was on display, Clinton claimed that 'Today we are learning the language in which God created life'.[41] Collins and Blair, both devout Christians, presumably concurred.

The draft sequence was published in 2001 in two versions: the Celera genome appeared in *Science*, and the publicly funded version was published in *Nature*.[42] The International Human Genome Sequencing Consortium sequence is now taken as the definitive version, and was initially a mosaic of information from more than 100 individuals who contributed DNA to the publicly funded project (one of whom was Jim Watson) and the five individuals used in the Celera effort (one of whom was Craig Venter). It continues to be updated as genes are more effectively annotated, and functions or similarities can be more reliably located to particular stretches of DNA.

However, there is no such thing as 'the' human genome. On average, each of our genomes differs by about one base pair in a thousand, so by about 3 million base pairs in total. Most of those differences are not in coding DNA, and those differences that are in coding sequences are generally silent – they do not alter our amino acid sequences. Nevertheless, the overall structure of the human genome, its mixture of coding and non-coding sequences, and the way in which the coding sequences are expressed in time and space, form part of what it is to be human. And as the publicly funded researchers intended, the human genome is a public good, open to all and freely accessible on the Internet, out of reach of the patent lawyers – in 2013, after years of argument, the US Supreme Court finally ruled that no human genes could be patented, striking down patents that had been awarded to Myriad Genetics for use in diagnostic testing for the *BRCA1* gene (mutations in this gene can increase the risk of breast cancer).[43] That situation may change: in 2014, the Australian courts supported Myriad Genetics's claim that human gene sequences could be patented.[44]

*

Since the beginning of the twenty-first century and the triumph of the Human Genome Project, genome sequencing has been transformed from a highly complex international affair, immensely costly in terms of people and money, into something that can be undertaken by relatively small groups of researchers, interested in the most obscure organisms. Behind this change has been the appearance of what are called next-generation sequencing techniques based on robotics and powerful computers that were developed after the human genome sequence was completed.[45]

The best sequencers available at the turn of the century used Sanger sequencing to simultaneously sequence about 100 stretches of DNA, each stretch producing reads that were up to 800 bases long. Next-generation sequencing is very different; it uses a variety of techniques to enable the machine to detect each base as it is incorporated into a new chain of DNA during DNA replication. The technology is continually being upgraded; as of 2014, hundreds of thousands of short strings of DNA – each 75–125 bases long – can be simultaneously sequenced, meaning that millions of bases can be detected in a second (when I was hand-sequencing in the 1990s, I was happy if I did 400 bases in a day). These fragments are randomly selected from the genome, and by carrying out this process millions of times, the entire genome can be covered. Computer algorithms are then used to assemble the sequence, meaning that next-generation sequencing is as much about mathematics as it is about molecular biology.

As the price of sequencing machines and computers has dropped, so too has the price of genomes. The human genome cost the public purse around $3bn – more or less a dollar a base pair – and used more than 1,000 sequencing machines. In 2010, the Chinese employed next-generation sequencing to analyse the 2.3 billion base pairs in the genome of the giant panda for a mere $900,000 – less than 0.04 cents per base, or 1/2,500 of the cost for the human genome. The whole project took less than a year, and used the equivalent of just thirty machines.[46]

Most of the genomes from multicellular organisms thus far completed have been published in one of the leading scientific journals. That will inevitably change. According to the Genomes Online Database, at the end of 2014 there were more than 700 projects to sequence non-human vertebrate genomes alone. The genomes of the rattlesnake, the turkey vulture and Nancy Ma's night monkey, along with hundreds of others, are all no doubt fascinating and will provide insight into evolution and medicine, but they – along with the hundreds of arthropods and the thousands of fungi that are being sequenced – are unlikely to get the same kind of attention as the platypus and the panda. The leading journals will focus on genomes with a high commercial or scientific impact, and which therefore promise a high rate of citation in the future. For the remainder of the natural world there will be electronic-only genome journals – already, most bacteria that are sequenced receive a brief one-page announcement with a link to the online database where the information is stored.[47]

More sequencing developments are just around the corner: in

2014, Oxford Nanopore Technologies delivered early models of its nanopore sequencer to researchers around the world for beta testing. The device is the size of a mobile phone and plugs into a computer via the USB port. Unlike next-generation sequencing, which relies on computing power and parallel processing, this technology is claimed to create continuous DNA sequences of up to 10,000 base pairs on your desktop. If it lives up to the hype, DNA sequencing will become commonplace, and could even be done in the field on wild-caught samples, to identify particular genetic variants. Already, next-generation sequencing is being used on oceanic research expeditions.[48]

Meanwhile, the market leader in next-generation sequencing machines, Illumina, has announced that its latest device will be able to sequence the equivalent of sixteen human genomes in three days, bringing the price for sequencing a whole human genome down to less than $1,000. There is a catch: the company insists that to get access to their technology, the user will have to buy ten machines at a total cost of at least $10m.[49] Whatever the coming years hold, the price of sequence data will continue to fall, and the number of sequences will continue to grow.

Eventually personalised medicine based on our individual genetic make-up will finally become widely available. The president of Illumina, Francis de Souza, has predicted that in 2015, an astonishing 228,000 human genomes will be sequenced in the name of medicine – the British government is currently supporting a project to sequence 100,000 genomes with the aim of improving the diagnosis, prevention and treatment of disease.[50] De Souza's ambition is to move Illumina technology into the hospitals and to carve out a chunk of a diagnostic market that he estimates at $20bn. Whatever the hype associated with such claims, our understanding of the significance of small genetic differences between individuals – what is known as intraspecific variation – is growing as governments and research agencies around the world realise that there will be health benefits, as well as insight into the history and demography of human populations.[51] Already, the analysis of the genomic variations found in particular cancers has opened the road to new, more precise treatments. For example, the breast cancer drug Herceptin is targeted solely at women with cancers that have a genetic profile called HER2-positive, while patients with lung cancer whose tumours show mutations in the *EGFR* gene can be treated with drugs called Iressa and Tarceva.[52]

Studies of genetic variation have led to radical new drug treatments that will transform the health of millions of people around the globe. At the beginning of the century, it was noticed that some families with extremely high levels of cholesterol had a particular form of a gene known as *PCSK9*. It then appeared that some people with very low levels of cholesterol had a mutation in this gene. In an extremely short period, drugs were developed to target the PCSK9 protein, and these should become available in 2015.[53] Scientists are now trawling through data from populations around the world, looking for genetic variants that correlate with particular health conditions and which could provide an insight into new drug development. Sequencing is beginning to transform medicine.

These technical developments highlight the ingenuity of molecular biologists, engineers and computer scientists, but they have created an intriguing new problem. We now have thousands of genomes sequenced, and the rate at which they are being completed has grown exponentially, outstripping our ability to analyse them.[54] In July 2011, only 36 eukaryotic genomes had been sequenced; a year later, another 140 had been added; by 2014, 5,628 eukaryotic genome sequences had been either begun or completed, and 36,000 prokaryotic genomes had been sequenced.[55] At the time of writing, the largest known genome is that of the loblolly pine tree, which comes in at a whacking 22 billion base pairs – about seven times the size of the human genome.[56] In contrast, the microbe *Nasuia deltocephalinicola* has a genome of just 112,000 base pairs – this organism is found uniquely in the guts of leafhopper insects, so that much of its metabolic work is done by its host.[57] With fewer chemical reactions to process, its genome has gradually become reduced in size over the 260 million years that the microbe has been living in the insect, losing unnecessary protein-coding genes much as a parasitic animal loses unnecessary anatomical structures. Scientists have calculated that such symbionts could get by with as few as 93 protein-coding genes, which would probably fit into a genome of merely 70,000 base pairs.[58]

Producing a genomic sequence is now relatively simple, at least compared with the effort involved in the pioneering studies. The problem begins when you try to understand what the genome actually does. One of the main tasks when a genome has been completed is to annotate it, identifying genes and their exons and introns, and above all finding genes that have equivalents in other organisms, preferably with some kind of known function. Often the only basis for identifying the function of a gene is because its DNA sequence is similar to a gene in a different organism where a function has been demonstrated. This has led to a new discipline called genomics, which involves obtaining genomes and understanding their nature and evolution. It includes a new set of techniques, collectively called bioinformatics, which combine computing and population genetics to make inferences about the patterns of evolution and enable us to determine which genes have a common origin or function. Training biologists in the techniques of computer science will be an important part of twenty-first-century scientific education.

One of the most far-reaching scientific consequences of sequencing came with the work of Carl Woese, who realised in the 1960s that he could use the RNA found in ribosomes (rRNA), which is common to every organism on the planet, to study patterns of evolution. Woese began studying variation in the nucleotide sequence of the 16s rRNA subunit in a range of bacteria, and by the mid-1970s he had sequenced part of this rRNA from around thirty species – the work was extremely slow and arduous. In 1977, Woese published two papers with George Fox in which they claimed that prokaryotes – single-celled organisms with no nucleus – were not a single group with a common evolutionary history. Basing their analysis on the rRNA sequences – a far more rigorous approach than the mixture of morphological, physiological and ecological data that had previously been employed – Woese and Fox proposed to split the bacteria into two groups: the Eubacteria (or true bacteria) and the Archaebacteria.[59] The data showed clearly that the Archaea, as they were later called, were no more closely related to bacteria than they were to eukaryotes, like you and me. Eventually this led Woese to propose that life evolved into what he called three domains: Bacteria, Archaea and Eukaryota.

For the past twenty years or so, this view has been widely accepted, and it appears in university-level textbooks. But it looks as though it is probably wrong. More extensive analyses of ribosomal RNA and of protein-encoding genes suggest that there are only two primary domains – Archaea and Bacteria, with the Eukaryota being formed when an Archaean microbe engulfed the bacterial ancestor of the mitochondria.[60] According to this view, the Archaea and the Bacteria make up the two great branches of life, with the Eukaryota positioned as a sudden genetic bridge between them.

Studies of evolution using genetic sequences have resolved fascinating issues that have troubled biologists for decades. For example, we now have confirmation that dogs are merely domesticated wolves, and we even have competing scenarios for where and when they were first domesticated, and why.[61] Meanwhile the genome of *Pleurobrachia bachei*, the Pacific sea gooseberry – a ctenophore, or box jelly – suggests that it evolved its neurons and its musculature separately from the rest of the animals, indicating that these two fundamental features of animal anatomy probably evolved at least twice.[62] Some of the results have been surprising: insects, it appears, are simply one form of crustacean. Despite appearances (insects have three pairs of legs and do not live in the sea; crustaceans generally have lots of appendages and are mostly marine), sequence analysis suggests that insects nestle as one very large branch in the middle of an evolutionary tree of crustaceans.[63]

Findings such as this, which are being published every week, are confirmation of Crick's far-seeing prediction, made in his 1957 'central dogma' lecture, that the study of amino acid sequences would reveal 'vast amounts of evolutionary information'.[64] The only thing wrong with Crick's vision was that he was not ambitious enough: we can now directly compare DNA sequences. There is even a web site and an accompanying free app called Timetree that allows ordinary members of the public to interrogate the sequence databases and discover how long ago different groups of organisms separated, enabling you to settle those annoying evolutionary arguments such as whether a hippo is more closely related to a rhino or to a whale.[65]*

---

*The answer, surprisingly, is whale.

*

One major development that has occurred over the past few years would have been dismissed as science fiction by Crick and by virtually every other twentieth-century scientist. We can now look back deep into evolutionary time: if the preservation conditions are favourable (preferably very cold, and not acidic), and extremely careful techniques are used to avoid contamination, then reliable DNA sequences can be obtained from samples that are up to 700,000 years old – that is the age of a horse bone from the Yukon permafrost that has been analysed.[66] It is quite possible that older samples will be sequenced in the future, although the prospect of extracting DNA from dinosaur fossils or from amber from the age of the dinosaurs will almost certainly remain the realm of science fiction.[67]

The advent of what is called palaeogenomics has led to a wave of evolutionary genetic studies of extinct organisms and above all a focus on our closest relatives in the human lineage, the Neanderthals – extinct members of the human lineage who lived in Europe until about 30,000 years ago. The results have been truly astounding.[68] The driving force behind much of this work has been the German-based Swedish–Estonian scientist, Svante Pääbo, who for nearly three decades has pioneered the study of ancient DNA.[69] In 1997, Pääbo, who is based at the Max Planck Institute for Evolutionary Anthropology in Leipzig, shocked the scientific world by sequencing mitochondrial DNA from Neanderthals; in 2010, Pääbo's grand ambition was realised when the draft sequence of the Neanderthal nuclear genome was published, all 3 billion base pairs of it.[70]

The sequencing of the Neanderthal genome was a technical tour de force and provided astonishing information about human history. To everyone's surprise, including Pääbo's, it revealed that, somewhere along the line, Neanderthals mated with humans, and vice versa. This sexual activity – which we now think first occurred up to 58,000 years ago – produced babies who survived and left offspring.[71] The genetic traces of these people can be found in the genomes of all humans in the world except Africans: the ancestors of non-African human populations left Africa and then met the Neanderthals; Africans never encountered Neanderthals. In their final phase, the Neanderthals lived in a mosaic of populations across Europe; over a period of at most 5,400 years, humans and Neanderthals overlapped and interacted.[72] The genetic consequences of some of those interactions live on in us.

Around 3 per cent of the genome of non-African human populations is composed of Neanderthal genes, with some of the characters involved being those to do with skin colour and the immune response, which apparently gave our ancestors an advantage; in contrast, some Neanderthal genes caused reduced male fertility in a human genetic background.[73] We are only beginning to explore this unexpected part of our past and its consequences for our understanding of what it is to be human. There are only about ninety genetic differences that lead to different amino acids in humans and Neanderthals – so at most ninety of our proteins are different.[74] Whatever differences there were between us were probably based more on variation in the regulatory parts of our genomes, which control how, where and when genes are active.

The most surprising proof of the power of palaeogenomics came in 2011, when Pääbo announced the existence of an entirely new and unsuspected group of extinct human relatives, known as the Denisovans who, like the Neanderthals, interbred with humans.[75] This discovery was based solely on the DNA analysis of a 40,000-year-old tiny finger bone from a young girl that had been found in a cave in Denisova in Siberia. The Denisovans branched off from the Neanderthals about 300,000 years ago but still interbred with humans, leaving their traces in today's Polynesian populations, who may have encountered them as the ancestors of the Polynesians slowly migrated through South-East Asia. We have no idea what the Denisovans looked like (all we have is the finger bone, a tooth and a toe bone), but we know that they interbred with humans, leaving traces of their DNA in us. One of the clearest examples of natural selection in the human genome – the existence of an adaptation to living at high altitude, seen in modern Tibetan populations – turns out to have originated with the Denisovans.[76] At some point, ancestors of the Tibetans mated with the Denisovans and acquired a gene that enabled their present-day descendants to survive in low oxygen conditions.[77] Intriguingly, genomic comparison of modern humans,

Denisovans and Neanderthals suggests that interbreeding may have also taken place in Asia and in sub-Saharan Africa with other, unknown members of our lineage.[78] This is a golden age in the study of human evolution, thanks to the breakthroughs produced by the advent of ancient DNA sequencing.*

*

For the general public, even more intriguing than the spectre of our ancestors mating with Neanderthals is how those billions of letters in our DNA sequence make us both human and therefore alike, and unique and therefore different. The study of the organisation of the genetic code has led to some unexpected and mysterious discoveries, as well as a great deal of scientific argument.

In 2000, Dr Ewan Birney set up a sweepstake inviting scientists to guess how many protein-encoding genes would be identified once the human genome was sequenced. The *Drosophila* genome, which had just been published, contained around 13,500 genes, and entrants in the sweepstake chose numbers between 26,000 and 140,000. The number announced in 2003 was, to everyone's surprise, around 21,000 (the prize was split three ways, although none of the entrants came within 5,000 genes of the actual figure).[79] Despite many claims at the time that this number would creep up, perhaps to as high as 40,000, the currently accepted number of protein-encoding human genes is around 19,000.[80] Alternative splicing may result in many different protein variants being produced, but even so, this is a counter-intuitively small figure. Around 10 per cent of our genome – twice as much – is composed of regulatory genes that control the activity of protein-encoding genes. In all species, biological richness resides not simply in the sheer number or variety of protein-encoding genes but above all in the way in which those genes are activated at different points in time, in different tissues and in response to different environmental stimuli.

---

*In 2004, the remains of another archaic human, *Homo floresiensis*, were discovered in a tropical cave on the island of Flores in Indonesia (Brown *et al.*, 2004; Morwood *et al.*, 2004). *H. floresiensis*, popularly known as 'the hobbit', inhabited the cave until about 18,000 years ago. For the moment, it is not possible to extract DNA from bones preserved in such bacteria-rich conditions, but this may change (Callaway, 2014a).

When Jacob and Monod put forward their operon model in 1961, they initially suggested that the *lac* operon repressor, the gene product that affects the activity of the *lac* gene, was an RNA molecule. Although it soon became apparent that the *lac* repressor was in fact a protein, their suggestion that RNA might be involved in gene regulation was extremely prescient. At the end of the 1960s, Roy Britten and Eric Davidson argued that networks of RNA-producing genes controlled the activity of genes in different tissues at different points in development.[81] Britten and Davidson suggested that most genes were what Jacob and Monod had called regulator or regulatory genes, producing RNA that would control the activity of protein-encoding structural genes. The activity of those genes is far more complex than Jacob and Monod's repressor.

We now know that there are many different forms of RNA, often consisting of very short sequences of less than a couple of dozen nucleotides, that are produced as part of the complex networks that control gene expression.[82] These RNA sequences bind to the DNA of the structural gene and are often produced by the complementary DNA strand in the double helix – this is called anti-sense RNA.[83] A large proportion of RNA transcripts produced by mammalian genomes have an anti-sense counterpart that seems to be involved in gene regulation.[84] Among the different kinds of regulatory sequences that are known to exist, promoters are sequences that are found just before the beginning of the coding gene; they allow an enzyme to begin the process of transcribing DNA into RNA – they effectively act as an 'on' switch.[85] Promoter sequences can also be targets for transcription factors – proteins produced by regulatory genes. In eukaryotes, some regulatory stretches of DNA called enhancer regions activate the promoter; these enhancers can be thousands of bases distant from the protein-encoding part of the gene. They exert their effect when the DNA forms a loop, bringing two distant parts of the molecule into relative proximity. So not only are eukaryotic genes in pieces, their constituent parts can also be spread far outside the area containing the protein-encoding exons.

The multiple roles of nucleic acids have expanded far beyond the initial definition of a gene as the fundamental unit of inheritance and show the inadequacy of Beadle and Tatum's 1941 suggestion

that each gene encodes an enzyme. As a consequence, some philosophers and scientists have suggested that we need a new definition of
'gene', and have come up with various complex alternatives.[86] Most
biologists have ignored these suggestions, just as they passed over
the argument by Pontecorvo and Lederberg in the 1950s that the
term 'gene' was obsolete.[87]

In 2006, a group of scientists came up with a cumbersome definition of 'gene' that sought to cover most of the meanings: 'A locatable
region of genomic sequence, corresponding to a unit of inheritance,
which is associated with regulatory regions, transcribed regions
and/or other functional sequence regions.'[88] In reality, definitions
such as 'a stretch of DNA that is transcribed into RNA', or 'a DNA
segment that contributes to phenotype/function', seem to work in
most circumstances.[89] There are exceptions, but biologists are used
to exceptions, which are found in every area of the study of life. The
chaotic varieties of elements in our genome resist simple definitions
because they have evolved over billions of years and have been continually sieved by natural selection. This explains why nucleic acids
and the cellular systems that are required for them to function do not
have the same strictly definable nature as the fundamental units of
physics or chemistry.

*

The fact that about 5 per cent of our genome is made up of protein-
encoding structural genes, whereas around 10 per cent is composed
of regulatory genes producing RNA or protein-based transcription
factors, raises the question of what the other 2.7 billion base pairs
are there for. Long before the human genome was sequenced, it was
obvious that our genome is not simply made up of protein-encoding
DNA. Apart from the stretches of apparently functionless introns
that break up most of our protein-encoding genes, scientists realised
that there were substantial parts of genomes that have no apparent
function and that represent genetic fossils, remnants of the evolutionary past.

Genes that no longer give an organism an advantage tend to
accumulate mutations and eventually cease to function altogether,

but their DNA ghost remains in the genome, a sign of what once was. The human genome, like all genomes, contains many of these non-functional elements, which are called pseudogenes. One of the clearest examples of how this process works can be seen in whales and dolphins (cetaceans). These aquatic mammals use their noses only to breathe every few minutes when they surface briefly; they therefore have very little use for the sense of smell that their terrestrial ancestors possessed 40 million years ago. The genomes of cetaceans contain genes that once coded for smell receptors (olfactory receptor genes) but which now no longer function because the animals spend virtually all their lives with their heads under the water, unable to smell in the air. Over millions of years these olfactory receptor genes have become riddled with mutations and are now pseudogenes – non-coding stretches of DNA that nonetheless retain certain sequences of their functional ancestors and can be identified by comparison with the intact olfactory receptor genes of terrestrial mammals.[90] These pseudogenes provide evidence of evolution by natural selection: the only explanation of their presence in the genome of cetaceans is that these animals were once terrestrial and had a use for their sense of smell.

Some parts of our genome seem to be pure selfish DNA – sequences that apparently have no function beyond to survive.[91] Some of these genetic elements, which riddle our genome, are the remnants of what are effectively genetic parasites – transposons. Transposons are sequences of DNA that can move about the genome, jumping from one location to another. They probably originated as RNA retroviruses that copied themselves into DNA and then became trapped in our genomes. They no longer produce viral RNA but retain the ability to move from place to place in our genome by producing an enzyme called transposase, which effectively unglues them from the DNA strand. Occasionally, the transposon may land in or next to a protein-encoding gene that then hijacks the transposon and converts its activity into a product that is delivered into the cell, thereby leading to the evolution of a new gene.[92] Transposons, and their potential regulatory functions, were first identified by Barbara McClintock at Cold Spring Harbor in the 1940s, to widespread disbelief in the scientific community. Not only was she right, in 1983 she was awarded the Nobel Prize in Physiology or Medicine for her discovery – the only woman to be the sole recipient of this prize.

Over evolutionary time, transposon sequences accumulate mutations in the part of their genome that codes for transposase, and cease to be able to move. They become frozen in our DNA, recognisable but immobile. The remnants of these invasive DNA sequences make up an astonishing 45 per cent of the human genome, with one element, known as Alu, leaving genetic traces that make up to 10 per cent of your DNA. Sometimes bits of these pseudogenes can even be transcribed, producing short bits of RNA that can regulate the activity of genes.[93] Apart from their potential transformation into transposons, retroviruses can play a direct evolutionary role – it is probable that the origin of the placenta in mammals was due to our distant ancestors being infected with a retrovirus that produced a protein called syncytin that is now essential for the development of the placenta. This infection seems to have occurred several times in the mammalian lineage, perhaps explaining the varying forms of this organ in different mammals.[94]

The long stretches of non-coding DNA lie at the heart of the most mysterious result that has been discovered since the beginning of widespread genomic sequencing. Different species can have substantial differences in the size of their genomes, which do not seem to be related to anything in their ecology or degree of apparent physiological complexity. For example, the genome of the 'primitive' lungfish is 350 times larger than that of the pufferfish. No one has been able to come up with any explanation for why this might be. This problem is called the 'C-value paradox' or 'C-value enigma' – 'C' is the amount of DNA in a genome.[95] Some of these differences may be due to a well-known phenomenon: chunks of genomes can be duplicated during evolution, particularly in plants, which can double their genome size in one generation when chromosome duplication goes slightly awry. Because of factors such as duplication, the variation in genomic size that we see between species resists any overall functional explanation. This is highlighted by what is known jocularly as the onion test: the onion genome contains around 16 billion base pairs, or five times that of a human. It is hard to explain this in terms of the contrasting physiology and behaviour of the two organisms, or to imagine that every one of these bases is necessary to the onion.[96]

In the late 1950s and early 1960s, researchers began to use the term 'junk DNA' to describe DNA that had no apparent function.[97] In 1972, Susumu Ohno defined junk DNA as a sequence that cannot be affected by a deleterious mutation. According to this definition, junk DNA is a sequence that, if it were changed, would have no effect on the organism's fitness (that is, on its success in passing its genes onto the next generation). Both pseudogenes and the remnants of tranposon activity would seem to be junk DNA, but scientists argue about this term, and some dispute whether any DNA can truly be considered junk.

In September 2012 this rather arcane debate erupted onto the pages of the press and on the Internet, focused on the question of what the human genome actually does. This was prompted by the publication of the findings of a large-scale project to study the cellular activity of the whole of the human genome, called ENCODE (Encyclopaedia of DNA Elements). The results of the ENCODE project were published in an unprecedented wave of thirty papers, signed by 442 authors, backed up by a web site and an iPad app. The leaders of the project claimed that 80 per cent of the human genome could be assigned a 'biochemical function'; the coordinator, Ewan Birney, went on to claim that the final figure would 'likely go to 100%'.[98] This led to great excitement in the press: *Science* proclaimed that ENCODE had written the 'eulogy' for junk DNA, the *New York Times* stated that ENCODE had shown that 80 per cent of the human genome was 'critical' and 'needed', while *The Guardian* trumpeted 'Breakthrough study overturns theory of "junk DNA" in genome'.[99] This hyperbole led to a backlash on the Internet and in scientific publications as scientists who had not been involved in the project disagreed with the suggestion that there was no 'junk DNA', or that 80 per cent of our genome is 'functional'.[100]

The argument turned on the meaning of the word 'function'. The ENCODE project deliberately cast its net wide by looking for a 'reproducible biochemical signature', which they defined as any consistent biochemical reaction induced by a given stretch of DNA, from mRNA production to protein binding.[101] That was where the

80 per cent figure came from. The computational biologist Sean Eddy pointed out that the ENCODE study lacked what scientists call a 'negative control' – a set of DNA sequences that did not have any function, by any definition, and should therefore have not been identified as functional by the biochemical criteria used by ENCODE.[102] Shortly afterwards, a paper appeared in which researchers carried out this experiment: they randomly generated 1,300 DNA sequences and found that most of these artificial sequences were 'functional' according to the ENCODE criteria. This suggested that the ENCODE definition could not systematically discriminate between random bases and DNA that has some kind of biochemical role in the cell.[103] The lead author of the study, Mike White, wrote:

> most DNA will look functional at the biochemical level. The inside of a cell nucleus is a chemically active place. The real puzzle is this: how does functional DNA manage to distinguish itself from the vast excess of dead transposable elements, pseudogenes, and other accumulated junk?[104]

That question remains unanswered.

In 2014, the ENCODE consortium published a second wave of papers and seemed to back away from their earlier headline claim of 80 per cent function, admitting that 'it is not at all simple to establish what fraction of the biochemically annotated genome should be regarded as functional'. Instead, they emphasised their indisputable finding that an important part of the human genome seems to induce reliable biochemical activity of some kind:

> The major contribution of ENCODE to date has been high-resolution, highly-reproducible maps of DNA segments with biochemical signatures associated with diverse molecular functions. We believe that this public resource is far more important than any interim estimate of the fraction of the human genome that is functional.[105]

For the moment, despite the initial claims of ENCODE, and despite the fact that much of the genome seems to be transcribed into RNA in one form or another, a substantial proportion of our DNA, and that of other organisms, seems to have no discernible role in our existence and could be deleted without causing any selective disadvantage. Future discoveries may change this view, but in those strict terms, much of our DNA still appears to be 'junk'.

THIRTEEN

THE CENTRAL DOGMA REVISITED

In his 1957 lecture, Francis Crick outlined what he called the central dogma of molecular genetics:

> once information (meaning here the determination of a sequence of units) has been passed into a protein molecule it cannot get out again, either to form a copy of the molecule or to affect the blueprint of a nucleic acid.[1]

Information can get out of DNA into RNA to determine the structure of a protein, but proteins cannot specify the sequence of new proteins, and the information in proteins cannot make the reverse journey back into your genes – your DNA cannot be rewritten by a protein. The central dogma has been the focus of repeated criticism over the past sixty years, partly because of the discovery of new facts, and partly because the unfortunate term 'dogma' tends to be a lightning rod for debate.

In 1970, *Nature* magazine trumpeted 'Central dogma reversed' when it was discovered that information can flow from RNA into DNA. *Nature*'s claim was prompted by a discovery that explained how RNA viruses can infect healthy cells and transform them into cancerous cells that produce viruses. In 1964, Howard Temin, a

30-year-old cancer researcher at the University of Wisconsin-Madison, had boldly suggested that the basis of this effect was that RNA viruses turned their RNA code back into DNA, which was then integrated into the host's chromosome, where it de-regulated cell growth and produced more virus RNA. There was no known mechanism whereby such a 'reverse transcription' could take place, so Temin was forced to hypothesise the existence of an enzyme that could carry out that task, transcribing the RNA virus into DNA. In 1970, Temin was proved right, when he, along with 32-year-old David Baltimore at MIT, reported the existence of an enzyme in RNA viruses that copies RNA into DNA. This enzyme, now called reverse transcriptase, enables information to flow from RNA back to DNA. *Nature* magazine, which published Baltimore's paper, editorialised somewhat pompously:

> The central dogma, enunciated by Crick in 1958 and the keystone of molecular biology ever since, is likely to prove a considerable over-simplification.[2]

Piqued by the tone of the editorial, Crick replied in the pages of the journal, graciously acknowledging Temin and Baltimore's 'very important work' and setting the record straight with regard to what he had argued thirteen years earlier. Crick's original hypothesis explored all the possible transfers of information between nucleic acids and proteins, and prohibited only those, such as protein → DNA, that either had been excluded experimentally or for which there was no conceivable mechanism. In subsequent years, this rich view tended to be replaced by the cruder DNA → RNA → protein, as summarised in Jim Watson's influential 1965 textbook, *Molecular Biology of the Gene*.[3] In 1957, Crick considered that the RNA → DNA step was 'rare or absent', but not impossible. As Crick pointed out in 1970, there was no 'good theoretical reason why the transfer RNA → DNA should not sometimes be used. I have never suggested that it cannot occur, nor, as far as I know, have any of my colleagues.'[4]

Crick's view of the significance of Temin and Baltimore's discovery was that the RNA → DNA transfer probably did not occur in most cells but might take place in special circumstances such as some viral infections. Temin was not so restrained, and within a year he was arguing that RNA → DNA information transfers were a fundamental part of normal development in the somatic cell line (that is, in all cells except the egg and sperm). As a result, he claimed, 'new DNA sequences are formed by this process during the lifetime of a single organism'.[5] According to Temin, reverse transcription was an everyday process, helping to shape how our cells develop – except that it is not, and reverse transcriptase is only ever found in cells infected by a particular class of RNA viruses called retroviruses. In this respect, Crick was right, and Temin – and the editorial writers at *Nature* – were wrong.

Although Temin's more extreme claims were misplaced, his discovery of reverse transcriptase was significant because it showed how viruses could cause cancer by altering the DNA of the cells they infect and by deregulating genes, leading to uncontrolled growth. The enzyme also went on to play an important role in the development of molecular genetics and of our ability to genetically modify organisms by introducing new sequences into DNA – it is used by scientists to make complementary DNA (cDNA) from mature mRNA. In 1975, Temin and Baltimore, along with Temin's PhD supervisor, Renato Delbucco, won the Nobel Prize in Physiology or Medicine for their work on how cancer viruses affect our genes.

*

In his 1970 clarification of exactly what he meant by the central dogma, Crick highlighted three kinds of information transfer that he postulated would never occur: protein → protein, protein → DNA and protein → RNA. However, even as he made such a clear prediction, Crick was cautious, underlining our ignorance and the fragility of the evidence upon which he based his slightly revised 'dogma':

> our knowledge of molecular biology, even in one cell – let alone for all organisms in nature – is still far too incomplete to allow us to assert dogmatically that it is correct.[6]

And in the very next sentence he highlighted a potential exception:

There is, for example, the problem of the chemical nature of the agent of the disease scrapie.

Scrapie is a neurodegenerative disease affecting sheep and goats that has been known for hundreds of years. In 1970, its cause was mysterious – the disease-causing agent was known to be resistant to heat, formalin, ultraviolet radiation and ionising radiation (all of which destroy nucleic acids and inactivate viruses) and it left no sign of infection in the animal's immune system. This curious set of facts led some scientists to argue that scrapie was in fact a genetic disorder rather than an infectious disease. Others daringly suggested that the scrapie infectious agent was a protein – this was what lay behind Crick's remark in 1970. At this time, all known infectious agents were based on nucleic acids and were either organisms or viruses. A protein-based infectious agent would be a truly radical discovery, and this suggestion was therefore treated with some scepticism.[7]

In 1982, Stanley Prusiner's group discovered that scrapie could be detected by the presence of a protein that was also a potential infectious agent – they called it the prion protein, and it seemed to act by altering the shape of non-infectious proteins that were otherwise identical to the prion.[8] This was utterly novel, both because it suggested that a protein could transmit a disease and because it implied that the prion might breach the central dogma by allowing the transmission of information from protein → protein. In the 1940s, Mirsky had suggested that there might be minute levels of protein contamination in Avery's purified DNA; in the 1980s, some of Prusiner's critics argued that there must be small amounts of nucleic acid in the apparently pure prion protein extracts. The prejudices that prevented some scientists from accepting Avery's discovery that DNA is the hereditary material reappeared in the case of this infectious agent that was apparently not based on nucleic acids.[9]

Interest in scrapie grew in the late 1980s and the 1990s with the horrific outbreak of variant Creutzfeldt–Jakob Disease (vCJD) in humans and its equivalent in animals, 'mad cow' disease (bovine spongiform encephalopathy, or BSE). These diseases infected millions of cattle and caused the deaths of hundreds of people, most of them teenagers and young adults. Both BSE and vCJD showed similarities

to scrapie, and again the evidence suggested that an infectious protein was involved. It was eventually shown that the same prion
protein causes all three diseases. Although it is still unclear how the
BSE outbreak began, it is possible that cows initially got the disease
from scrapie-infected sheep, the remains of which were fed to cows
as meat and bonemeal. Whatever the original source of BSE, people
caught the disease by eating diseased sections of the bovine nervous
system that had been included in processed meat such as burgers.

It is now accepted that the aberrant prion protein alters the conformation of normal prion proteins, thereby producing the brain
pathology in sheep, goats, cows and humans.[10] In 1997, Prusiner
was awarded the Nobel Prize in Physiology or Medicine for his discovery, but despite the widespread acceptance of the prion hypothesis, a few scientists continue to argue that virus-like particles are
involved in scrapie and similar diseases.[11] Although it is known that
yeast prion transmission involves only proteins, there remains the
slim possibility that unknown nucleic acid-based cofactors may be
involved in mammalian prion diseases.[12]

In 1982, Prusiner suggested that the prion codes directly for the
synthesis of another prion, not merely for its shape. This would have
completely destroyed one fundamental point of the central dogma,
that protein does not code for protein. Prusiner was wrong. Prion
proteins are produced by the action of the prion gene, encoded in
DNA and transcribed into RNA and then translated into a chain of
amino acids – the normal prion protein plays a role in producing
myelin, which protects nerves.[13] In both the benign and the pathogenic forms of the prion, the amino acid sequence remains the
same, so there is no transfer of information as defined by the central dogma, which referred solely to the sequence, not the structure.
Although it can be argued that three-dimensional conformation is a
form of information – indeed, Crick accepted as much – the change
induced by the prion protein is probably more similar to the action
of a crystal growing by assembling identical copies of itself than it
is to that of a DNA molecule, which can produce a correspondence
with a sequence in a different kind of molecule.[14] Despite the highly
unusual and pathological conditions that produce prion disease, the
central dogma remains fundamentally intact.[15]

*

It was 1977, and I was an undergraduate at Sheffield University listening to a lecture by Professor Kevin Connolly, a world expert on child development and behaviour genetics with whom I eventually studied for my PhD. Kevin was describing the effects of social deprivation, and he highlighted a 1967 study that showed that if a female rat pup was removed from its mother for only three minutes a day, her sons and daughters and even *their offspring* would show pathological changes in their activity and weight, even if they and their parents had been reared under normal conditions. For my excited 20-year-old brain, this study had two implications. First, it suggested that the effects of social deprivation in humans might continue to echo down the generations, even if people were subsequently provided with an excellent environment. Second, and more fundamentally, as the title of the paper put it, the effect involved a 'nongenetic transmission of information'.[16] Not all hereditary information is made of DNA, the result suggested. Almost speechless with surprise, I went up to Kevin after the lecture to confirm that I had understood correctly. I had. I went away, thinking hard about what this effect might mean, and above all how it might work.

Nearly forty years on, I am still interested in the non-genetic transmission of information down the generations and the underlying mechanisms – one of my PhD students, Becky Lockyer, has recently studied them in *Drosophila*.[17] The existence of intergenerational transmission of environmentally induced changes is now well established, and it is known that these effects can buffer populations of organisms against rapid environmental change before new adaptations evolve.[18] They form part of the complex route from DNA to phenotype, presenting biologists with fascinating examples of plasticity – how a given DNA sequence can generate a variety of different phenotypes.

Some of these effects can be thought-provoking, like the study I heard about in 1977. For example, in 2009 Larry Feig's group at Tufts University in Boston reported that if female rats were given an enriched environment during their adolescence, their offspring – conceived after the enrichment had ceased – showed an increased learning ability. The effect was even strong enough to overcome genetic defects in learning.[19] There is no evidence that such memory effects occur in humans, nor is it known how this particular trans-generational effect works, but it does not necessarily involve genes. In the case of deprivation, poor parenting can cause behavioural and hormonal changes in offspring that lead to those individuals' being poor parents in turn, and so on.

This kind of phenomenon is often described as an 'epigen-etic' effect, even where no effect on genes has been demonstrated. Strictly speaking, 'epigenetic' refers to any way in which the genetic code is modulated on its route from DNA sequence in a cell into an expressed character; that is, how genes are regulated.[20] However, the term is increasingly being used primarily to describe rare cases in which changes in gene regulation are transmitted down the gen-erations. Journalists, philosophers and scientists have claimed that transgenerational epigenetics radically alters our understanding of inheritance and evolution, and even marks 'victory over the genes' as the German magazine *Der Spiegel* put it in 2010.[21] The truth is somewhat less dramatic.

In their most widespread form, epigenetic effects explain how genes are turned off and on in our cells, enabling each specific cell type to appear, allowing a complex organism with various kinds of tissues to develop from a single-celled embryo – precisely the mys-tery highlighted by Jacob and Monod when they discovered the first example of a regulator gene. In other words, epigenetic effects, whether they are transgenerational or occur in a single organism, are examples of gene regulation.*

Epigenetic regulation often involves the activity of small RNA molecules, which are produced by genes in complex regulatory networks.[22] One of the most widely studied forms of epigenetic control is the placing of epigenetic marks on genes, which occurs when the cell adds a methyl group (CH_3) onto a cytosine base of a DNA sequence. This process, known as methylation, does not alter the sequence but can result in the gene being silenced – the gene is ignored, as though the transcription machinery no longer recognises

---

* 'Epigenetics' sounds much more exciting than 'gene regulation', which is no doubt why the term is increasingly being used.

the sequence. Methylation is relatively common in plants, but it is rare in animals. Where methylation does occur in animals, it is overwhelmingly in somatic cells, which form the organism's body, not in the germ cells that pass genes to the next generation. Methylation marks that may occur in germ cells are mostly removed during the formation of eggs and sperm; any that survive this process are generally wiped immediately after fertilisation.[23] In a recent example, the undernourishment of a female mouse affected DNA methylation in the sperm of her sons, although this was not then transmitted to the son's offspring.[24]

Epigenetic chemical marks can also be left in histone – the proteins that package DNA. It is widely assumed that changes to histones are involved in gene regulation, but the evidence is unclear, and in *Drosophila* one kind of histone can be deleted completely without any effect on gene transcription.[25] For the moment, there is no evidence that histone modifications are directly passed on during ordinary cell division, never mind to the next generation. Instead, in some circumstances enzymes may re-introduce histone marks in daughter cells from which these marks have been wiped.[26]

Epigenetic effects are particularly important in the development of certain cancers: genes that normally silence genes that can lead to uncontrolled growth can be themselves silenced by epigenetic effects from the environment, leading to disease. In the late 1950s, Szilárd, Jacob and Monod called this effect the 'derepression of repression', and we now know that it can lead to some forms of cancer that, in certain rare circumstances, can be passed from one generation to another.[27] For example, an increased susceptibility to a genetic form of testicular cancer in mice was found to be transmitted down several generations, suggesting that expression (or silencing) of a gene in the parent can lead to that gene being expressed (or silenced) in the offspring.[28] It is this kind of transgenerational effect that captures the imagination of journalists, scientists and the general public, because it apparently contradicts the basic teachings of genetics.

One of the most widely recounted examples of apparent transgenerational epigenetic effects is the terrible 'Dutch famine' that took place during the winter of 1944–45. Women in the Netherlands who were pregnant at the time had smaller children; now that these children are adults, they show poor glucose tolerance and are more likely to suffer from diabetes. They also show differences in their levels of DNA methylation.[29] But it is unclear whether the methylation differences are the cause of the abnormalities or are the consequence of them, and above all there is no evidence that the germ line in these people has been affected.[30]

Mammals often show a particular form of transgenerational epigenetic effect known as genomic imprinting. We inherit two copies of each gene, one from each parent; in some cases, either the maternal or paternal copy is silenced (or, more rarely, enhanced) by the action of epigenetic marks or imprinting, and the other parental form reappears in the offspring.[31] Although genomic imprinting affects only a small proportion of mammalian genes, in females it lies at the basis of the inactivation of one of their two X chromosomes that occurs shortly after fertilisation and is essential for normal development. Like other epigenetic changes, genomic imprinting effects can be reversed in the next generation – this is not a permanent change that alters the genes that are carried by a population.

Plants are much more likely than mammals to show epigenetic inheritance, partly because they do not have strictly separated germ and somatic cell lines. However, the division of germ and soma in animals is not quite as clear-cut as might be thought (indeed, not all animals have this division). Egg and sperm are cells that contain half your genomic DNA, but they also carry other material that may affect the fitness of your offspring. For example, there are 'maternal effects', whereby characters are differentially expressed depending on the mother, generally due to the presence of certain mitochondrial genes in the egg – we inherit our mitochondria from our mother, and this separate genome can exert specific effects before and during embryogenesis. Other molecules can have long-lasting effects down the generations, so for example if the nematode worm *C. elegans* becomes resistant to a virus, that resistance can be transmitted for several generations through the presence of small protective RNA molecules in the sperm cells; these molecules are directly involved in resistance to the virus, not in gene regulation.[32]

Even in plants there are limits to the significance of the inheritance of epigenetic factors. First, the environment can only alter gene

regulation: there is no evidence that it can lead to any direct alteration of the genetic code. Second, for the moment, despite some tantalising hints that DNA methylation may be involved in a plant's
response to bacterial infection, there is not one clear example of an
epigenetically based adaptation that is of a character that increases
the fitness of the organism.[33] However, epigenetic manipulation may
have important consequences for plant breeding in the future. In the
laboratory plant *Arabidopsis*, heritable epigenetic factors have been
found to affect flowering time and root length, and can be subject to
artificial selection. Heritable gene silencing in this species can occur
through the activity of small RNA molecules, which may be transmitted in the gametes. For the moment these gene silencing effects
have been found mainly in characters associated with the activity of
transposons; one of the leaders in the field, David Baulcombe, has
suggested that epigenetic effects in plants are probably associated
with variation in particular traits, rather than in determining major
characters.[34]

A recent review in the journal *Cell*, entitled 'Transgenerational
epigenetic inheritance: myths and mechanisms', summarised our
understanding of the contrasting situations in plants and animals in
appropriately sober fashion:

> although transmission of acquired states can occur in some
> animals (such as nematodes), proof that transgenerational
> inheritance has an epigenetic basis is generally lacking in
> mammals. Indeed, evolution appears to have gone to great
> lengths to ensure the efficient undoing of any potentially del-
> eterious bookmarking that a parent's lifetime experience may
> have undergone.[35]

Nevertheless, some scientists have continued to suggest that
evolutionary biology needs a major theoretical rethink; others have
responded correctly that our current views already include the existence of epigenetic effects.[36] Epigenetics remains fascinating, but it is
an adjunct to our understanding of the complexities of gene regulation and the origin of plasticity, not a radical new model of inheritance and evolution. The fundamental claim of the central dogma is

that once information has gone out of the DNA sequence into protein, there is no way for it to get back into the genome. Despite the existence of epigenetic inheritance in certain unusual circumstances, this statement remains true. We know of no way in which the information expressed in proteins can alter the DNA sequence.

The idea that an organism's experience affects the characters shown by its offspring was an early contender for explaining evolution, and is associated with the name of the nineteenth century French naturalist Jean-Baptiste Lamarck. In fact, Lamarck never used the phrase by which he is now most widely known – 'the inheritance of acquired characteristics' – and according to his view only the youngest organisms could acquire new characters.[37] The most widely-known example given by Lamarck was his suggestion that the giraffe acquired its long neck by gradual increases in length over many generations, caused by the act of stretching to reach higher branches to eat the leaves.

In its nineteenth-century context, Lamarck's suggestion that acquired characteristics were the driver of evolutionary change provided a mechanism for a process that was still widely contested.[38] The mechanism that Lamarck described was widely believed – Darwin also assumed that characters acquired during the organism's lifetime affected the hereditary particles that he hypothesised existed in every tissue of the body and which were involved in reproducing those tissues in the next generation.[39] In this sense, Darwin was also a Lamarckian. By the beginning of the twentieth century, with Weismann's discovery of the germ and somatic cell lines in animals and the development of genetics, it soon became apparent that heredity was based on particles that were transmitted with no detectable effect of experience. Lamarck's hypothesis was stripped from its context and became a caricature for the condescending amusement of students, while the fact that Darwin shared this concept was often passed over in silence.

The most notorious advocate of the inheritance of acquired characteristics was the Soviet agronomist Trofim Denisovich Lysenko. In the 1930s, Lysenko claimed that he was able to alter the make-up of wheat by environmental manipulation, and that genetics was a fantasy invented by the capitalist West. In the Cold War years,

Lysenkoism led to the persecution of a whole generation of Soviet geneticists as their science was effectively outlawed. When Lysenko's star waned in the late 1950s and 1960s, the Stalinist leaders of the USSR gradually allowed Soviet genetics to re-emerge, but even half a century on, the damage is still evident in terms of the relatively weak status of Russian genetics on a world scale as compared to other areas of science.

With the recent interest in transgenerational epigenetic effects, some science journalists have suggested that Lamarckian evolution is making a comeback (the name of Lysenko is rarely mentioned in such articles).[40] Lamarckism is not on the rise, because there is no cellular pathway from protein to DNA. In the absence of such a route, this mode of evolution would require epigenetic effects to be reliably transmitted over many generations. There would then have to be some way of fixing the epigenetically determined character in the population in the absence of the environmental factor that initially induced it. This is a fundamental challenge for the idea that epigenetics plays a major role in evolution: Darwinian natural selection on protein-encoding genes leads to stable adaptations that persist even in the absence of the selection pressure that shaped them – a polar bear born in a southern European zoo still has thick white fur, even though it has never been near ice or cold. All its descendants will also show this character. This would not be the case with epigenetic inheritance, unless we were to discover new mechanisms for permanently fixing the epigenetic marks.

There is an even more fundamental problem: although Lamarckian inheritance apparently provides a common-sense explanation of adaptation – that is, of the appearance of characters such as a giraffe's long neck – if there were a route from acquired characteristic to DNA, that route would carry information both for good things such as longer necks and for bad things such as high cholesterol or type 2 diabetes, which also occur during an individual's lifetime. Unless the cell could know which changes would be advantageous in the future, both kinds of change would end up being translated back into DNA, or tagged epigenetically, and the next generation would show a confusing mixture of advantageous and disadvantageous characters.

*

Virtually all of the evolutionary adaptations we see around us are consequences of the sifting of DNA sequences by natural selection over many generations. However, other factors also affect the frequencies of sequences of DNA in a population. For example, shifts in apparently non-functional DNA sequences generally occur through completely random processes because they are not subject to natural selection. Even sequences that are subject to natural selection, which include both protein-coding genes and lots of sequences involved in gene regulation, can show changes in a population through random sampling, producing an effect called genetic drift. One of the easiest effects to understand occurs where a sudden reduction in the number of organisms in a population can affect gene frequencies and therefore the raw material that natural selection has to work on.

As well as being the primary motor that produces the fantastic adaptations we can see in the world around us, natural selection is also the only conceivable way in which it can be argued that information encoding adaptations can flow back from a character – be it a protein or whatever – into the gene sequence shared by a population. That is how natural selection works: the characters produced by different genotypes lead the organisms that carry those genes to leave different numbers of offspring in the next generation. The DNA sequences in the population therefore gradually change over time as random effects change the range of sequences available in a population, and the environment sifts the genetic composition of the population, with those individuals that can survive and reproduce in particular environmental conditions leaving more copies of their genes in the next generation. Natural selection is not a challenge to the central dogma, because the two phenomena function at different physical and temporal levels. Natural selection is a force that operates in populations over many generations, whereas the central dogma describes processes taking place in a single cell – or even in a single molecule – during the lifetime of an individual.[41]

When Crick enunciated the central dogma, his aim was not to reframe Weismann's division of cells into the somatic line and the germ line, or to defend the modern understanding of evolution by natural selection against the idea of the inheritance of acquired characteristics. The central dogma was based on known or assumed patterns of biochemical information transfer in the cell rather than any dogmatic position. As such it was vulnerable to being invalidated by future discoveries. Nevertheless, in its fundamentals it has been shown to be correct. Real or apparent exceptions to this rule, such as retrotranscription, prion disease or transgenerational epigenetic effects, have not undermined its basic truth.

Crick's operational definition of genetic information was 'the determination of a sequence of units' either in a nucleic acid (DNA or RNA) or in the amino acid chain of a protein. A separate part of Crick's argument was the sequence hypothesis – the idea that the three-dimensional structure of proteins is in some way inherent in the sequence and emerges as the amino acid chain is assembled. Crick admitted that it was 'possible that there is a special mechanism for folding up the chain' but felt that 'the more likely hypothesis is that the *folding is simply a function of the order of the amino acids.*'[42] He therefore hypothesised that there is no separate genetic code for protein conformation. Nearly half a century later, this still seems to be correct. By comparing a novel DNA sequence with known amino acid sequences and three dimensional protein structures, it is possible to make predictions about the three-dimensional conformation of the protein that could be produced by the DNA sequence. However, despite our deep understanding of the physicochemical rules underlying the shape that proteins take, there is as yet no way of predicting with absolute certainty the three-dimensional structure of a protein from a DNA sequence. Our predictions are becoming increasingly accurate because of novel computer algorithms and because the process is cumulative: as the number of correlations between sequence and structure increases, so, too, does accuracy.

There is something missing from this description of protein synthesis. In many cases – perhaps most, perhaps almost all – protein synthesis involves the presence of molecular chaperones, which alter the rate of folding.[43] These chaperones can be molecules such as the heat shock proteins that prevent polypeptide chains from clumping together, or they may block competing biochemical reactions during the folding process. Chaperones can equally be hollow molecules that provide a favourable chemical micro-environment for the protein to be correctly folded, within a space known by the sci-fi-sounding term Anfinsen's cage – named after Christian Anfinsen, who won the 1972 Nobel Prize in Chemistry for his work showing the link between the one-dimensional amino acid sequence and the three-dimensional protein structure.[44]

In a 1970 letter to Howard Temin, Crick, open-minded as ever, showed that he recognised the possibility that other factors apart from the amino acid sequence might determine three-dimensional protein structure: 'I do not subscribe to the view that all "information" is *necessarily* located in nucleic acids. The central dogma only applies to residue-by-residue sequence information.'[45]

Crick would no doubt have been intrigued but unworried by some aspects of protein synthesis that might appear to contradict the central dogma. In a large number of bacteria and some eukaryotes such as fungi, some peptides are assembled without the direct involvement of DNA, mRNA or ribosomes. The existence of these oddities does not go against the key argument of the central dogma, because there is no route for information to flow from the protein into DNA. Instead, non-ribosomal peptides raise the question of how the information they contain is represented and transmitted genetically, if it is not contained in a nucleic acid sequence. There are two linked answers. Non-ribosomal peptides are synthesised using an assembly line of enzymes and the order of the amino acids is determined by the order in which the enzymes act on the growing chain – each functional part of each enzyme adds a particular amino acid. However, these enzymes are themselves encoded by genes, so ultimately the information that represents the peptide's amino acid sequence is in fact contained in DNA sequences, albeit indirectly.[46]

At the beginning of 2015, a protein called Rqc2p was discovered that actually gets involved in ribosomal protein synthesis, and recruits tRNA molecules to add two kinds of amino acid – alanine and threonine – to the end of a protein chain when synthesis gets stalled.[47] By adding a number of these two amino acids in what seems to be a random order, Rqc2p appears to mark the protein (or perhaps the ribosome) for imminent destruction by the cell's house-keeping machinery. The order of the two amino acids does not seem to be important, and they are not added in any consistent sequence, so, strictly speaking, this example does not contradict Crick's hypothesis that a protein cannot determine the amino acid sequence of another protein. However, it represents a step towards that possibility. Other odd examples may yet be discovered – as Crick emphasised in 1970, our knowledge is still far too incomplete for us to assert that our current understanding is completely correct. It explains what we have so far discovered, but we may find there are further surprises.

For some philosophers of science, the role of chaperones and the potential existence of information outside the genetic code undermines Crick's 1957 assumption that protein folding is a spontaneous, self-directed phenomenon. Some even argue that proteins are an agency of heredity, opening the door to the inheritance of acquired characteristics.[48] These are very much minority views among philosophers – and even more so among scientists. The role of chaperones is simply what their metaphorical name suggests: they protect and facilitate interactions that lead to three-dimensional protein structure; they do not actively guide and structure them. And even if it eventually turns out that some proteins do directly form the three-dimensional structure of certain proteins (as with prions), the wealth of existing data about protein synthesis indicates that these will be minor curiosities, exceptions that prove the rule.[49]

*

Some readers – and in particular any philosophers out there – may be uneasy with the way in which findings that do not conform to the central dogma seem to have been dismissed as exceptions or the products of pathology, thereby apparently leaving the fundamental argument intact when in reality it has been severely weakened. Apart from the fact that none of these examples provides evidence for the transfer of information from protein $\rightarrow$ DNA, this relaxed attitude, which I share with the vast majority of biologists, underlines a difference between general statements or hypotheses in biology and axioms or laws in mathematics or physics. A single example of a particle travelling faster than light would require a great deal of work by theoretical physicists in order to reshape our understanding of the Universe. In contrast, a solid example of information flowing directly from protein → DNA would not cause a radical revision of our concepts of how genetics and evolution work, unless it was discovered that such transfers take place systematically and on a wide scale.

Were such an example to be discovered, that part of the central dogma would no longer be true, and it is possible that new technologies would become available for manipulating organisms. But virtually all of our existing results and experimental protocols would almost certainly emerge unscathed, because they have been shown to function perfectly well in the absence of such an additional mode of information transfer. The challenge would be for scientists to put the new exception into the existing framework, explaining it in the historical and evolutionary context of the central dogma. If that were not possible, then a radically new explanation would be necessary, and the central dogma would be relegated to the status of an abandoned fruitful hypothesis, an idea that led to successful and informative experimental work, but which was ultimately shown to be wrong.

This would not constitute some kind of moral or philosophical victory for the epigenetic revolutionaries: the reason that scientists accept the central dogma is not because it is a dogma but because the evidence supports it. If new evidence were to arise, then, as the French phrase puts it: *Il n'y a que les imbéciles qui ne changent pas d'avis* – only fools do not change their mind.

FOURTEEN

BRAVE NEW WORLD

In 2010, the molecular geneticist and entrepreneur Craig Venter hit the headlines. In an article published in *Science*, his group claimed they had created the world's first synthetic organism.[1] More than a decade earlier, Venter's group began to study the bacterium *Mycoplasma mycoides*, which causes lung disease in ruminants. Over years of painstaking work, they succeeded in creating a synthetic version of the *M. mycoides* genome, having disarmed various pathogenic genes. They then introduced the synthetic DNA chromosome – over a million base-pairs long – into a cell of a related species from which the genomic DNA had been removed. Once installed in its new host, the *M. mycoides* genome was able to function successfully, controlling the cell and reproducing. A new life-form had appeared, created through the work of scientists.

This feat had two important limitations. First, the cell they used was not empty: it contained all the natural cellular machinery, such as ribosomes, metabolites and enzymes, needed for the synthetic DNA to make the new organism function. These vital ingredients had not been touched by Venter's group. Second, the DNA they introduced into that cell had not been written from scratch; it was copied from the genome of an existing organism. Despite all the human ingenuity involved, the success of the project relied fundamentally on work that had already been done over hundreds of millions of years by natural selection in creating the cell and its contents and in encoding the genome.

Nevertheless, in typical entrepreneurial fashion, the researchers from the J. Craig Venter Institute stamped their ownership on the new bacterium – called *Mycoplasma mycoides* JCVI-syn1.0 – in the shape of genetic watermarks. Using a complex code made up of combinations of letters from the genetic code, the Venter group hid several identifying marks in the DNA sequence of their creation. These included three quotations (one from *A Portrait of the Artist as a Young Man* was briefly the subject of a humourless legal action by the James Joyce estate), the names of forty-six people who were involved in the work, and an address to which e-mails could be sent by anyone able to crack the code. The first correct solution was received a little more than three hours after the watermark sequence went on line.[2]

In a 2012 lecture in Dublin to commemorate Schrödinger's *What is Life?*, Venter suggested that he could use his technique to teleport life-forms from the surface of Mars. He proposed sending a robot to the Red Planet that could sequence Martian DNA (assuming that Martians contain DNA) and then transmit the sequence back to Earth. We could then reassemble the Martian in a laboratory, using the technique employed to create *Mycoplasma mycoides* JCVI-syn1.0.[3] The idea of transmitting Martians caused some excitement in the press ('Geneticist aims to teleport Mars life back to Earth', said *The Boston Globe*) even though Venter did not even invent the idea of teleporting genes – Norbert Wiener first proposed this method for transmitting an organism through the ether in his 1950 book *The Human Use of Human Beings*.

As Venter points out, recreating a Martian in a maximum-containment facility on Earth would be safer than bringing it hurtling back through the atmosphere, with the potential of a crash and the pollution of the planet. There are some difficulties, however. If there were life on Mars, it would be very surprising if a Martian genome were able to pop into an Earthling cell and just start working – the cellular context would almost certainly be utterly different from that required by the Martian DNA. In the extremely unlikely event that a Martian was found, that it was based on DNA and that it could kickstart itself into life in an Earthling cell, recreating it on Earth would show that the Earthling and Martian branches of life shared a common ancestor. The most probable explanation would be that the Martian microbe came from Earth, blasted into space on a lump of rock after a meteorite strike and eventually plummeting onto the Red Planet. If there is life on Mars that is truly Martian, it seems highly unlikely it is similar to Earth life. There is no reason to imagine that DNA is the only possible informational molecule; in fact, our deep evolutionary past, and the ingenuity of today's scientists, both show that is not the case.

*

Biotechnology is not a recent development. For thousands of years, humanity has used the power of microbes to produce two food-stuffs that are seen as an essential part of everyday life for much of the planet: bread and beer. Both rely on harnessing the respiratory mechanisms of yeast to produce carbon dioxide (which makes bread rise) and alcohol (which makes beer intoxicating). What was initially a blind process has been utterly transformed over the past four decades, as modern biotechnology has exploited our ability to manipulate the genetic code to create organisms containing new genes, including genes from other species.[4] New terms have been coined – biotechnology, genetic engineering, synthetic biology – but they all ultimately describe the use of genetic manipulation to alter living organisms.[5]

Many drugs, including hormones, are now produced by harnessing the power of genetic engineering, involving the insertion of the relevant gene into a microbe that then churns out the desired material. Some examples are frankly bizarre, such as the goats that express a spider gene for producing silk, and excrete the stuff in their milk.[6] If the spider-goats can produce sufficient quantities of the strong and flexible silk, new products such as stab-proof jackets could be created. Looking to the future, research groups around the world are trying to address the two central problems facing our species – energy and food supplies – by manipulating cells to produce fuel and meat.

Over the past couple of decades, genetically modified (GM)

plants have become widespread in agriculture, in particular in the US. In 2014, 94 per cent of US soybean crops were GM, as were 93 per cent of corn crops, 95 per cent of sugar beet crops and 96 per cent of cotton.[7] Some of these crops are pest-resistant because they have been engineered to produce a natural insecticide that is normally produced by the soil bacterium *Bacillus thuringiensis* (these are therefore known as *Bt* crops). Other crops are herbicide-resistant and enable farmers to increase yield by reducing the need to leave space for weeding – more plants can be grown per acre.

The best-known GM crop is the Roundup Ready soybean, produced by the agrichemical company Monsanto, which resists Monsanto's own brand of herbicide, Roundup. Despite the very real benefits of these crops in terms of higher productivity, the increased use of herbicides reinforces the bleak monoculture of much of modern industrial farming, reducing biodiversity in the immediate vicinity of the farm. Herbicides can also pollute local water sources, with unintended consequences for wildlife, in particular for amphibians.[8] The safety and reliability of GM technology is not at issue here; it is the aims and consequences of its use that need to be addressed.

When GM food crops were first introduced into the UK, the tabloid press described them as 'Frankenfood', and there was widespread hostility to scientific trials of new GM crops, including direct action by activists who trashed the fields. Health fears relating to the consumption of GM food have been widespread but are entirely unjustified: there is no evidence that consumption of GM organisms will do you any harm at all. Vague unease about 'manipulating nature' is similarly mistaken – all the food we eat has been genetically manipulated over thousands of years through artificial selection by our ancestors. The difference is simply one of method: artificial selection of our foodstuffs is merely slower and generally less effective than direct genetic manipulation.

Even the widespread feeling that there is something unnatural about transferring genes from one species to another is unfounded. Exchange of genes between species – known as horizontal gene transfer – occurs quite readily in microbes, and sometimes in animals and plants. Very specific adaptations have appeared through horizontal gene transfer. For example, the pea aphid is the only animal in the world that can synthesise red pigments known as carotenoids (all other animals have to acquire these compounds from the environment, by eating plants). The aphid gained this ability by incorporating genes from a fungus into its genome; when, how and why this happened is unknown, but it demonstrates that horizontal gene transfer, a form of inadvertent natural genetic engineering, can also explain some adaptations of multicellular organisms.[9]

A recent experimental study showed that horizontal gene transfer could help ordinary bacteria to become plant symbionts by simultaneously transferring genes involved in symbiosis and genes that led temporarily to an increased mutation rate in the bacteria. As a result, the bacteria responded rapidly to selection pressure, accelerating their transformation into symbionts.[10] Horizontal transfer effects may not be limited to genes – the parasitic plant dodder exchanges not genes but mRNA with its plant host. In some circumstances even gene products can jump the species barrier.[11]

Even more spectacularly, horizontal gene transfer was at the heart of the evolution of the eukaryotic lineage, which includes all multicellular organisms, along with some single-celled organisms such as yeast. As well as containing the DNA of the ancestor of the mitochondrion, our genome also plays host to many genes from prokaryotic organisms, which were transferred into our ancestors by horizontal gene transfer. We are all the product of gene transfer between species. Faced with the evidence that gene transfer between species is so widespread, no one can argue against GM technology on the basis that it is unnatural.

*

Some applications of genetic engineering relate to areas of technology that are neither controversial nor bizarre. By manipulating the genetic sequence, it is possible to use DNA to store information generated by humans, perhaps providing an efficient, compact and future-proofed storage system. DNA, unlike cassettes, floppy discs or VHS tapes, will not go out of date. In 2013, Ewan Birney's group in Cambridge announced that they had written 739 kilobytes of computer data into DNA, using a code made out of groups of nucleotides.[12] They synthesised DNA containing this encoded information, sequenced it and reconstructed the original files – these included a text file containing all of Shakespeare's sonnets, a PDF version of Watson and Crick's description of the double helix, an MP3 extract of Martin Luther King's 'I have a dream' speech, and a JPEG image of the team's laboratory. There was not a single error, and all the files were functional.

The idea behind this proof of principle was to find a method that could guarantee future data storage in systems where huge amounts of data are being produced, such as at CERN, where the amount of data from the Large Hadron Collider and other experiments currently stands at more than 80 petabytes (1 petabyte (PB) = 1,000,000,000,000,000 bytes, or 10^{15} bytes) and is growing by at least 15 PB per year. At the moment, these data are stored on magnetic tape; DNA storage would be ideal for archiving data that rarely need to be accessed, in particular with the inevitable future decline in costs for reading and encoding DNA. Information storage in DNA is far more space-efficient than in other materials: a single 1-gram drop of DNA could store as much information as hard drives weighing around 150 kilograms. In November 2014, New York rock band OK Go announced that their new record, *Hungry Ghosts*, would be released on DNA as well as in the usual formats.[13] Although this was clearly a stunt, it may point the way to the future: DNA-based data cannot be read as quickly as a magnetic tape, but it can be safely stored for thousands of years if kept in the right conditions. One joker has facetiously suggested that the double helix would be particularly useful for storing the wave of genetic sequence data that is being generated at an exponential rate in laboratories all over the planet.*

Perhaps the most exciting recent development in DNA information storage has been the announcement by Fahim Farzadfard and Timothy Lu of MIT that they were able to engineer a population of bacterial cells to record real-time data as the cells were treated with a particular chemical inducer – or, as Farzadfard and Lu put it, they created 'genomic "tape recorders" for the analog and distributed recording of long-term event histories'.[14] This information was

_____

*I'm afraid it was me.

stored in a special form of DNA that only has one strand; it could be written and rewritten, and could be recovered after several days. This breakthrough brings the development of organic sensors, for both environmental and medical uses, much closer.

*

Although we generally describe the structure of DNA as a double helix, it is in fact a bit more complicated than this. Unlike a screw thread, which has a constant pitch or interval between each turn, the double helix has two different intervals, which alternate as the molecule spirals round. These are known as the major and minor grooves – the major groove, which is larger, tends to be the point at which DNA-binding proteins affect gene activity, as the sequence of bases is physically more accessible there. Above all, the DNA double helix spirals in only one direction – anticlockwise as seen from the top, or right-handed, like a normal screw. It is easy to get confused about which way the double helix should spiral, and in many representations of DNA the molecule spirals the wrong way. In 1996 Tom Schneider began posting images of leftward spiralling double helices on his web site, but he was soon overwhelmed by the number of examples.[15] I will not cast the first stone over mistaken representations of DNA – I once managed to put a wrong double helix on the side of a building: it was not even left-handed, it was geometrically impossible.

The DNA double helix comes in several shapes. The two forms studied by Wilkins and Franklin – A-DNA and B-DNA – both have right-handed helices; B-DNA is the iconic version that exists in your cells, and A-DNA occurs under conditions of low humidity and can be found in organisms although its biological function (if any) is unclear. In 1961, a group including Wilkins observed a third right-handed form of DNA, known as C-DNA, which appears in the presence of particular salts and has a slightly different structure again.[16] Left-handed DNA, known as Z-DNA, can be found in our cells. In one of the ironies of history, the first structure of a DNA molecule to be determined precisely was of the Z form, in 1979.[17] For complex chemical reasons, the Z form lacks the minor groove, and it turns in a looser helix than the B form – it has twelve base pairs per twist, whereas the B form has ten base pairs.[18] It is not simply an elegant left-handed version of the B form, but a kind of twisted Bizarro-DNA, with bases turned upside down relative to the B form, so the phosphate backbone of the molecule forms a zigzag rather than a smooth spiral.[19] The function of Z-DNA is still being explored – early hopes that it would turn out to be of importance in gene regulation or a useful tool in biotechnology have yet to be fulfilled.[20] DNA can also form other structures, including the four-stranded G-quadruplex and a cruciform shape; although it is assumed that these non-helical structures have some functional role, probably in regulating transcription, the evidence is still unclear.[21]

DNA is not the only molecule that can form a double helix. In 1961, Watson and Crick, together with Alexander Rich and David Davies, suggested that in certain circumstances, RNA, which is normally single-stranded, could double up.[22] Over half a century later, researchers were able to crystallise double stranded RNA and to describe its structure. It, too, has a right-handed spiral.[23] There is no evidence that the RNA double helix has any biological function in normal cells, but it may be possible that biotechnology will be able to employ this novel molecular structure. Although RNA is generally presented on diagrams as a single strand, in fact RNA molecules often bend around on themselves, forming complex double-stranded stems with a single-stranded loop at the top, like a hairpin. Such secondary structures may be important in the various functions of RNA – they give tRNA its distinctive shape, for example.

Both DNA and RNA are made of a common ribose-phosphate backbone, onto which are linked the bases (A, T, C and G) that carry the genetic information. In 2012 a collaborative project led by Philipp Holliger at Cambridge described the creation of six new kinds of informational molecule that did not use ribose, dramatically called xeno-nucleic acids or XNA. In place of ribose, these weird molecules each have different forms of sugar in their backbones onto which the usual bases can be attached. As well as creating these six forms of XNA, the group also engineered the enzymes necessary to enable DNA to be copied into XNA, and for XNA to be copied into DNA – this was a truly remarkable feat.[24] In a series of experiments the group demonstrated that genetic information (that is, the sequence of bases) could be successfully copied from DNA into XNA and back again. They were even able to subject one of the XNAs to selection and to show that its sequence evolved as a result. In principle, DNA and RNA are not the only potential informational molecules. Alien life-forms – if there are any – may well use non-DNA or non-RNA information.

The biotechnological potential for XNA is immense. As Holliger's group concluded:

> 'synthetic genetics' – that is, the exploration of the informational, structural, and catalytic potential of synthetic genetic polymers – should advance our understanding of the parameters of chemical information encoding and provide a source of ligands, catalysts, and nanostructures with tailormade chemistries for applications in biotechnology and medicine.

Commenting on the creation of XNA, the veteran biochemist Gerald Joyce recognised that it opened the road to what he called an alternative biology, but he also sounded a warning.[25] Use of DNA-based and RNA-based synthetic molecules carries a fail-safe mechanism in that they are susceptible to degradation by enzymes that have evolved over billions of years – indeed, this is one of the obstacles that restricts their widespread use. Furthermore, all DNA-based life-forms are susceptible to attack by other organisms and the enzymes they contain. This would not necessarily apply to XNAs. As Joyce put it:

> XNAs are unnatural and would pass through the biosphere unscathed. The benefits of their unusual chemical properties must be weighed against their greater cost, both literally and with regard to operating in the uncharted waters of XNA biochemistry. … Synthetic biologists are beginning to frolic on the worlds of alternative genetics but must not tread into areas that have the potential to harm our biology.

For the moment, no one has been able to create an informational molecule that does not contain phosphate. In December 2010 a twelve-person team including the NASA astrobiologist Felisa Wolfe-Simon published an article online in *Science*, suggesting that bacteria found in Mono Lake in California, which has high levels of arsenic, naturally replace the phosphorus in their DNA by arsenic.[26] This claim, which was announced at a high-profile NASA news briefing to the excitement of the world's press, was immediately contested on social media, such as Dr Rosie Redfield's blog RRResearch. What became known on Twitter as #arseniclife took on the proportions of a major scientific row and also showed the power of social media to act as a form of peer review. In 2011, when the original paper was finally published in *Science*, it was accompanied by an unprecedented seven short articles that were critical of the paper's claims. Redfield and others eventually published papers in *Science* that showed that, in this case, there was no integration of arsenic into DNA.[27] The original Wolfe-Simon paper has not been retracted, and it remains possible that, perhaps on another world or eventually in an Earth laboratory, other forms of informational molecule without phosphate may exist.

The #arseniclife debacle was partly driven by an idea put forward in 2005 by Carol Cleland and Shelley Copley of the University of Colorado, who published a speculative paper exploring the possibility that our planet hosts microorganisms that do not use DNA or RNA or our set of amino acids, and which are therefore undetectable by traditional methods such as the polymerase chain reaction.[28] This hypothesis, which has been given the dramatic names 'the shadow biosphere' or 'weird life', has attracted some interest from those prone to theoretical speculation, but has not been treated with any degree of seriousness by the scientific community. As Cleland and Copley put it, 'the fact that we have not discovered any alternative life forms cannot be taken as evidence that they do not exist'. That is logically correct, but it is hardly an enticing starting point for a research programme and would not be taken seriously by any funding agency.

Physicists have realised that 95 per cent of the Universe is made of stuff we cannot directly detect – dark matter and dark energy – because calculations of the amount of matter in the Universe based on gravitational effects revealed a substantial discrepancy between observed and expected values. For the shadow biosphere to be more than day-dreaming, a similarly overwhelming signature of its existence would be necessary. Those intrigued by the remote possibility of weird life have come up with some potential indicators of its existence, such as the supposedly anomalous varnish that is found on desert rocks.[29] However, something as impressive as the indirect evidence of the existence of dark matter and dark energy would be required for this hypothesis to be taken seriously.

*

The potential for synthetic biology is enormous. Scientists are already able to integrate unnatural amino acids into proteins, for example by manipulating enzymatic machinery associated with an 'amber' stop codon (UAG) that has been introduced into bacteria, yeast, the nematode worm *Caenorhabditis elegans* and even mammalian cells.[30] A synthetic form of transfer RNA is used to allow the UAG codon to code for an unnatural amino acid, thereby producing a novel protein. Some of these engineered proteins produce pulses of light when they are subjected to particular forms of biochemical activity in the cell, acting as an exquisitely sensitive marker. Some produce modifications in the three-dimensional structure of the protein, allowing greater understanding of the precise organisation of the molecule; others enable light to be used to activate molecules in the cell, giving an insight into the roles of key components of the cellular machinery. For the moment, many of these novel proteins are aimed at increasing our understanding of basic processes. But it is only a matter of time before they will be used to develop new forms of biotechnology with potentially massive implications for our future.

The ability to manipulate DNA has recently been extended to altering the genetic code itself. Although only two base pairs occur in nature (A binds with T and C binds with G), unnatural base pairs can be used to make novel forms of DNA in test tube reactions. More than twenty-five years ago, Steven Benner's research group extended the alphabet of the genetic code by introducing two new base pairs into DNA and RNA molecules – one pair is known as κ and Π, the other as iso-G and iso-C.[31] In 2011, Benner's group was able to amplify and sequence DNA containing the four usual bases and an unnatural base pair (Z and P), which they called GACTZP DNA.[32] For decades, the manipulation of base pairs has been used in some forms of everyday antiviral medicine, such as those regularly used by cold sore sufferers. The cold sore virus is persuaded to replace the G bases in its sequence by one of several proprietary molecules that cannot be copied by the infected cell's machinery, thereby blocking reproduction of the virus.[33]

In 2014, a group of researchers led by Floyd Romesberg of the Scripps Research Institute in California took a giant step towards the creation of a truly synthetic organism when they were able to make *Escherichia coli* bacteria copy a piece of DNA containing a sequence that involved an unnatural base pair (the two new bases go by the unfriendly names of d5SICS and dNaM), which had previously been created in a test tube.[34] Where the *E. coli* DNA replication machinery found a d5SICS in the artificial molecule, it inserted a dNaM on the new complementary strand, and vice versa. The bacteria seemed quite happy with this new alphabet, showing no serious problems, and the classic DNA repair pathways in the cell, which normally snip out and repair errors, did not make any move against the intruding pairs of bases.

For the moment, this remains at the level of a technical breakthrough – as the title of the paper put it, the researchers had created 'A semi-synthetic organism with an expanded genetic alphabet'. The two extra letters that were introduced into the DNA are currently mute; the expanded genetic alphabet does not yet form new words. But the authors made clear that their aim is to create a system in which unnatural base pairs will code for unnatural amino acids. The possibilities for synthetic biology – using living cells to produce new molecules – are almost endless.

It is inevitable that these astonishing developments in our ability to manipulate the essential elements of life will soon come together. Someone will eventually create a system in which XNA carries unnatural base pairs that code for unnatural amino acids that are assembled into bizarre proteins, and an utterly novel form of life, created entirely through human ingenuity, will be only a step away.

It does not seem too outlandish to imagine that by the end of the century, entirely synthetic life-forms will exist, able to produce drugs, foods and novel compounds at the service of humanity, perhaps in the most inhospitable of environments. Synthetic organisms able to survive in low levels of oxygen and cold temperatures might be one way in which we could terraform Mars, should it prove to be barren, and should we consider it ethical to destroy such a pristine environment. In this respect, as others, science and technology pose questions; they do not necessarily provide the answers.

*

Despite the optimism that surrounds synthetic biology and genetic engineering, ever since the first appearance of these techniques in the 1970s scientists have consistently expressed concern about the potential dangers. With the development of restriction enzymes (proteins that will snip a piece of DNA in two at a defined sequence) it became possible to create what is known as recombinant DNA – DNA from more than one organism, generally from two different species. This led scientists, including those involved in pioneering the approach, to be concerned that the introduction of new genes into organisms could have unforeseen consequences. They were particularly worried about what would happen if the organisms escaped and transferred their genes into the wild or if the new genes were inherently dangerous to humans or the ecosystem. This was not just an abstract concern: the era of genetic engineering was heralded by the introduction of *SV40*, a viral gene that can cause cancer in rodents, into the DNA of a bacteriophage virus, which was then used to transform *E. coli*.[35] Faced with the novelty of this technique, it was legitimate to worry that the transformed *E. coli* might end up inducing cancer in humans. The experiment, by Paul Berg, David Jackson and Robert Symons, was published in 1972. Eight years later, Berg won the Nobel Prize in Chemistry for this feat, together with Wally Gilbert and Fred Sanger, who were recognised for their work on DNA sequencing. Within a year of publication, Berg, along with other scientists, was arguing for a partial moratorium on recombinant DNA research because of the potential dangers.[36]

In February 1975, a conference took place at Asilomar, on the edge of Monterey Bay in California, to discuss the risks associated with the new technique and above all how to minimise the dangers. The conference, which included journalists and lawyers among the attendees, adopted a set of laboratory procedures, including strict containment facilities and biosecurity measures, which would enable research to continue safely. Many of these are still in force, but others have been abandoned as it has been realised that the dangers are far less than was originally feared.[37] In my own field, the study of behaviour in *Drosophila*, the introduction of DNA from other species into the fly's genome has become widespread in order to mark and manipulate tissues, enabling us to turn genes on and off, simply by allowing two flies to mate. The technique is perceived as entirely risk-free, and recombinant fly stocks, which may contain genes from yeast, jellyfish or bacteria, are routinely sent around the world by ordinary post and are used in ordinary laboratories, with no restrictive containment procedures.

However, genetic engineering can pose very real dangers. In June 2014, a group of US and Japanese scientists, led by Yoshihiro Kawaoka of the School of Veterinary Medicine at University of Wisconsin-Madison, attempted to recreate the Spanish Flu virus, which killed millions of people after the First World War.[38] As is well known, we are in danger of another global flu pandemic, with avian flu being the most likely source because it seems also to have been the source of the Spanish Flu. Kawaoka and his colleagues took bits of avian flu virus that were similar to the Spanish Flu infectious agent and put them together in a new DNA sequence, which proved to be highly infectious, just as the Spanish Flu virus was.

They justified their study by arguing that it would help identify which are the most dangerous parts of the viral genome and would therefore increase our preparation to meet any future outbreak. Although the US National Institute of Allergy and Infectious Diseases, which funded the study, defended the research both in terms of the information it provided about the potential dangers of newly emerging flu strains and the stringent biosecurity measures that were applied, there are clear dangers. The newly created virus could escape, or it could conceivably be used by a hypothetical group of bioterrorists, although this would require them to breach the stringent security procedures around such facilities and to be highly trained microbiologists.

Despite these very slim risks, researchers around the world were aghast at the news of the recreation of the Spanish Flu virus.[39] Lord May, a former president of the Royal Society, said:

> The work they are doing is absolutely crazy. The whole thing is exceedingly dangerous. Yes, there is a danger, but it's not arising from the viruses out there in the animals, it's arising from the labs of grossly ambitious people.

Perhaps the strongest reaction was from Simon Wain-Hobson, a virologist at the Institut Pasteur:

> It's madness, folly. It shows profound lack of respect for the collective decision-making process we've always shown in fighting infections. If society, the intelligent layperson, understood what was going on, they would say 'What the F are you doing?'

This was precisely the kind of response that Berg and his colleagues feared would become widespread with the development of the new technology, and which led them to propose first the moratorium and then the adoption of stringent biosecurity measures. In response to such concerns, in October 2014 the US government introduced a temporary moratorium on funding experiments that would increase the pathogenicity of viruses. This in turn met with criticism from researchers and pharmaceutical companies who argued that our ability to respond to future pandemics might be damaged by this policy.[40] Whatever the eventual outcome of this debate, the way in which this issue has been handled is in striking contrast to the self-regulation embodied by the Asilomar conference.

*

Berg's 1972 paper on genetic engineering in *E. coli* raised the possibility of altering humans suffering from genetic diseases, by introducing a correct copy of a faulty gene, in a process known as gene therapy (the term was coined before Berg's paper appeared).[41] These procedures are generally directed at the affected tissues, not the germ line (eggs and sperm), so they do not alter the genes that are passed on to the next generation – several European countries have banned germline gene therapy because of uncertainty about its long-term consequences. Gene therapy was first used in 1990, and interest grew after the launch of the Human Genome Project, although renewed doubts about the safety and effectiveness of the procedure surfaced in 1999 after the death of 18-year-old Jessie Gelsinger, a patient who had received treatment for liver disease. In recent years there has been a resurgence of interest in the technique, with hundreds of clinical trials of a range of therapies, including treatments of various forms of leukaemia and retinal disease and of Parkinson's disease, many of which have been successful. In 2012, the European Union licensed gene therapy as a treatment for a rare defect in fat metabolism.[42] Although the pipeline from concept to therapy is long, complex and expensive, the future looks promising. That is certainly the view of venture capitalists, who have begun pouring hundreds of millions of dollars into the field.[43]

One technique that is being widely touted as a game-changer for both science and medicine is a method for directly editing the genetic code, generally known as CRISPR. The technique takes its name from the full title of the enzyme that does the work, which goes by the mouthful of 'Clustered Regularly Interspaced Short Palindromic Repeats (CRISPR)-associated RNA-guided endonuclease Cas9'. There are several similar enzymes found in bacteria, where they serve the function of a defence molecule, attacking and chopping up bits of invading viruses. The bacterial genome contains short palindromic repeats of twenty-four to forty-eight base pairs, which are separated by other sequences of DNA of about the same length, called spacers. The enzyme is activated by RNA transcribed from stretches of spacer DNA, which correspond to the genetic code of an invading virus that the bacterial strain has encountered in the past and which the bacteria have incorporated into their genome as a kind of memory. When a virus enters the bacterial cell and tries to hijack the bacterium's machinery to reproduce itself, Cas9 (or a similar enzyme) is activated and attacks the virus, snipping out the bit of DNA that it recognises, thereby disabling the invader.

In 2012, Emmanuelle Charpentier and Jennifer Doudna, then based at Umeå University in Sweden and the Howard Hughes Medical Institute at Berkeley, announced that they had found out how to harness this system to change any DNA sequence.[44] Within a year, CRISPR was being used to genetically manipulate DNA from a wide range of organisms, including humans.[45] The principle is straightforward: the Cas9 enzyme is introduced into a cell along with a piece of synthetic RNA containing CRISPR sequences interspersed with a sequence from a gene that you are interested in rather than a bit of viral DNA. The Cas9 enzyme looks for that sequence, finds it in the genomic DNA of your organism, and snips it out. The gene of interest has either been disabled, or, if you combine CRISPR with other techniques, altered in some way. This approach is called directed mutation – targeting a particular gene in a predetermined way – and will apparently be available in virtually any organism; it is even possible to correct mistakes in the DNA sequence, such as occur in genetic diseases.[46] Although the technique is in its early days, it is clearly going to revolutionise scientific discovery and may lead to the development of new gene therapies.

CRISPR looks like it will be far more effective and flexible than the previous tool of choice, RNAi (RNA interference). RNAi is based on a naturally-occurring mechanism of gene regulation that is of fundamental importance in our cells, in which short strands of RNA that complement the mRNA from a particular gene, together with a complex of proteins, block the activity of the gene by binding to its mRNA. On hearing of the CRISPR breakthrough, Craig Mello, who with Andrew Fire won the 2006 Nobel Prize in Physiology or Medicine for the discovery of RNAi, described his reaction:

> CRISPR is absolutely huge. It's incredibly powerful and it has many applications, from agriculture to potential gene therapy in humans … It's one of those things that you have to see to believe. I read the scientific papers like everyone else but when I saw it working in my own lab, my jaw dropped. A total novice in my lab got it to work.[47]

If patent issues can be overcome – the Broad Institute, jointly run by MIT and Harvard, has successfully obtained a patent on CRISPR, and the two main inventors of the technique, Charpentier and Doudna, have also filed patents – then this technique could transform biology and medicine.[48] Already, Doudna has extended the power of CRISPR to be able to alter RNA, thereby enabling finely tuned detection and manipulation of mRNA.[49] Whatever happens next, I would bet that Charpentier and Doudna will eventually receive that telephone call from Stockholm.

A glimpse of the radical implications of CRISPR is given by the suggestion from a group of Harvard researchers that CRISPR could be used to potentially 'prevent the spread of disease, support agriculture by reversing pesticide and herbicide resistance in insects and weeds, and control damaging invasive species'.[50] None of the researchers were ecologists, but they sounded the alarm about potential side-effects, and simultaneously published a call for discussion about how to regulate the new technology, coming up with criteria that should be adhered to before the implementation of any such programme, and also identifying regulatory gaps that need to be filled by legislators around the globe.[51]

In January 2015, the same group of Harvard researchers came up with an ingenious technofix for ensuring that GMOs with potentially problematic modifications do not cause havoc in the environment – a group from Yale simultaneously published a similar report. Both groups used a 'genetically recoded organism' – a special strain of *E. coli* in which certain codons had been manipulated to code for synthetic amino acids that are not available in the environment. These synthetic amino acids are essential to the functioning of key proteins in these organisms, which are therefore effectively restricted to living in artificial conditions. Were the bacteria to escape, they would die. Furthermore, the authors claim that the alternative genetic code used by these organisms effectively prevents horizontal gene flow. This would seem to open the road to creating potentially hazardous GMOs in the knowledge that they would be contained by their engineered physiological requirements. However, a great deal of further work will be needed before this approach can be applied in the real world, and I suspect few scientists – or readers – would want to rely solely on this technique to ensure biosecurity.[52]

These responsible approaches to the potential impact of a new technique of unprecedented power are a direct descendant of the Asilomar conference on recombinant DNA that so successfully guided science as it was catapulted into the new world of genetic manipulation. In 2008, Paul Berg reflected on the impact of the Asilomar conference:

> In the 33 years since Asilomar, researchers around the world have carried out countless experiments with recombinant DNA without reported incident. Many of these experiments were inconceivable in 1975, yet as far as we know, none has been a hazard to public health. Moreover, the fear among scientists that artificially moving DNA among species would have profound effects on natural processes has substantially disappeared with the discovery that such exchanges occur in nature. ... That said, there is a lesson in Asilomar for all of science: the best way to respond to concerns created by emerging knowledge or early-stage technologies is for scientists from publicly-funded institutions to find common cause with the wider public about the best way to regulate – as early as possible. Once scientists from corporations begin to dominate the research enterprise, it will simply be too late.[53]

Faced with a future potentially populated by CRISPR-modulated DNA-based organisms and full of bizarre synthetic life-forms that use XNA and unnatural base pairs and can record what is happening to them in their genetic material, Berg's view, from a man who has looked at the question from both sides, is a salutary reminder for us all. His approach was to recognise the potential dangers and to find ways of countering them in conjunction with the public and regulators. The implication is that science is too important to be left to the corporations – or to the scientists.

FIFTEEN

ORIGINS AND MEANINGS

In May 1953, a week before Watson and Crick's second *Nature* paper introduced the world to the concept of genetic information, an article appeared in *Science*, signed by a 23-year-old PhD student, Stanley Miller.[1] Together with his supervisor, Harold Urey, Miller had attempted to discover how life might have begun on Earth. Using two connected flasks, they replicated the conditions of about 3.5 billion years ago: one flask represented the primitive ocean (sea water), the other represented Earth's early atmosphere, and contained hydrogen, ammonia and methane (oxygen appeared in large quantities much later, and reached modern levels only about 600 million years ago). Pulses of electricity were periodically sent through the apparatus to mimic the effect of lightning. To Miller's surprise, within a few days he could detect amino acids, in particular glycine. A simple chemical process, with no direct human guidance, had produced the components of a protein. Glycine has since been detected on a comet, showing that amino acids exist elsewhere in the Universe and could have been brought to Earth by comets shortly after the formation of the planet.[2]

Although the Miller–Urey experiment shows that amino acids can be formed relatively simply, it does not shed light on how life arose – we are more than just bags of amino acids. There are several scenarios for the origin of life – we do not know which is correct, and it is possible that we will never know. Here I will describe one hypothesis that is being explored by Nick Lane at University College London and Bill Martin at the Heinrich-Heine-Universität in Düsseldorf.[3] According to this view, the first replicating molecules appeared perhaps 4 billion years ago in the microscopic pores of rock around a deep-ocean hydrothermal vent.* Experimental evidence shows that such pores can act as a cell, containing and constraining molecular interactions, including the accumulation of nucleotides, and also allowing compounds to be exchanged with the outside world.[4]

Today every cell on the planet uses electrochemical gradients to move energy around and power its activities – known as proton gradients, they are also found around hydrothermal vents, where alkaline water bubbling up from under the sea bed meets acidic sea water. According to Lane and Martin, early life, which would just have consisted of a small number of types of replicating molecule, could have used these proton gradients to gain energy. Deep in the sea, and encased in rock, these molecules would also have been protected from the destructive effects of the powerful ultraviolet radiation that bombarded the surface of the planet at that time.[5]

Other scenarios are available. In his 1981 book *Life Itself*, Francis Crick put forward a theory he developed with Leslie Orgel, in which he argued that life was the result of what they called directed panspermia. Their surprising suggestion was that life on Earth originated with microorganisms that 'travelled in the head of an unmanned spaceship sent to earth by a higher civilisation which had developed elsewhere some billions of years ago.'[6] Aside from the distinct lack of proof, this does not explain the origin of life at all – it simply puts the problem back a long time ago in a galaxy far, far away.[7] It is possible that life originated elsewhere in the Universe and came to Earth on a meteorite or a comet. However, that hypothesis does not seem to be necessary – we seem to be within touching

---

*These first reactions may not have occurred near deep-sea vents, but instead in small vesicles made of fatty acids. This is the view of Jack Szostak, who has been able to create such an artificial protocell and get RNA to replicate spontaneously within it (Adamala and Szostak, 2013).

distance of understanding the chemical dynamics that created life spontaneously.

Proteins and DNA, which are so important to life today, have not always been present. The RNA machinery that exists in every cell of every organism on the planet, and the ability of RNA molecules to act as enzymes, catalysing biochemical reactions without the involvement of proteins, all indicate that another form of life existed before DNA-based life-forms: the RNA world.[8] Exactly what the first replicating molecules were, and how they made the transition from merely replicating to also interacting with the world and therefore truly becoming alive, we do not know – they may have been RNA molecules, or simpler compounds, such as peptides.[9] One essential feature of those early replicating systems would have been that they were able to speed up the chemical reactions that define life. The ability of molecules like RNA to act as enzymes and to catalyse reactions was discovered in the early 1980s by Sidney Altman and Thomas Cech, who won the 1989 Nobel Prize in Chemistry for their work. If left to their own devices, the kind of reactions that take place in our cells would need billions of years to occur spontaneously; in the presence of RNA they take a fraction of a second.[10]

At some point, perhaps after a period of evolution and competition between various biochemical types of life, the RNA world came into being.[11] There are no direct traces of this world, so our views are based on strong suppositions rather than physical evidence. This was a very different kind of life. In the RNA world, RNA molecules were the basis both for reproduction and for biochemical interaction. In a world without DNA or proteins, the genetic information contained in an RNA molecule coded simply for that piece of RNA. There was therefore no code, in terms of the genetic material containing a representation of another molecule – the earliest RNA genes coded themselves and that was it. Reproduction involved the copying of RNA molecules that acted as enzymes to direct chemical reactions. These RNA molecules provided the raw material for natural selection to begin its long work of sifting between variants, eventually leading to the DNA-based life that now covers the planet.

The idea of the RNA world seems to have first been put forward by Oswald Avery's colleague, Rollin Hotchkiss, at a symposium organised by the New York Academy of Sciences in 1957. Struck by the fact that some viruses use RNA and others use DNA, Hotchkiss suggested that

> as a genetic determinant, RNA was replaced during biochemical evolution by the more molecularly and metabolically stable DNA. Cell lines have preserved the RNA entities which, evolutionwise, were primary to DNA and may have allowed them to store their information in DNA and thereby become subservient to it metabolically.[12]

For many years it was difficult to see how RNA could have appeared spontaneously, because the biosynthetic pathways involved in its creation seemed to be too complex. But in 2009, John Sutherland's group, then at the University of Manchester, showed that the RNA pyrimidines (U and C nucleotides) could appear through a relatively simple series of reactions, using as their starting point the kind of chemicals that could have been floating about in early Earth conditions.[13] We are getting closer to understanding how life might have appeared spontaneously. Already, researchers have been able to create artificial systems in which pairs of short RNA enzymes can grow and evolve in a self-sustained manner, each catalysing the growth of the other.[14]

Although the RNA world no longer exists (but who knows what secrets lurk in the deep ocean?), we all carry its legacy within our cells. When our DNA-based life appeared, evolution did not redesign life from scratch: it used what was to hand, adapting existing RNA biochemical pathways and turning them into something new and strange. This explains why RNA is not simply a passive messenger between the two apparently fundamental components of life – DNA and proteins. It plays many roles, shuttling genetic information around the cell and shaping how it is expressed, just as it did in the RNA world. As the RNA biochemist Michael Yarus has put it: 'Without RNA, a cell would be all archive and no action.'[15]

RNA is involved in almost all of the cell's machinery for getting the genetic information out of DNA and either creating proteins or controlling the activity of genes. In its many forms, RNA performs essential functions within the cell, even if it has lost its role as the embodiment of genetic information, replaced by the semi-inert double helix of DNA. The double helix – iconic, rigid and fixed – contrasts with the many physical forms that RNA can take, enabling it to carry out such a wide range of functions, which would have been such an important feature of the RNA world.

Just as we do not know when the RNA world appeared, so we also do not know when it finally disappeared. All we can do is trace the ancestry of modern, DNA-based organisms back to the Last Universal Common Ancestor (LUCA), a population of single-celled DNA organisms that lived perhaps 3.8 billion years ago. LUCA evolved out of the RNA world, eventually – perhaps rapidly – outcompeting and replacing it.

The replacement of RNA as the repository of genetic information by its more stable cousin, DNA, provided a more reliable way of transmitting information down the generations. This explains why DNA uses thymine (T) as one of its four informational bases, whereas RNA uses uracil (U) in its place. The problem is that cytosine (C), one of the two other bases, can easily turn into U, through a simple reaction called deamination. This takes place spontaneously dozens of times a day in each of your cells but is easily corrected by cellular machinery because, in DNA, U is meaningless. However, in RNA such a change would be significant – the cell would not be able to tell the difference between a U that was supposed to be there and needed to be acted upon, and a U that was a spontaneous mutation from C and needed to be corrected. This does not cause your cells any difficulty, because most RNA is so transient that it does not have time to mutate – in the case of messenger RNA it is copied from DNA immediately before being used. Thymine is much more stable and does not spontaneously change so easily. The adoption of DNA as the genetic material, with its built-in error-correction mechanism in the shape of the two complementary strands in the double helix, and the use of thymine in the sequence, provided a more reliable information store and slowed the rate of potentially damaging mutations.

These kind of mechanisms also provide one answer to Schrödinger's concern about how gene-molecules are able to remain apparently constant down the generations, despite the existence of quantum effects that should alter their structure. Life is even stranger than Schrödinger imagined: it has evolved ways of checking the stability of genetic information and of reducing errors.

The new DNA life-forms would have had a substantial advantage because they involved proteins in all their cellular activities. Although we do not know when or why protein synthesis developed, it seems unlikely that it occurred instantaneously – there was probably no protein revolution.[16] Instead, amino acids present in the primitive cell would have interacted spontaneously with pieces of RNA that were acting as enzymes, with small nucleotide sequences binding with a few amino acids – this was the first glimmer of the genetic code. Initially the interaction of RNA and amino acids would have enabled RNA life-forms to gain some additional metabolic property, before eventually the appearance of strings of amino acids – proteins – created the world of protein-based life. At some point DNA supplanted RNA as the informational molecule, keeping the genetic sequence safe, using RNA to produce rapid translations of that sequence into the patterned production of proteins, as the RNA enzymes were co-opted and turned into bits of cellular machinery such as transfer RNAs and ribosomes.

Proteins can carry out an almost infinite range of biological functions, both as structural components and as enzymes. In both respects, they far surpass RNA. The appearance of proteins therefore opened new niches to life, spreading DNA and protein across the planet, creating and continually altering the biosphere. These new DNA-based life-forms would have out-competed the RNA world organisms in terms of their flexibility and the range of niches they could occupy. They would also have been able to grow much more quickly: a modern DNA-based cell can replicate itself in about 20 minutes. Experiments on RNA enzymes involved in replication suggest that it would have taken days for an RNA-based life-form to reproduce.[17] The RNA world was slow, limited and probably confined to the ocean depths.

The evolutionary and ecological advantages gained through the use of proteins by DNA-based life show that the appearance of translation from a sequence of RNA bases into a sequence of amino acids was a decisive evolutionary step. The evolution of the genetic code was therefore essential for life as we know it. It truly is life's greatest secret. This raises the obvious questions of how the code evolved and why it is the way it is. There is a simple but frustrating answer to both these questions: we do not know.

*

In December 1966, shortly before the final word in the genetic code was read, Francis Crick explored the origin of the code. Speaking at a meeting of the British Biophysical Society in London, he developed ideas that still dominate scientists' thinking about this difficult question.[18] Crick's first suggestion, which was also being explored by Carl Woese and Leslie Orgel, was that there is a physical link between each codon and the amino acid it codes for, and that the code is therefore in some way inevitable.[19] This assumption lay behind many of the theoretical attempts to break the code, and could trace its intellectual origins right back to Gamow's letter to Watson and Crick in 1953. However, Crick was unable to fully explain the code in these terms, and there is still no physical explanation of the relation between all RNA codons and their amino acids.

Part of the problem is that there is obviously something going on – the distribution of codons and amino acids is clearly not random. As Crick pointed out, in many cases the final base in an RNA codon is irrelevant if the first two bases are alike: XYU and XYC always code for the same amino acid, and XYA and XYG often do so. In half of the cases it does not matter what base follows XY – all those combinations code for the same amino acid. There seems to be some link with the physical nature of the amino acid: for example, if the second base in the codon is a U, then the amino acid has a particular chemical characteristic called hydrophobicity, and the more acidic and the more alkaline groups of amino acids each have similar codons.

These tantalising patterns have led several scientists, including Crick, to suggest that at first the code would have enabled the protocell to process a relatively small number of amino acids, on the basis of the physicochemical interactions between RNA molecules and amino acids, and indeed there is some evidence for this.[20] These would initially have been coded by one or two bases alone, but very quickly the triplet code was established, with the number of amino acids subsequently being expanded to the current twenty. Unable to see any coherence in the code, Crick described its modern structure as a frozen accident.[21] He argued that the Last Universal Common Ancestor of all existing life just happened to use the current system of translation from DNA to protein, and that it has stuck because any deviation from the universal code would be disadvantageous. Despite the fact that we know that minor variations of the code are possible, this intellectually unsatisfactory explanation remains as a rather limp conclusion to most explanations of the origin of the modern genetic code.

Attempts to explain why the code is the way it is are generally divided into three types: Crick's physicochemical hypothesis, which seeks to explain the code in terms of the links between codons and amino acids; the co-evolution hypothesis, which suggests that the pattern of codon assignment reflects the evolution and expansion of the code in terms of function; and the adaptive hypothesis, which sees the code as a reflection of processes that reduce the number of errors. This final suggestion was first set out by Stephen Freeland and Laurence Hurst in 1998. They compared the actual genetic code with all the possible alternatives and found that in terms of the errors that could occur in the chemical binding between a codon and an amino acid, only one alternative in a million outperformed the genetic code we currently use.[22]

In the 1950s, theoreticians assumed that there must be some physicochemical explanation of the link between a DNA codon and its amino acid. Once it was realised that proteins were assembled with the help of RNA, it was the RNA codon that became the focus. But this is not correct, either: in fact, the amino acid does not actually attach to a codon at all. It fits onto the open end of the tRNA molecule, which is a long loop in the shape of a clover leaf. On the opposite side of the tRNA molecule sits the anticodon, which is what the mRNA molecule recognises because it is composed of three complementary bases. There is no known link between the amino acid attachment site on the tRNA and the anticodon, which are on separate sides of the tRNA molecule. All those theoretical explanations were barking up the wrong tree.

In 2005, RNA biochemist Michael Yarus and his colleagues reviewed various explanations for the origin of the code, together with the evidence for them, and attempted to integrate all three types of hypothesis.[23] They suggested that the initial allocation of codons was based on physicochemical relations between a few RNA enzymes (either tRNA molecules or their predecessors) and the small number of amino acids they processed. These primitive molecules were then subjected to natural selection to optimise their ability to recognise different amino acids, and these were then selected to minimise error in their recognition by mRNA, leading to the current code. Although this is an attractive compromise, it has by no means settled the argument. Despite fifty years of research on the origin and evolution of the genetic code, there is still no consensus as to which of these hypotheses, or which combination of them, is correct. A recent review of the topic gloomily predicted that the answer – if there is one – might elude us for another half-century.[24]

*

Francis Crick's starting point in his 1957 lecture on protein synthesis, which contained his description of the central dogma, was what he called the sequence hypothesis: the sequence of nucleotides on the DNA or RNA molecules enables the cell to produce a corresponding sequence of amino acids in a protein, or nucleotides in another molecule of nucleic acid. That is all the genetic code is. Crick could see no need for any other explanation of the way in which proteins fold, and, with the exception of the protective role of chaperones, this seems to be the case. Nonetheless, there have been repeated suggestions that the genetic code might contain more than sequence information – there may be a code within the code. This in turn may shed light on the origin of the particular version of the code that all life now uses.

The simplest example of this kind of hidden genetic information is to be found in the pattern of codon usage. For those amino acids that are coded by more than one codon, the frequency with which the alternative codons are used is not equal. For example, leucine can be encoded by six DNA codons – TTA, TTG, CTT, CTC, CTA and CTG. In human genes, these six codons, each of which do the same thing, are found at varying frequencies: CTA makes up 0.7 per cent of your DNA, whereas CTG is 4.1 per cent of your DNA. Broadly similar results are found for the mouse and for *Drosophila*, however in yeast CTA and CTG make up around 1.3 and 1.0 per cent of the DNA, respectively, whereas TTG is the most frequently found leucine codon, at 2.7 per cent.

For the moment there is no agreement about why this effect – called codon bias – exists, nor what it tells us about evolution. It seems to be related to selection, in that codon bias is more easily detected in genes that are highly expressed. Among the factors that may produce codon bias are the possibility of mutation leading to a switch between the various redundant codons, the number of genes coding for the various tRNAs in a given species, and selection pressure to use one form of codon rather than another to avoid potential errors. Genome-wide analyses of codon bias in twelve *Drosophila* species have shown that codon bias can even extend across codons: the most frequent pairs of codons in these flies are XXG-CXX (so any codon ending in G, followed by any codon starting in C), whereas the least frequent are XXT-TXX.[25] This is telling us something, although it is not clear what. A similar effect has been observed in yeast, where there is a tendency for the codons using the same tRNA to follow each other in the sequence, perhaps because tRNA molecules that have released their amino acid diffuse away from the ribosome at a slower rate than that with which other copies of these tRNAs are recruited by the translation process. To avoid a molecular traffic-jam, natural selection may have tended to favour the sequential involvement of the same tRNAs, leaving a signature in the genome.[26]

There is even information in the frequency that the four bases occur in our genomes. In humans, the GC pair of bases (guanine on one DNA strand, cytosine on the complementary DNA strand) is not found at the same frequency as the other pair of bases (AT – adenine and thymine). Again, the reasons for these effects are not known: longer genes tend to have a higher proportion of GC than AT pairs, and there are substantial differences between species – for example, in large stretches of mammalian genomes, the proportion of GC pairs varies between 35 per cent and 55 per cent (under random variation you would expect both types of pair to be at 50 per cent).[27] Furthermore, GC pairs tend to be more frequent in some parts of chromosomes than others – these GC-rich zones, called iso-chores, have been known about for more than forty years, but there is still no agreement about their origins or significance.[28] Although in mammals there are links between GC content and both body mass and genome size, many researchers argue that the effect is not due to selection and is instead produced by neutral changes that flow from gene duplication and mutations in non-selected parts of the genome.[29] This apparent code within the code may contain nothing except noise.

The boldest suggestion that there is more than a sequence code in our DNA emerged at the end of 2013 from the University of Washington, after the publication of a paper in *Science*.[30] The researchers claimed that they had identified a second layer of information in the human genetic code, overlying the sixty-four triplet codons. Around 14 per cent of the normal codons both specified amino acids and allowed transcription factors that control gene regulation to bind to the sequence.[31] Despite the high-profile publication and the claims of the university's communications agency, this effect had been described a few years earlier in a wide range of organisms, including mammals.[32] The researchers suggested that mutations in the codons that also act as binding sites for transcription factors could lead to genetic diseases. However, they did not show any consequences (good or bad) of the existence of these binding sites, nor did they show that these sites actually affected the regulation of even one gene.

The paper provoked irritation on social media, mainly because of the overblown claims about a phenomenon that had already been described.[33] The research was part of the ENCODE project, which in 2012 claimed that most of our genomes are functional because detectable levels of biochemical activity are associated with virtually all of our DNA, even though the biological significance of that activity was not clear. These dramatic claims of a second genetic code may be a consequence of the criteria initially used by the ENCODE consortium to interpret its data. The findings may turn out to be correct, but it will require experimentation to prove that such a high proportion of our genome is the focus of gene regulation, and a great deal of work will be needed to convince the scientific community that this is indeed an example of a code within the code.

We now know that our genomes contain information that enables the cell to process the DNA sequence in various ways, most of which are related to gene regulation.[34] The simplest form of this additional information can be found in the untranslated regions upstream and downstream of the gene, which appear in mature mRNA and help direct gene expression. The region at the end of the gene consists of a long sequence of adenine bases called a poly(A) tail, which can be up to 200 bases long and is involved in the stability of the mRNA molecule. The beginning of the mRNA molecule has a chemical 'cap' and a series of bases that control how the mRNA is processed by the cell.[35] None of these forms of information are a systematic code like the genetic code. Called auxiliary or complementary genetic information by some researchers, this form of information, which is dispersed throughout our genomes, is more like a set of additional, particular and precise instructions, which has still to reveal all its secrets.[36] Rather than an alternative code of life, it is instead a sign of our deep evolutionary history, revealing ways in which our far-distant ancestors discovered new ways of manipulating genes and their outputs.[37] It helps shape our DNA into something more than simply a set of codons that produce sequences of proteins or nucleic acids: it reveals our genome as a palimpsest, overlaid with other forms of information that do not obscure or invalidate the original sequence-encoding signal but enrich our understanding of our present and of our past.

*

Ever since 1953, when Watson and Crick wrote those apparently simple words 'the precise sequence of the bases is the code which carries the genetical information', biologists have considered that the idea that genes contain information is intuitively obvious. Philosophers have not been so easily convinced, and over the past two decades a debate about genetic information has taken place, away from the gaze of biologists. The main issue that has preoccupied the philosophers is the exact nature of the kind of information that is in genes, and, indeed, whether there is something there that can strictly be called information. The fact that most scientists are unaware of these arguments is due partly to the divisions between academic disciplines and partly, I suspect, to the fact that many of my colleagues take a dim view of philosophy. This is unfortunate, because one of the jobs of philosophers is to explore the complexity that lurks in apparently straightforward concepts such as information. Indeed, it is possible that had philosophers paid more attention to the issue in the 1950s, they might have been able to persuade the theoreticians not to view the code literally as a code or as a language, and less time might have been wasted on fruitless speculation.

Many scientists would probably agree with Michael Apter and Lewis Wolpert, who argued in 1965 that genetic information is simply a metaphor or an analogy, a way of describing what genes contain and how they exert their effects.[38] Apter and Wolpert claimed that the most thorough definition of information, as described in Shannon's communication theory, does not apply to genetic information because the whole point of genetic information is that it does something, it has a function, a meaning, whereas Shannon's view of information has no place for meaning. The difficulty involved in expressing the content of DNA in Shannon's terms can be shown by trying to calculate the information content of a genome. The problems begin at the beginning: it is unclear whether the fundamental unit should be a single base, with four alternative states (and therefore two bits of information), or a codon – three bases, with sixty-four alternative states (and therefore eight bits of information) – or the output of the system, with twenty-one alternative states (twenty amino acids and 'stop'), and therefore five bits of information. Calculations based on each of these approaches would produce different answers, and in all cases it is not clear what the outcome would mean.

Two other possibilities highlight the problem associated with using Shannon's measure of information on data from molecular genetics. First, imagine two stretches of DNA of identical lengths, containing the same proportions of the four bases but in different orders. According to Shannon, the information content of those stretches of DNA, if calculated using each base, would be identical, and yet they would almost certainly have differing gene products that would affect the fitness of the organism in various ways – the biological content of their information would not be alike. Second, there is no agreed answer as to whether most of the DNA sequences in our genome, which have no apparent function and do not seem to be subject to natural selection, contain information or not. Most biologists would probably say not, because they would link information with function, whereas a mathematician would probably argue that they do. Although from Shannon's point of view a sequence of junk DNA contains as much information as a sequence of codons from a protein-encoding gene, that is clearly not the case from the point of view of the cell, the organism or natural selection. Despite these obstacles, some scientists and philosophers continue to claim that DNA does contain Shannon information and have applied information theory to data from molecular genetics.[39] None of these attempts has yet convinced the scientific community as a whole.

Towards the end of his life, the theoretical population geneticist John Maynard Smith (1920–2004) began to explore the role of information in biology. In 1997 he wrote a book with Eörs Szathmáry entitled *The Major Transitions in Evolution*, in which they described the evolution of life as a set of changes in the way in which information is stored and transmitted.[40] For example, the evolution of multicellular organisms altered how information is transmitted and stored, with the appearance of differentiation between cells, underpinned by spatially and temporally modulated gene regulation. The most recent evolution of an information transfer system is the one we are using at this very moment – the appearance of language in humans.

In 2000, Maynard Smith wrote the first of a series of articles in which he explored the nature of genetic information and exchanged views with philosophers of biology.[41] Maynard Smith put evolution at the heart of his view of genetic information and where it comes from: 'DNA contains information that has been programmed by natural selection', he stated, and as a consequence the quantity and quality of genetic information has increased over the past 3.8 billion years.[42] From this point of view, natural selection is the coder that has given the DNA sequence meaning: 'genomic information is "meaningful" in that it generates an organism able to survive in the environment in which selection has acted', he wrote.[43] In other words, genes provide the cell with instructions that have been encoded through natural selection: that is the nature of genetic information.

Genetic information is not like the effect of the environment – most scientists and philosophers consider that environmental factors, although they have shaped genetic information through natural selection and form the conditions that allow genes to be expressed, do not themselves contain information (some philosophers disagree).[44] For example, although changes in temperature can alter the expression of sex-determining genes in crocodiles thereby changing the sex-ratio of a population, the meaning of increased temperature is not the production of more male crocodiles. 'It is for this reason that we speak of genes carrying information during development, and of environmental fluctuations not doing so', argued Maynard Smith.[45]

For Ulrich Stegmann, a biologist turned philosopher at the University of Aberdeen, DNA contains information that is conditionally expressed. One way of thinking about this is that DNA sequences act a bit like a recipe. Protein synthesis proceeds in a step-by-step fashion, where each step depends on an external factor (codons in DNA and then in mRNA), in the same way that a recipe determines the order in which a cook puts together the ingredients and uses the utensils.[46] The idea of a gene as a computer program is another popular metaphor, according to which the program responds to input conditions in various ways and, depending on those inputs, produces various consistent outputs.[47] However, these are only metaphors. Genes are not programs or recipes, and organisms are not computers or cakes.

The first systematic critic of the concept of genetic information was the philosopher of biology Sahotra Sarkar, of the University of Texas at Austin. For Sarkar, like Wolpert and Apter in 1965, genetic information is 'little more than a metaphor that masquerades as a theoretical concept'.[48] Sarkar's critique rests partly on the fact that in eukaryotes, with their complex system of gene splicing, the DNA sequence does not correspond to the amino acid sequence. Strictly speaking, our genes therefore do not correspond to Crick's definition of genetic information, because the DNA sequence has to be processed and mediated before it appears as an amino acid sequence. Sarkar also points out that genetic information differs from artificial codes because it is impossible to back-translate a protein sequence into a DNA sequence, owing to the redundancy of the genetic code, the presence of introns in eukaryotes and the existing of multiple splicing. For Sarkar, genetic information therefore fails what he calls the test of reverse differential specificity, and, he argues, the concept has ceased to be a useful tool for discovery.[49] To my mind, Sarkar's critique does not invalidate the use of the term information when discussing the content of genes. Instead, it underlines that genetic information is not like other kinds of information. Neither does this critique undermine the existence of a genetic code: a particular codon will produce a particular amino acid – the triplet of bases represents and encodes that amino acid. That is a code. The fact that you cannot reliably back-translate from amino acid into DNA may disqualify the use of the word code for a philosopher, but it does not for a scientist, or for a member of the public.

As the philosopher Peter Godfrey-Smith has pointed out, part of the problem flows from the fact that the meaning of the word code as used to describe the content of genes is not strictly identical to the word code as used in other contexts (Godfrey-Smith nevertheless thinks that it is legitimate to use the term code in molecular genetics).[50] The genetic code is not an artificially designed system, it is a phrase that describes the sixty-four ways in which a part of one molecule (messenger RNA) binds with part of another (a tRNA), which in turn binds with another (an amino acid), the detail of which can only be fully understood in an evolutionary context. Sarkar put it pithily: 'DNA is, ultimately, a molecule and not a language'.[51] DNA is a replicating molecule that, in the right context, leads to the production of certain chemical sequences through the information it contains.

Despite these philosophical clarifications, at first glance the genetic code does indeed look like an artificial code, and the initial assumption was that it therefore came with the associated baggage of such an artefact, such as strictly logical rules and the ability to back-translate. This apparent similarity between the genetic code and artificial codes beguiled many scientists in the 1950s as they tried to crack the code using mathematical principles. Interpreting the genetic code in terms of precise analogies, strict definitions and exact parallels to artificial systems will almost certainly fail, because the genetic code, like every other aspect of biology, has not been designed. It is part of life, and has evolved. It can be properly understood only in its historical, biological context. That was the lesson of the doomed attempts to break the code in the 1950s, and it should guide us today in trying to understand what is in our genes.

For some philosophers, describing the content of genes as information suggests that DNA determines all the characters of an organism in an absolute and unmediated fashion. This critique is misplaced, because in reality few, if any, scientists hold such extreme views. There is a rule of thumb in reading popular science reporting (or, indeed, a scientific paper): if an article describes 'the gene for' something, you are almost certainly reading an over-simplistic account. Genes rarely do just one thing; even if a gene produces only one kind of protein, that protein can have different consequences in different contexts.

The gene that got me interested in studying the effects of genes on behaviour, back in 1976, was a *Drosophila* gene called *dunce* that was identified in Seymour Benzer's lab – flies with a mutation in this gene show defects in learning and memory.[52] *Dunce* might seem to be a gene 'for' learning or memory, and it primarily codes for an enzyme that affects the level of an intracellular signalling molecule called cAMP, which has been implicated in learning in a wide range of organisms. But through multiple splicing *dunce* can produce seventeen separate proteins, varying in length from 521 to 1,209 amino acids. Mutations in this gene can affect a wide range of characters apart from learning and memory, including female fertility and the insect's responses to organophosphates.[53] In the light of this knowledge, what exactly *dunce* is 'for' escapes easy definition. Although we know what it does under some circumstances, and what happens when specific parts of the gene are mutated, that does not mean that the gene has a single function. And remember, *dunce* is nothing special, it is just one gene out of billions that exist throughout nature.

Many of those philosophers who criticise the idea that genes contain information rightly point out that DNA can do nothing on its own, emphasising the role that proteins play in life.[54] This is hardly a major criticism – it is true of all representations, codes or languages. The printed symbols that you are looking at represent words and ultimately concepts that I have encoded onto paper, but they mean nothing until they are read. That does not stop them from being part of a language, and does not undermine their fundamental importance in communication. As to the essential role of proteins, Crick said basically the same thing in his 1957 lecture:

> the main function of the genetic material is to control (not necessarily directly) the synthesis of proteins. ... Once the central and unique role of proteins is admitted there seems little point in genes doing anything else.[55]

Some of these critics argue that DNA is merely one of many factors, including the environment, that equally determine the life-cycles of organisms – this is called the parity thesis.[56] There are a handful of scientists who agree with this extreme position and argue that proteins, the environment, or the cell's metabolism, play a role that is equal to, or greater than, DNA in determining the characteristics of organisms.[57] These scientists remain in a very small minority, because the overwhelming evidence is against their view. It does not correspond to what happens in our laboratories, where DNA is manipulated, altered and transferred according to gene-centred experimental protocols, and where the expected outcome occurs. When students in my laboratory take genes from three separate organisms and combine them, using a regulatory gene from yeast to drive the expression of a jellyfish gene that encodes fluorescent protein so that a single cell in a maggot's nose glows, the determining causal factor is the genes. The environment, the cell, the maggot, and the ingenious humans who designed the experiment are all permissive factors that had to be in the correct state for the genes to produce their desired effect, but the contribution of these peripheral conditions to the outcome is qualitatively unlike the contribution of the DNA. In this case, the genes function exactly as if they contained information that determines the outcome, because they do.

Although philosophers tend to be interested in the majority of cases where genes are not destiny, it is worth remembering that in some situations they most definitely are. If you have two copies of the sickling version of the haemoglobin gene, you will suffer from terrible anaemia and other debilitating symptoms. Nothing in your environment or upbringing seems to be able to alter that. Even more tragically, if you carry a single copy of the *Huntingtin* gene containing a CAG trinucleotide that is repeated potentially sixty times over, then you will eventually suffer from Huntington's disease, a neurodegenerative disorder. This genetic disease shows varying symptoms in different individuals, partly as a result of differences in the number of CAG repeats, but it is always fatal.[58]

In many cases, however, genes are not the ultimate determiner or cause of biological phenomena. In the example of sex determination in crocodiles given earlier, the genes are constant, and the proportion of male and female crocodiles is determined not directly by the genes but by the way in which the temperature affects the activity of those genes. In that case, the decisive causal factor is temperature, but it does not act alone. Temperature exerts its effects by altering the activity of sex-determining genes, through the production of proteins and RNA molecules that are themselves the product of other genes. Genes need cells, which they create, to realise the conditional instructions that they contain, and the environment has to be permissive. However, in similar conditions, similar effects will tend to be produced. The way in which those effects percolate out into the anatomy, physiology and behaviour of a whole organism can be unpredictable, making it hard to draw a direct line between a particular gene and a particular character.

The behaviour geneticists Doug Wahlsten and John Crabbe explored this problem in 1999 when they got separate laboratories to carry out the same behavioural experiments on the same inbred strains of mice. There were systematic differences in the behaviour of the mice in different laboratories, indicating that the route from gene to behaviour depends on many complex factors, including the experimental set-up and the immediate environment.[59] That does not mean that it is impossible to test reliably for genetic effects on behaviour: in 2006, Wahlsten and Crabbe reported that inbred mice strains can show very high levels of behavioural consistency over

time (for example in locomotor activity or in preference for etha-
nol), even when the experiments were conducted with a gap of fifty
years.[60]

These results are not particularly surprising to anyone who
has done an experiment on the genetics of behaviour. Organisms
are not robots, and their continual interaction with the environment throughout their development and during the experiment creates genetic, physiological and behavioural noise that can affect the
results. That does not mean to say that genes are not involved in
determining anatomy, physiology and behaviour; it simply means
that it is sometimes extremely hard to study these effects.

Attempts to detect genetic factors underlying intelligence have
proved particularly problematic. There are clear genetic effects on
cognitive ability: no chimpanzee will ever be able to act, speak and
think like the average human. That flows from the relatively small
differences in our DNA – our genes produce two species with different levels of intellectual ability. The problems begin when it comes to
studying the differences in intelligence (whatever that might be) that
can be observed between humans: pinning down what part is due to
our slightly different sets of genes is very difficult. In 2014, a study of
more than 100,000 people sought to correlate genetic variability with
variations in cognitive ability and educational attainment.[61] The
authors found just three genetic variations across the whole genome
that might be implicated in the cognitive differences they were measuring, and these all had extremely small effects. There are undoubtedly genetic differences between humans that affect our intelligence,
but it seems probable there are very many such genetic factors, each
contributing a tiny amount, with any individual having a mixture
of a wide range of these genes. The lesson of such studies is that if
the character that is being investigated is largely determined by the
environment, as seems reasonable to imagine is the case for educational attainment, then it will be difficult to detect genetic effects.

If it turns out that there are important genetic factors underlying individual differences in human cognitive ability, the challenge
would be for society to decide how to use – or not – that information.
However, I would be very surprised if this were the case. The fact that
genes contain information that determines the sequence of nucleic

acids and proteins does not imply that all characters are genetically determined. Some are, many are not. Biology is complicated.

*

Describing the content of genes as information, and viewing the activity of cells and organisms as involving the movement of information, puts all levels of life into a single framework. As Crick put it, life is characterised by the flow of energy, the flow of matter and the flow of information. Information flow involves the activity of specific molecules and, at the level of a whole organism, of cells or groups of cells. Conceptualising the whole process as having an underlying unity in terms of information provides a context that helps explain how molecules, cells, organs and organisms interact and are coordinated.

This reflects one of the central conceptual approaches in the history of the genetic code and of gene function, the cybernetic vision. Cybernetics – the study of control and apparently purposive behaviour in animals and machines – exerted tremendous influence in the late 1940s and throughout the 1950s, because it appeared that it would form a new science, providing a way of uniting all levels of biology with engineering and mathematics. That did not happen, and the tide of enthusiasm for cybernetics gradually ebbed when it became evident that, beyond its emphasis on control and the existence of negative feedback loops to produce apparently purposive behaviour, cybernetics did not provide a predictive framework for future discoveries.

Nonetheless, cybernetics was important in helping Jacob and Monod understand the data from their operon experiments and thereby contributed to our understanding of gene regulation. In 1970, Monod attempted to explain the organisation of living systems in a book entitled *Chance and Necessity*. Even though Monod was writing after the fashion for cybernetics had begun to wane, he still argued that organisms were quite literally cybernetic structures, consisting of patterns of control and feedback that were embodied by the action of specific molecules. He also argued that the components of many cellular molecular networks interact in a way that is based on information, not on chemical structure. For example, when enzymes are induced by the presence of their substrate, this does not occur through a direct chemical link but through the activity of an intermediary protein and the gene that codes for it. The only way of understanding this process is as a flow of information that passes through each component, taking different physical forms as it goes.[62]

Modern analytical techniques enable scientists to understand such biochemical interactions in exquisite detail. This has given rise to the field known as systems biology, which studies patterns of chemical interaction and gene regulation. Some claim that systems biology will embrace all the levels of life, right up to the ecosystem.[63] For the moment, research under the systems biology label focuses on the chemical processes taking place in the cell. The vast data sets produced by such analyses, and the ability of modern computers to process and model those data, have inevitably led to a resurgence of interest in cybernetic approaches to biochemical processes, with a focus on the importance of feedback.[64] However, despite the confidence of both Jacob and Monod that cybernetics would provide a way of understanding how the genetic code turns into instructions, the influence of cybernetics on modern science remains at the level of broad effects rather than any precise detail. That is even true in the field of neurobiology: although neural networks are clearly processing information and control, for most students and scientists cybernetics is a dimly remembered ancestor, rather than an essential part of their experimental approach.[65]

Science, like other parts of human culture, can be influenced by fashion, and by our apparently endless appetite for novelty. When fashions change in science, it is not simply because people become bored and crave change, but because the old approach or technique has at best proved disappointing, at worst a failure. The influence of cybernetics and information theory on genetics can be seen in this way. In the 1940s and 1950s these two related approaches had a massive impact on the development of biology as a whole and on molecular genetics in particular. In the end, their influence waned as they failed to provide a framework that could stimulate further discovery. Both views ended up influencing genetics as vital metaphors and ways of viewing the world, not as essential theoretical foundations. This metaphorical role remains today, and it explains why scientists are so comfortable in saying that genes contain information and that they exert control over cellular networks.

CONCLUSION

In his book *Ways of Knowing*, the historian John Pickstone pointed out something that might seem obvious: science is a form of work. He argued that changes in how scientists gain their knowledge of the world can be interpreted in terms of changes in the organisation of work that have also occurred in manufacturing, which in different phases has been dominated in turn by what he called craft, rationalised production and systematic invention.[1] The race to crack the genetic code was mostly a matter of craft. Individuals or small groups were struggling with ideas and concepts as much as they were with facts; they were not only trying to understand what would be the right experiment to answer a question, they had to work out what the question was. Only in its final phases, after the breakthrough of Nirenberg and Matthaei, did craft partially cede pride of place to something like rationalised production, as the answer became visible and knowable, although it had not yet been attained. During those years from 1961 to 1967, cracking the code gradually became as much about biochemical technology as it was about imagination, even if the development and application of that technology required a great deal of craft and insight.

These discoveries created a revolution in our understanding and in our ways of thinking about life, a revolution that changed how science is done, shaping both our present and our future. In many respects we are now in a phase of systematic invention, in which new discoveries are being made in a more coordinated way, often involving large teams. Through the development of technology, we are now able to sequence the genomes of whole organisms in a matter of weeks – and soon even more quickly than that. Norbert Wiener, the founder of cybernetics, was concerned about how automation would alter factory work. It has most certainly transformed how science is done: robots can now decipher our genes, turning our genetic code into digital data that can then be explored anywhere in the world.

In 1991, just as the genome projects were being dreamt up, Wally Gilbert published an article in *Nature* in which he looked to the future.[2] Quite remarkably, he pretty much described the world we live in, suggesting that computers around the world would be hooked into databases, and that biologists would need to learn computing techniques to cope with the tide of data, investigating gene function first through a comparison of genes in different species rather than in an experiment. Gilbert pointed to skills that had already been lost in the brief history of molecular genetics, such as the ability to isolate restriction enzymes in the lab, which had been rendered obsolete by the availability of commercial products, and he rightly predicted that this process would continue. He also recognised that this rolling change was nothing new – once upon a time scientists blew their own glassware; later they bought it from a catalogue. The advent of automated sequencing of whole genomes is a huge step forwards – few scientists who went through the drudgery of hand-sequencing genes would want to return to those days. Scientists can now think about the biology instead of struggling with the chemistry.

Who those scientists are and how they work together have also changed dramatically. The work that resulted in the cracking of the genetic code was virtually entirely carried out by men, with a few exceptions – in chronological order, the women featured here were Harriet Ephrussi-Taylor, Martha Chase, Rosalind Franklin, Marianne Grunberg-Manago, Maxine Singer, Leslie Barnett and Norma Heaton. Some of these women were leading scientists, others were mid-level researchers, still others were technicians. Women now have a far more significant role: most fields of biology include leading female scientists, and it is quite usual for women to run laboratories. Nevertheless, although there are generally more women studying biology at university, at PhD level this becomes approximately equal numbers of male and female students, and there is then a growing proportion of men as you go up the academic scale, culminating in an overwhelmingly male professoriate. We are still far from equality between the sexes.

Most of the scientists described here were from the US, the UK and France. Science is now a truly international activity; even if the main contributors to the pages of the leading journals are still based in the richest and most developed countries, those researchers are often from all around the planet, with an increasing number coming from China. The current route for training a scientist involves not only a PhD but also a period of several years working in different laboratories, preferably in other countries, gaining experience and techniques. Most leading laboratories are now mini-United Nations. However, it is a striking fact that, even in the US, men and women of Afro-Caribbean origin are still substantially under-represented. In each country, the recruitment into science is biased by the multiple effects of race and class on educational attainment and on what is seen as being possible. Increasing the number of scientists from ethnic minorities and from the working class is a complex issue that science cannot solve on its own, but it needs to be addressed – at the moment there is a substantial pool of talent that we are not accessing because of inherent inequalities in our education system.

The way in which these multinational teams work has also altered in comparison with the 1950s and 1960s. Although papers in genetics are still published by small groups and, very occasionally, by single individuals, there is a clear tendency for research to be produced by large multidisciplinary teams. This is especially the case in genomics, in which many groups from around the world may be involved in obtaining and analysing the data. In 2014, a paper appeared in *Nature Genetics* describing how hundreds of variants in the human genome contribute to differences in height between individuals; the article was signed by more than 440 authors.[3] Big Science, typical of particle physics and astronomy, had not been seen in biology until the major genome sequencing projects. It is now becoming commonplace, changing the relationship of individual scientists to the work they produce, rendering each person's contribution relatively minor and highly specific. The increasingly tight budgets of funding organisations encourage large teams by promoting multidisciplinarity and often require the probable outcomes to be clear before the experiments have begun. It seems unlikely that the small, curiosity-driven teams that led to the cracking of the genetic code would survive in today's climate.

How we think about genes and what they do has also been transformed. In the 1830s, when the word heredity was first applied to biological characteristics, they were said to be 'passed down', just like more worldly inheritances such as money, land or furniture. Once the electronic age began, characteristics were said to be transmitted; after the growth in interest in codes and computing during and following the Second World War, it seemed obvious to suggest that genes contain a code and transmit information. The most powerful metaphors in science are often those that flow from new technological developments. The summit of the current phase of technology is the computer – this is also the richest metaphor that science currently employs. Not all the metaphors we use to describe the genetic code and the way it functions are so complex – 'transcription' and 'translation' suggest that the code is a language that is written down, and is either copied from DNA into RNA ('transcription'), or is turned into another language completely, that of proteins ('translation'). These metaphors weigh heavily on how we think about the nature of the genetic code and what it does. The complex linguistic and computational metaphors wrapped up in the seemingly simple idea of a genetic code frame our ideas about heredity.

But a 'frame' means two things – it both enables and limits how we think. We understand the nature of heredity with a far greater richness than people a century ago because of the wealth of research that has been done and also, because of the way in which we think about this research, the context in which we interpret it. But we are unable to conceive of other ways of viewing these phenomena because we do not yet have the appropriate metaphors. The frame is also a cage.

Nirenberg and Matthaei, the first to crack the code, were

outsiders, unaware of the debates of the previous decade that had led theoreticians to think that a repetitive sequence of bases would be meaningless. Their imaginations were free of the shutters that seem to have operated in thinking at other laboratories around the world. Ideas can help scientists understand data and can also prevent them from seeing what is under their nose: either way, they are essential to how science works.

Metaphors and analogies carry a risk. It is easy to forget that a particular term is a figure of speech, a way of viewing a given phenomenon, rather than being literally true. A gene is like a computer program, but it is not a program and does not function according to the same rules, even though it may be usefully understood in this way. Organisms are not machines, even if they work on physical principles and share some features with devices we have invented. The genetic code is not literally a code and it is not a language. It is a process that enables organisms to carry out particular functions by turning stored information into structures or actions, using evolved systems of control.

As became clear after the failed attempts to apply the strict mathematical view of information to genetic data, our way of describing information in genetics is primarily metaphorical. Although experimentation is generally the most powerful way of obtaining evidence that can test a hypothesis, to interpret this evidence we need theories and conceptual frameworks, which in turn are made up of words, metaphors and analogies. Understanding the power and limits of such metaphors will help us prepare for the breakthroughs of tomorrow, when we will reinterpret what we know and discover what we have yet to imagine.

New technological and scientific developments will provide us with new metaphors, new ways of understanding how life works, and new approaches to manipulating molecules. That future will inevitably contain opportunities and challenges. Synthetic life may enable us to resolve major economic and ecological problems, or it may inadvertently threaten the human race and the ecosystem. We may find ourselves able to manipulate aspects of our behaviour or anatomy by deliberately and precisely changing our genes and those of our offspring. This might open the road to health, fulfilment and pleasure, but it will also pose major ethical dilemmas. By revealing and cracking the genetic code, science has shown itself capable of revealing life's greatest secret. But science cannot tell us what to do with that secret, nor ensure that the knowledge and technology that flow from it are used for the greatest good of the many and with the least damage to the planet. Such a positive outcome will require the active involvement of the populations of all countries, as well as a clear understanding of the scientific and political issues raised by the amazing discoveries we have made and by the yet more amazing discoveries that are to come.

UGA

GLOSSARY AND ACRONYMS

Amino acid. A small molecule containing amine (-NH$_2$) and carboxylic acid (-COOH) groups. There are hundreds of different amino acids, but only twenty of them generally occur in organisms. They are strung together to make proteins.

Anticodon. A sequence of three bases of RNA found on the small tRNA molecule, which bind with a codon on the mRNA molecule.

AUG. The opening 'word' of a gene, this mRNA codon instructs the cell's protein synthesis machinery to 'start here', thereby also setting the reading frame for the gene. When AUG occurs in the middle of a gene, it codes for methionine.

Base. A molecule – adensoine, cytosine, guanine, thymine or uracil – that forms part of a nucleotide in DNA or RNA.

Chromosome. Cellular structures composed of DNA and proteins that contain genes.

Codon. A sequence of three bases in a DNA or RNA molecule that codes for an amino acid.

CRISPR. A new technique for editing genes in organisms, using a method derived from bacteria. The name comes from the kind of sequences where the phenomenon was first observed: Clustered Regularly Interspaced Short Palindromic Repeats. The technique has enormous scientific and medical potential.

Crystallography. The study of the molecular structure of crystals.

Cybernetics. The study of control and information flow in organic, mechanical or electronic systems, with an emphasis on the ability of negative feedback to produce apparently purposeful behaviour.

DNA. Deoxyribonucleic acid, a double helical molecule composed of a sugar/phosphate backbone and four bases: adenine, cytosine, guanine and thymine (A, C, G and T). The genetic material in all organisms and some viruses.

Enzyme. A large biological molecule – made of either protein or RNA – that catalyses (speeds up) a particular chemical reaction. Essential for life to exist.

mRNA. Messenger RNA. These molecules are copied from the gene and move from the chromosome to the ribosome, where they bind with a series of transfer RNA molecules, each of which is attached to an amino acid.

Nucleic acid. RNA or DNA.

Nucleoproteins. The mixture of proteins and nucleic acids that make up chromosomes.

Nucleotide. A molecule that combines a base with a five-carbon sugar (ribose or deoxyribose) plus phosphate; forms the basis of the nucleic acid sequence.

Operon. A group of genes that act under the concerted control of a single genetic element.

PCR. Polymerase chain reaction. Technique developed in the 1980s for amplifying small sequences of identified DNA. Now routinely used in science, in medicine and in the legal system.

Phage. Short for bacteriophage. These are viruses that attack bacteria.

Protein. A large molecule consisting of chains of amino acids. Proteins come in a vast variety of forms and carry out many biological functions.

Purines. Ring-shaped molecules, rich in nitrogen, larger than pyrimidines. In DNA and RNA, adenine and guanine are purine bases; each pairs with a particular pyrimidine (A with C, G with T or U).

Pyrimidines. Ring-shaped molecules, rich in nitrogen, smaller than purines. In DNA, cytosine and thymine are pyrimidine bases; in RNA, thymine is replaced by uracil. Each pairs with a particular purine (C with A, T or U with G).

Reading frame. In a DNA or RNA sequence, the correct order in which the bases should be read.

Repression. Inhibition of gene function.

Ribosome. Complex RNA structure found in all cells that is the primary site of protein synthesis.

RNA. Ribonucleic acid. A helical molecule composed of a sugar/phosphate backbone and four bases: adenine, cytosine, guanine and uracil (A, C, G and U). The genetic material in some viruses; carries out a wide range of regulatory functions in all cells.

Specificity. A term widely used until the 1960s to describe the various qualities of molecules and in particular the ability of proteins to carry out many functions.

Transcription. Copying of the genetic message from DNA to RNA.

Transcription factor. RNA or protein molecule that binds to a particular DNA sequence and regulates the activity of a gene.

Translation. The process whereby the genetic message in RNA is turned into an amino acid sequence; part of the protein synthesis process.

tRNA. Transfer RNA. Small cloverleaf-shaped piece of RNA, predicted to exist by Crick and Brenner. Each tRNA attaches to a particular amino acid and also carries an anticodon that enables it to bind with the relevant codon on the mRNA molecule.

UGA. The final word (or codon) in the genetic code to be deciphered, in 1967. Known as the opal codon, this mRNA sequence instructs the cell's protein synthesis machinery to 'stop here'.

FURTHER READING

Some of the research articles cited here are available as open access articles on the Internet; sadly, that is not true of all of them. You can generally find at least the abstract or summary of the article on line by putting its title into a search engine. Archival material covering the life and work of Avery, Crick, Nirenberg and others is available at http://profiles.nlm.nih.gov. The Wellcome Trust Codebreakers web site also holds many original documents: http://wellcomelibrary.org/using-the-library/subject-guides/genetics/makers-of-modern-genetics. Many informal photos of the key figures in this story can be found at http://www.estherlederberg.com.

Several academic works cover the material presented here and provide excellent additional sources: Lily E. Kay's *Who Wrote the Book of Life?*, Evelyn Fox Keller's *Refiguring Life: Metaphors of Twentieth-Century Biology*, *The Century of the Gene* and *Making Sense of Life*, and the articles and chapters by Sahotra Sarkar (see the reference list). Michel Morange's *A History of Molecular Biology* provides the scientific context (declaration: I translated it; a second edition is apparently in the works), while H. Freeman Judson has written a huge and fascinating oral history of the subject *The Eighth Day of Creation: Makers of the Revolution in Biology*. If you want to explore the history of information without much mathematics, James Gleick's *The Information* is for you, while for those interested in the scientific importance of metaphors, Theodore Brown's readable *Making Truth: Metaphor in Science* is a great place to start. Above all, I recommend reading the memoirs of four of the central people involved in this work: the inevitable *The Double Helix* by James Watson, Francis Crick's *What Mad Pursuit*, Maclyn McCarty's account of work in the Avery lab, *The Transforming Principle*, and François Jacob's marvellous but little-known *The Statue Within*.

If you want to know what happened before this story begins, you should read my earlier book, *The Egg and Sperm Race: The Seventeenth Century Scientists Who Unravelled the Secrets of Sex, Life and Growth* (published in the US as *Generation*).

ACKNOWLEDGEMENTS

This book owes a debt to the scholarship of the late Lily E. Kay, whose *Who Wrote the Book of Life?* provided me with inspiration and acted as a pathfinder. As the dedication indicates, my friend and colleague Professor John Pickstone died before he could subject the manuscript to his incisive criticism. We talked about the book several times over coffee or a pint as I was planning and writing it, and John's insight and good humour always cheered and helped me.

My agent, Peter Tallack, and my London publisher, John Davey, were enthusiastic and attentive during the pitching and the writing, respectively. John's careful edits have improved the manuscript no end, and at his suggestion we visited the King's College Archive and were able to handle the camera that took the notorious photo 51 (see Chapter 6). Bruce Goatly edited the copy with great efficiency and also rescued me from some howlers that will remain our secret. Picture researcher Lesley Hodgson gathered the illustrations with aplomb. Penny Daniel ensured that the passage from manuscript to printed page went smoothly, and was tolerant of my changes to the proofs.

My thanks go to my friends, colleagues, folk on Twitter and people whom I contacted out of the blue by e-mail, all of whom helped me in all sorts of ways, providing information, encouragement, articles, and in the case of Jerry Hurwitz an eye-witness account of the moment that Marshall Nirenberg told the world that the genetic code had been cracked: Tom Avery, Stuart Bennett, Casey Bergman, Sam Berry, Dave Briggs, Thony Christie, Dan Davis, Jerry Hurwitz, Nick Lane, Richard Lenski, Florian Maderspacher, Bjorn Poonen, Brian Sutton, Alex Wellerstein, Michael Wells and

Vivian Wyatt. Alok Jha (then of *The Guardian*), Steve Mao of *Cell* and Geoff North of *Current Biology* were all generous enough to allow me to sketch out my ideas through articles in their publications. Jerry Coyne encouraged me to post material on http://whyevolutionistrue.com, and the readers' comments helped me clarify my ideas. Similarly, the students on Carsten Timmerman's University of Manchester course *A History of Biology in 20 Objects* have been guinea pigs for some of my arguments. When it came to reviewing the manuscript, Jerry Coyne, Stephen Curry, Larry Moran, Michel Morange, Adam Rutherford, Ulrich Stegmann and Leslie Vosshall all generously provided extremely useful comments on chapter drafts. The errors and omissions that remain are my fault, of course.

While I wrote this book, my close family had to face a variety of life-changing events – PhD completion, university entry, major depression and vascular dementia. I am sure that when I was researching and writing, I was not as attentive to the needs of my loved ones as I ought to have been. My apologies to you all. Books have a price for writers' families, too.

REFERENCES

Adamala, K. and Szostak, J. W., 'Nonenzymatic template-directed RNA synthesis inside model protocells', *Science*, vol. 342, 2013, pp. 1098–100.

Administrative Framework of OSRD, *Organizing Scientific Research for War*, New York, Little, Brown, 1948.

Ageno, M., 'Deoxyribonucleic acid code', *Nature*, vol. 195, 1962, pp. 998–9.

Allison, A. C., 'Two lessons from the interface of genetics and medicine', *Genetics*, vol. 166, 2004, pp. 1591–9.

Anderson, T. F., 'Electron microscopy of phages', in J. Cairns, G. S. Stent and J. D. Watson (eds), *Phage and the Origins of Molecular Biology*, Cold Spring Harbor, Cold Spring Harbor Laboratory of Quantitative Biology, 1966, pp. 63–78.

Annett, R., Habibi, H. R. and Hontela, A., 'Impact of glyphosate and glyphosate-based herbicides on the freshwater environment', *Journal of Applied Toxicology*, vol. 34, 2014, pp. 458–79.

Anonymous, 'Award of the Gold Medal of the New York Academy of Medicine', *Science*, vol. 100, 1944, pp. 328–9.

Anonymous, *Symposium on Information Theory*, London, Ministry of Supply, 1950.

Anonymous, 'Biochemistry in Russia: Impressions gained at the Fifth International Congress of Biochemistry in Moscow', *British Medical Journal*, vol. 5253, 1961, pp. 701–3.

Anonymous, 'NIH researchers crack the genetic code', *Medical World News*, vol. 3, 1962, pp. 18–19.

Anonymous, 'Central dogma reversed', *Nature*, vol. 226, 1970, pp. 1198–9.

Anonymous, 'Max Delbrück – How it was (Part 2)', *Engineering and Science*, 43 (5), 1980, pp. 21–7.

Anonymous, 'Anfisen's cage', *Nature Structural Biology*, vol. 4, 1997, p. 675.

Apter, M. J. and Wolpert, L., 'Cybernetics and development: I. Information theory', *Journal of Theoretical Biology*, vol. 8, 1965, pp. 244–57.

Arai, J. A., Li, S., Hartley, D. M. and Feig, L. A., 'Transgenerational rescue of a genetic defect in long-term potentiation and memory formation by juvenile enrichment', *Journal of Neuroscience*, vol. 29, 2009, pp. 1496–502.

Archibald, J., *One Plus One Equals One: Symbiosis and the Evolution of Complex Life*, Oxford, Oxford University Press, 2014.

Ashburner, M., *Won for All: How the* Drosophila *Genome Was Sequenced*, Cold Spring Harbor, Cold Spring Harbor Laboratory Press, 2006.

Astbury, W. T., 'X-ray studies of nucleic acids', *Symposia of the Society for Experimental Biology*, vol. 1, 1947, pp. 66–76.

Attar, N., 'Raymond Gosling: the man who crystallized genes', *Genome Biology*, vol. 14, 2013, p. 402.

Augenstine, L. G., 'Protein structure and information content', in H. P. Yockey, R. L. Platzman and H. Quastler (eds), *Symposium on Information Theory in Biology*, London, Pergamon, 1958, pp. 103–23.

Avery, O. T., MacLeod, C. M. and McCarty, M., 'Studies on the chemical nature of the substance inducing transformation of pneumococcal types. Induction of transformation by a desoxyribosenucleic acid fraction isolated from pneumococcus type III', *Journal of Experimental Medicine*, vol. 79, 1944, pp. 137–58.

Axelsson, E., Ratnakumar, A., Arendt, M.-L. *et al.*, 'The genomic signature of dog domestication reveals adaptation to a starch-rich diet', *Nature*, vol. 495, 2013, pp. 360–4.

Baaske, P., Weinert, F. M., Duhr, S. *et al.*, 'Extreme accumulation of nucleotides in simulated hydrothermal pore systems', *Proceedings of the National Academy of Sciences USA*, vol. 104, 2007, pp. 9346–51.

Bada, J. L. and Lazcano, A., 'Stanley Miller's 70th birthday', *Origins of Life and Evolution of the Biosphere*, vol. 30, 2000, pp. 107–12.

Bar-Hillel, Y., 'Semantic information and its measures', in H. von Foerster, M. Mead and H. L. Teuber (eds), *Cybernetics: Circular Causal and Feedback Mechanisms in Biology and Social Systems*, New York, Josiah Macy, Jr Foundation, 1953, pp. 33–48.

Barrell, B. G., Bankier, A. T. and Drouin, J., 'A different genetic code in human mitochondria', *Nature*, vol. 282, 1979, pp. 189–94.

Basilio, C., Wahba, A. J., Lengyel, P. *et al.*, 'Synthetic polynucleotides and the amino acid code, V', *Proceedings of the National Academy of Sciences USA*, vol. 48, 1962, pp. 613–16.

Beadle, G., 'The role of the nucleus in heredity', in W. D. McElroy and B. Glass (eds), *A Symposium on the Chemical Basis of Heredity*, Baltimore, The Johns Hopkins Press, 1957, pp. 3–22.

Beadle, G. W. and Tatum, E. L., 'Genetic control of biochemical reactions in *Neurospora*', *Proceedings of the National Academy of Sciences USA*, vol. 27, 1941, pp. 499–506.

Bearn, A. G., 'Oswald T. Avery and the Copley Medal of the Royal Society', *Perspectives in Biology and Medicine*, vol. 39, 1996, pp. 550–5.

Beckwith, J., 'The operon as paradigm: Normal science and the beginning of biological complexity', *Journal of Molecular Biology*, vol. 409, 2011, pp. 7–13.

Behura, S. K. and Severson, D. W., 'Codon usage bias: causative factors, quantification methods and genome-wide patterns: with emphasis on insect genomes', *Biological Reviews*, vol. 88, 2013, pp. 49–61.

Belozersky, A. N. and Spirin, A. S., 'A correlation between the compositions of deoxyribonucleic and ribonucleic acids', *Nature*, vol. 182, 1958, pp. 1–2.

Bennett, G. M. and Moran, N. A., 'Small, smaller, smallest: the origins and evolution of ancient dual symbioses in a phloem-feeding insect', *Genome Biology and Evolution*, vol. 5, 2013, pp. 1675–88.

Bennett, S., 'Norbert Wiener and control of anti-aircraft guns', *IEEE Control Systems*, December 1994, pp. 58–62.

Bennett, S., 'A brief history of automatic control', *IEEE Control Systems*, June 1996, pp. 17–24

Benzer, S., 'The elementary units of heredity', in W. D. McElroy and B. Glass (eds), *A Symposium on the Chemical Basis of Heredity*, Baltimore, The Johns Hopkins Press, 1957, pp. 70–93.

Benzer, S., 'On the topology of the genetic fine structure', *Proceedings of the National Academy of Sciences USA*, vol. 45, 1959, pp. 1607–20.

Benzer, S., 'On the topography of the genetic fine structure', *Proceedings of the National Academy of Sciences USA*, vol. 47, 1961, pp. 403–15.

Benzer, S., 'Adventures in the rII region', in J. Cairns, G. S. Stent and J. D. Watson (eds), *Phage and the Origins of Molecular Biology*, Cold Spring Harbor, Cold Spring Harbor Laboratory of Quantitative Biology, 1966, pp. 157–65.

Benzer, S., Interview by Heidi Aspaturian. Pasadena, California, September 1990–February 1991. Oral History Project, California Institute of Technology Archives. http://oralhistories.library.caltech.edu/27/1/OH_Benzer_S.pdf, 1991.

Berg, P., 'Meetings that changed the world. Asilomar 1975: DNA modification secured', *Nature*, vol. 455, 2008, pp. 290–1.

Berg, P. and Singer, M., *George Beadle, an Uncommon Farmer: The Emergence of Genetics in the Twentieth Century*, Cold Spring Harbor, Cold Spring Harbor Laboratory Press, 2003.

Berg, P., Baltimore, D., Boyer, H. W. *et al.*, 'Potential biohazards of recombinant DNA molecules', *Science*, vol. 185, 1974, p. 303.

Berg, P., Baltimore, D., Brenner, S. *et al.*, 'Asilomar conference on recombinant DNA molecules', *Science*, vol. 188, 1975, pp. 991–4.

Berget, S. M., Moore, C. and Sharp, P. A., 'Spliced segments at the 5' terminus of adenovirus 2 late mRNA', *Proceedings of the National Academy of Sciences USA*, vol. 74, 1977, pp. 3171–5.

Bergstrom, C. T. and Rosvall, M., 'The transmission sense of information', *Biology and Philosophy*, vol. 26, 2011a, pp. 159–76.

Bergstrom, C. T. and Rosvall, M., 'Response to commentaries on "The transmission sense of information"', *Biology and Philosophy*, vol. 26, 2011b, pp. 195–200.

Berk, A. J. and Sharp, P. A., 'Ultraviolet mapping of the adenovirus 2 early promoters', *Cell*, vol. 12, 1977, pp. 45–55.

Bernardi, G., 'Isochores and the evolutionary genomics of vertebrates', *Gene*, vol. 241, 2000, pp. 3–17.

Berry, M. J., Banu, L., Harney, J. W. and Larsen, P. R., 'Functional characterization of the eukaryotic SECIS elements which direct selenocysteine insertion at UGA codons', *The EMBO Journal*, vol. 12, 1993, pp. 3315–22.

Beurton, P., Falk, F. and Rheinberger, H.-J., *The Concept of the Gene in Development and Evolution*, Cambridge, Cambridge University Press, 2000.

Bhattacharjee, Y., 'The vigilante', *Science*, vol. 343, 2014, pp. 1306–9.

Birnbaum, R. Y., Clowney, E. J., Agamy, O. *et al.*, 'Coding exons function as tissue-specific enhancers of nearby genes', *Genome Research*, vol. 22, 2012, pp. 1059–68.

Biscoe, J., Pickels, E. G. and Wyckoff, R. W. G., 'An air-driven ultracentrifuge for molecular sedimentation', *Journal of Experimental Medicine*, vol. 64, 1936, pp. 39–45.

Bohlin, G., *Arvstvisten. Om hur DNA-molekylen blev accepterad som bärare av genetisk information i Sverige och om ett uteblivet Nobelpris*, Stockholm, Nobel Museum, 2009.

Boivin, A., 'Directed mutation in colon bacilli, by an inducing principle of desoxyribonucleic nature: its meaning for the general biochemistry of heredity', *Cold Spring Harbor Symposia on Quantitative Biology*, vol. 12, 1947, pp. 7–17.

Boivin, A. and Vendrely, R., 'Sur le rôle possible des deux acides nucléiques dans la cellule vivante', *Experientia*, vol. 3, 1947, pp. 32–4.

Boivin, A., Vendrely, R. and Lehoult, Y., 'L'acide thymonucléique hautement polymerisé, principe capable de conditionner la specificité sérologique et l'équipement enzymatique des Bactéries. Conséquences pour la biochimie de l'hérédité', *Comptes Rendus de l'Académie des Sciences de Paris*, vol. 221, 1945a, pp. 646–8.

Boivin, A., Delaunay, Vendrely, R. and Lehoult, Y., 'L'acide thymonucléique polymerisé, principe paraissant susceptible de déterminer la specificité sérologique et l'équipement enzymatique des bactéries. Signification pour la biochimie de l'hérédité', *Experientia*, vol. 1, 1945b, pp. 334–5.

Boivin, A., Vendrely, R. and Tulasne, R., 'La spécificité des acides nucléiques ches les êtres vivants, spécialement chez les Bactéries', *Colloques Internationaux du Centre National de la Recherche Scientifique*, vol. 8, 1949, pp. 67–78.

Bond, D. M. and Baulcombe, D. C., 'Small RNAs and heritable epigenetic variation in plants', *Trends in Cell Biology*, vol. 24, 2014, pp. 100–107.

Bond, D. M. and Baulcombe, D. C., 'Epigenetic transitions leading to heritable, RNA-mediated de novo silencing in *Arabidopsis thaliana*', *Proceedings of the National Academy of Sciences USA*, vol. 112, 2015, pp. 917–22.

Boto, L., 'Horizontal gene transfer in the acquisition of novel traits by metazoans', *Proceedings of the Royal Society: Biological Sciences*, vol. 281, 2014, article 20132450.

Botstein, D., White, R. L., Skolnick, M. and Davis, R. W., 'Construction of a genetic linkage map in man using restriction fragment length polymorphisms', *American Journal of Human Genetics*, vol. 32, 1980, pp. 314–31.

Bowler, P. J., *The Mendelian Revolution: The Emergence of Hereditarian Concepts in Modern Science and Society*, London, Athlone, 1989.

Boyce, F. M., Beggs, A. H., Feener, C. and Kunkel, L. M., 'Dystrophin is transcribed in brain from a distant upstream promoter', *Proceedings of the National Academy of Sciences USA*, vol. 88, 1991, pp. 1276–80.

Brachet, J., 'La localisation des acides pentosenucléiques dans les tissus animaux et les oeufs d'Amphibiens en voie de développement', *Archives de biologie*, vol. 53, 1942, pp. 207–57.

Brannigan, A., 'The reification of Mendel', *Social Studies of Science*, vol. 9, 1979, pp. 423–54.

Bremer, J., Baumann, F., Tiberi, C. *et al.*, 'Axonal prion protein is required for peripheral myelin maintenance', *Nature Neuroscience*, vol. 13, 2010, pp. 310–18.

Brenner, S., 'On the impossibility of all overlapping triplet codes' Unpublished note, RNA Tie Club, Wellcome Trust Library, SB/2/1/106, 1956.

Brenner, S., 'On the impossibility of all overlapping triplet codes in information transfer from nucleic acid to proteins', *Proceedings of the National Academy of Sciences USA*, vol. 43, 1957, pp. 687–94.

Brenner, S., 'RNA, ribosomes and protein synthesis', *Cold Spring Harbor Symposia on Quantitative Biology*, vol. 26, 1961, pp. 101–10.

Brenner, S., *My Life in Science*, London, BioMedCentral, 2001.

Brenner, S., Jacob, F. and Meselson, M., 'An unstable intermediate for carrying information from genes to ribosomes for protein synthesis', *Nature*, vol. 190, 1961, pp. 576–81.

Brenner, S., Barnett, L., Katz, E. R. and Crick, F. H., 'UGA: a third nonsense triplet in the genetic code', *Nature*, vol. 213, 1967, pp. 449–50.

Bretscher, M. S. and Grunberg-Manago, M., 'Polyribonucleotide-directed protein synthesis using an *E. coli* cell-free system', *Nature*, vol. 195, 1962, pp. 283–4.

Brillouin, L., 'Life, thermodynamics, and cybernetics', *American Scientist*, vol. 37, 1949, pp. 554–68.

Brillouin, L., *Science and Information Theory*, New York, Academic Press, 1956.

Britten, R. J. and Davidson, E. H., 'Gene regulation for higher cells: a theory', *Science*, vol. 165, 1969, pp. 349–57.

Brown, P., Sutikna, T., Morwood, M. J. *et al.*, 'A new small-bodied hominin from the Late Pleistocene of Flores, Indonesia', *Nature*, vol. 431, 2004, pp. 1055–61.

Brown, T. L., *Making Truth: Metaphor in Science*, Urbana, University of Illinois Press, 2003.

Burian, R. M. and Gayon, J., 'The French school of genetics: From physiological and population genetics to regulatory molecular genetics', *Annual Review of Genetics*, vol. 33, 1999, pp. 313–49.

Burkeman, O., 'Why everything you've been told about evolution is wrong', *The Guardian*, 19 March 2010.

Burnet, M., *Enzyme, Antigen and Virus: A Study of Macromolecular Pattern in Action*, Cambridge, Cambridge University Press, 1956.

Burnet, M., *Changing Patterns: An Atypical Autobiography*, London, Heinemann, 1968.

Caetano-Anollés, G. and Seufferheld, M. J., 'The coevolutionary roots of biochemistry and cellular organization challenge the RNA world paradigm', *Journal of Molecular Microbiology and Biotechnology*, vol. 23, 2013, pp. 152–77.

Cairns, J., 'The autoradiography of DNA', in Cairns, J., Stent, G. S. and Watson, J. D. (eds), *Phage and the Origins of Molecular Biology*, Cold Spring Harbor, Cold Spring Harbor Laboratory of Quantitative Biology, 1966, pp. 252–7.

Cairns, J., Stent, G. S. and Watson, J. D. (eds), *Phage and the Origins of Molecular Biology*, Cold Spring Harbor, Cold Spring Harbor Laboratory of Quantitative Biology, 1966.

Caldwell, P. C. and Hinshelwood, C., 'Some considerations on autosynthesis in bacteria', *Journal of the Chemical Society*, vol. 4, 1950, pp. 3156–9.

Callaway, E., 'The Neanderthal in the family', *Nature*, vol. 507, 2014a, pp. 415–16.

Callaway, E., 'Geneticists tap human knockouts', *Nature*, vol. 514, 2014b, p. 548.

Cannarozzi, G., Schraudolph, N. N., Faty, M. *et al.*, 'A role for codon order in translation dynamics', *Cell*, vol. 141, 2010, pp. 355–67.

Carey, N., *The Epigenetics Revolution: How Modern Biology is Rewriting Our Understanding of Genetics, Disease and Inheritance*, London, Icon, 2011.

Carlson, E. A., *The Gene: A Critical History*, New York, W. B. Saunders, 1966.

Carlson, E. A., *Genes, Radiation and Society: The Life and Work of H. J. Muller*, London, Cornell University Press, 1981.

Carlson, E. A., *Mendel's Legacy: The Origin of Classical Genetics*, Cold Spring Harbor, Cold Spring Harbor Laboratory Press, 2004.

Carroll, S. B., *Brave Genius: A Scientist, a Philosopher, and Their Daring Adventures from the French Resistance to the Nobel Prize*, New York, Crown, 2013.

Caspersson, T., 'The relations between nucleic acid and protein synthesis', *Symposia of the Society for Experimental Biology*, vol. 1, 1947, pp. 127–51.

Caspersson, T. and Schultz, J., 'Pentose nucleotides in the cytoplasm of growing tissues', *Nature*, vol. 143, 1939, pp. 602–3.

Caspersson, T., Hammarsten, E. and Hammarsten, H., 'Interactions of proteins and nucleic acid', *Transactions of the Faraday Society*, vol. 31, 1935, pp. 367–89.

Cavalcanti, A. R. O. and Landweber, L. F., 'Genetic code', *Current Biology*, vol. 14, 2004, p. R147.

Cavalli-Sforza, L. L., 'Bacterial genetics', *Annual Review of Microbiology*, vol. 11, 1957, pp. 391–418.

Chantrenne, H., 'Information in biology', *Nature*, vol. 197, 1963, pp. 27–30.

Chargaff, E., 'On the nucleoproteins and nucleic acids of microorganisms', *Cold Spring Harbor Symposia in Quantitative Biology*, vol. 12, 1947, pp. 28–34.

Chargaff, E., 'Chemical specificity of nucleic acids and mechanism of their enzymatic degradation', *Experientia*, vol. 6, 1950, pp. 201–9.

Chargaff, E., 'Some recent studies of the composition and structure of nucleic acids', *Journal of Cellular and Comparative Physiology*, vol. 38 (Suppl. 1), 1951, pp. 41–59.

Chargaff, E., 'Base composition of desoxypentose and pentose nucleic acids in various species', in W. D. McElroy and B. Glass (eds), *A Symposium on the Chemical Basis of Heredity*, Baltimore, The Johns Hopkins Press, 1957, pp. 521–7.

Chargaff, E., *Heraclitean Fire: Sketches from a Life before Nature*, New York, Rockefeller University Press, 1978.

Chargaff, E. and Vischer, E., 'Nucleoproteins, nucleic acids, and related substances', *Annual Review of Biochemistry*, vol. 17, 1948, pp. 201–26.

Chargaff, E., Lipschitz, R., Green, C. and Hodes, M. E., 'The composition of the desoxyribonucleic acid of salmon sperm', *Journal of Biological Chemistry*, vol. 192, 1951, pp. 223–30.

Chien, A., Edgar, D. B. and Trela, J. M., 'Deoxyribonucleic acid polymerase from the extreme thermophile *Thermus aquaticus*', *Journal of Bacteriology*, vol. 127, 1976, pp. 1550–7.

Chow, L. T., Gelinas, R. E., Broker, T. R. and Roberts, R. J., 'An amazing sequence arrangement at the 5′ ends of adenovirus 2 messenger RNA', *Cell*, vol. 12, 1977, pp. 1–8.

Church, G. M., Gao, Y. and Kosuri, S., 'Next-generation digital information storage in DNA', *Science*, vol. 337, 2012, p. 1628.

Clark, B. F. and Marcker, K. A., 'The role of N-formyl-methionyl-sRNA in protein biosynthesis', *Journal of Molecular Biology*, vol. 17, 1966, pp. 394–406.

Cleland, C. E. and Copley, S. D., 'The possibility of alternative microbial life on Earth', *International Journal of Astrobiology*, vol. 4, 2005, pp. 165–73.

Cobb, M., 'Heredity before genetics: A history', *Nature Reviews: Genetics*, vol. 7, 2006a, pp. 953–8.

Cobb, M., *The Egg and Sperm Race: The Seventeenth Century Scientists Who Unravelled the Secrets of Sex, Life and Growth*, London, Free Press, 2006b. (Published in the US as *Generation*, New York: Bloomsbury.)

Cobb, M., 'The unclosed loop – ESF minibrains report', *EMBO Reports*, vol. 12, 2011, pp. 389–91.

Cobb, M., 'Oswald Avery, DNA, and the transformation of biology', *Current Biology*, vol. 24, 2014, pp. R55–R60.

Cochran, W. and Crick, F. H. C., 'Evidence for the Pauling-Corey α-helix in synthetic polypeptides', *Nature*, vol. 169, 1952, pp. 234–5.

Cochran, W., Crick, F. H. and Vand, V., 'The structure of synthetic polypeptides. I. The transform of atoms on a helix', *Acta Crystallographica*, vol. 5, 1952, pp. 581–6.

Cohen, S. S., 'Streptomycin and desoxyribonuclease in the study of variations in the properties of a bacterial virus', *Journal of Biological Chemistry*, vol. 168, 1947, pp. 511–26.

Cohen, S. S., 'Alfred Ezra Mirsky, 1900–1974', *National Academy of Sciences Biographical Memoir*, Washington DC, National Academies Press, 1998.

Cohn, M., Cohen, G. N. and Monod, J., 'L'effet inhibiteur spécifique de la methionine dans la formation de la methionine-synthase chez Escherichia coli', *Comptes rendus hebdomodaires des séances de l'Académie des sciences*, vol. 236, 1953a, pp. 746–8.

Cohn, M., Monod, J., Pollock, M. R. *et al.*, 'Terminology of enzyme formation', *Nature*, vol. 172, 1953b, pp. 1096–7.

Collier, J., 'Information in biological systems', in P. Adriaans and J. v. Benthem (eds), *Handbook of the Philosophy of Science. Volume 8: Philosophy of Information*, Amsterdam, Elsevier, 2008, pp. 736–87.

Colnort-Bodet, S., 'Pierre de Latil: La pensée artificielle', *Revue d'histoire des sciences et de leurs applications*, vol. 7, 1954, p. 196.

Comfort, N., 'Recombinant gold', *Nature*, vol. 508, 2014, pp. 176–7.

Commoner, B., 'Failure of the Watson-Crick theory as a chemical explanation of inheritance' *Nature*, vol. 220, 1968, pp. 334–40.

Cong, L., Ran, F. A., Cox, D. *et al.*, 'Multiplex genome engineering using CRISPR/Cas systems', *Science*, vol. 339, 2013, pp. 819–23.

Conway, F. and Siegelman, J., *Dark Hero of the Information Age: In Search of Norbert Wiener, the Father of Cybernetics*, New York, Basic Books, 2005.

Corden, J., Wasylyk, B., Buchwalder, A. *et al.*, 'Promoter sequences of eukaryotic protein-coding genes', *Science*, vol. 209, 1980, pp. 1406–14.

Cornelis, G., Vernochet, C., Malicorne, S. *et al.*, 'Retroviral envelope syncytin capture in an ancestrally diverged mammalian clade for placentation in the primitive Afrotherian tenrecs', *Proceedings of the National Academy of Sciences USA*, vol. 111, 2014, pp. E4332–41.

Corjito, S., Wardenaar, R., Colomé-Tatché, M. *et al.*, 'Mapping the epigenetic basis of complex traits', *Science*, vol. 343, 2014, pp. 1145–8.

Cosentino, C. and Bates, D., *Feedback Control in Systems Biology*, Boca Raton, CRC Press, 2012.

Couffignal, L. (ed.), *Le Concept de l'information dans la science contemporaine*, Paris, Gauthier-Villars, 1965.

Coyne, J. A., 'The gene is dead; long live the gene', *Nature*, vol. 408, 2000, pp. 26–7.

Crabbe, J. C., Wahlsten, D. and Dudek, B. C., 'Genetics of mouse behavior: Interactions with laboratory environment', *Science*, vol. 284, 1999, pp. 1670–2.

Creager, A. N. H., 'Phosphorus-32 in the Phage Group: radioisotopes as historical tracers of molecular biology', *Studies in History and Philosophy of Biological and Biomedical Sciences*, vol. 40, 2009, pp. 29–42.

Creager, A. N. H. and Morgan, G. J., 'After the double helix: Rosalind Franklin's research on tobacco mosaic virus', *Isis*, vol. 99, 2008, pp. 239–72.

Creeth, J. M., Gulland, J. M. and Jordan, D. O., 'Deoxypentose nucleic acids. Part III. Viscosity and streaming birefringence of solutions of the sodium salt of the deoxypentose nucleic acid of calf thymus', *Journal of the Chemical Society*, vol. 1947, 1947, pp. 1141–5.

Crick, F. H. C., 'On degenerate templates and the adaptor hypothesis' Unpublished note, RNA Tie Club, Wellcome Trust Library, SB/2/1/106, http://genome.wellcome.ac.uk/assets/wtx030893.pdf, 1955.

Crick, F. H. C., 'Nucleic acids', *Scientific American*, vol. 197, 1957, pp. 188–200.

Crick, F. H. C., 'On protein synthesis', *Symposia of the Society for Experimental Biology*, vol. 12, 1958, pp. 138–63.

Crick, F. H. C., 'The present position of the coding problem', *Brookhaven Symposia in Biology*, vol. 12, 1959, pp. 35–9.

Crick, F., 'Towards the genetic code', *Scientific American*, vol. 207 (3), 1962, pp. 8–16.

Crick, F. H. C., 'The recent excitement in the coding problem', *Progress in Nucleic Acids Research*, vol. 1, 1963a, pp. 163–217.

Crick, F. H. C., 'On the genetic code', *Science*, vol. 139, 1963b, pp. 461–4.

Crick, F. H. C., 'The genetic code – yesterday, today, and tomorrow', *Cold Spring Harbor Symposia on Quantitative Biology*, vol. 31, 1966a, pp. 3–9.

Crick, F. H. C., 'Codon-anticodon pairing: the wobble hypothesis', *Journal of Molecular Biology*, vol. 19, 1966b, pp. 548–55.

Crick, F. H. C., 'The origin of the genetic code', *Journal of Molecular Biology*, vol. 38, 1968, pp. 367–79.

Crick, F., 'Central dogma of molecular biology' *Nature*, vol. 227, 1970, pp. 561–3.

Crick, F., 'Split genes and RNA splicing', *Science*, vol. 204, 1979, pp. 264–71.

Crick, F., *Life Itself: Its Origin and Nature*, London, Macdonald, 1981.

Crick, F., *What Mad Pursuit: A Personal View of Scientific Discovery*, Cambridge, MA, Basic, 1988.

Crick, F. H. C. and Watson, J. D., 'The complementary structure of deoxyribonucleic acid', *Proceedings of the Royal Society of London. Series A, Mathematical and Physical Sciences*, vol. 223, 1954, pp. 80–96.

Crick, F. H. C., Griffith, J. S. and Orgel, L. E., 'Codes without commas', *Proceedings of the National Academy of Sciences USA*, vol. 43, 1957, pp. 416–21.

Crick, F. H. C., Barnett, L., Brenner, S. and Watts-Tobin, R. J., 'General nature of the genetic code for proteins', *Nature*, vol. 192, 1961, pp. 1227–32.

Crow, J. F., 'Erwin Schrödinger and the hornless cattle problem', *Genetics*, vol. 130, 1992, pp. 83–6.

Crow, E. W. and Crow, J. F., '100 years ago: Walter Sutton and the chromosome theory of heredity', *Genetics*, vol. 160, 2002, pp. 1–4.

Daly, M. M., Allfrey, V. G. and Mirsky, A. E., 'Purine and pyrimidine contents of some desoxypentose nucleic acids', *The Journal of General Physiology*, vol. 133, 1950, pp. 497–510.

Danchin, E., 'Avatars of information: towards an inclusive evolutionary synthesis', *Trends in Ecology and Evolution*, vol. 28, 2013, pp. 351–8.

Dancoff, S. M. and Quastler, H., 'The information content and error rate of living things', in H. Quastler (ed.), *Essays on the Use of Information Theory in Biology*, Urbana, University of Illinois Press, 1953, pp. 263–73.

Davies, K., *The Sequence: Inside the Race for the Human Genome*, London, Phoenix, 2002.

Davies, M., 'W. T. Astbury, Rosie Franklin, and DNA: A memoir', *Annals of Science*, vol. 47, 1990, pp. 607–18.

Davies, P. C. W., Benner, S. A., Cleland, C. A. *et al.*, 'Signatures of a shadow biosphere', *Astrobiology*, vol. 9, 2009, pp. 241–9.

Davis, D. M., *The Compatibility Gene*, London, Allen Lane, 2013.

Davis, J., 'Microvenus', *Art Journal*, spring 1996, pp. 70–4.

Davis, L. and Chin, J. W., 'Designer proteins: applications of genetic code expansion in cell biology', *Nature Reviews Molecular Cell Biology*, vol. 13, 2013, pp. 168–82.

Davis, T. H., 'Meselsohn and Stahl: The art of DNA replication', *Proceedings of the National Academy of Sciences USA*, vol. 101, 2004, pp. 17895–6.

Daxinger, L. and Whitelaw, E., 'Understanding transgenerational epigenetic inheritance via the gametes in mammals', *Nature Reviews Genetics*, vol. 13, 2012, pp. 153–62.

Debré, P., *Jacques Monod*, Paris, Flammarion, 1996.

De Chadarevian, S., 'Portrait of a discovery: Watson, Crick, and the Double Helix', *Isis*, vol. 94, 2003, pp. 90–105.

Deichmann, U., 'Early responses to Avery *et al.*'s paper on DNA as hereditary material', *Historical Studies in the Physical and Biological Sciences*, vol. 34, 2004, pp. 207–32.

Deichmann, U., 'Different methods and metaphysics in early molecular genetics – a case of disparity of research?', *History and Philosophy of the Life Sciences*, vol. 30, 2008, pp. 53–78.

de Latil, P., *La Pensée artificielle: Introduction à la cybernétique*, Paris, Gallimard, 1953.

de Latil, P., *Thinking by Machine: A Study of Cybernetics*, Boston, Houghton Mifflin, 1957.

Delbrück, M., 'Génétique du bactériophage', *Colloques Internationaux du Centre National de la Recherche Scientifique*, vol. 8, 1949, pp. 92–103.

Delbrück, M. and Stent, G., 'On the mechanism of DNA replication', in W. D. McElroy and B. Glass (eds), *A Symposium on the Chemical Basis of Heredity*, Baltimore, The Johns Hopkins Press, 1957, pp. 699–736.

De Lorenzo, V., 'From the selfish gene to selfish metabolism: Revisiting the central dogma', *BioEssays*, vol. 36, 2014, pp. 226–35.

Denenberg, V. H. and Rosenberg, K. M., 'Nongenetic transmission of information', *Nature*, vol. 216, 1967, pp. 549–50.

Di Giulio, M., 'The origin of the genetic code in the ocean abysses: new comparisons confirm old observations', *Journal of Theoretical Biology*, vol. 333, 2013a, pp. 109–16.

Di Giulio, M., 'The origin of the genetic code: Matter of metabolism or physicochemical determinism?', *Journal of Molecular Evolution*, vol. 77, 2013b, pp. 131–3.

Dobzhansky, T., *Genetics and the Origin of Species*, New York, Columbia University Press, 1941.

Donohue, J., 'Fragments of Chargaff', *Nature*, vol. 276, 1978, pp. 133–5.

Doolittle, W. F. and Sapienza, C., 'Selfish genes, the phenotype paradigm and genome evolution', *Nature*, vol. 284, 1980, pp. 601–3.

Doolittle, W. F., Brunet, T. D. P., Linquist, S. and Gregory, T. R., 'Distinguishing between "function" and "effect" in genome biology', *Genome Biology and Evolution*, vol. 6, 2014, pp. 1234–7.

Dounce, A. L., 'Duplicating mechanism for peptide chain and nucleic acid synthesis', *Enzymologia*, vol. 15, 1952, pp. 251–8.

Dounce, A. L., Morrison, M. and Monty, K. J., 'Role of nucleic acid and enzymes in peptide chain synthesis', *Nature*, vol. 176, 1955, pp. 597–8.

Doyle, R., *On Beyond Living: Rhetorical Transformations of the Life Sciences*, Stanford, Stanford University Press, 1997.

Dromanraju, K. R., 'Erwin Schrödinger and the origins of molecular biology', *Genetics*, vol. 153, 1999, pp. 1071–6.

Du, X., Wojtowicz, D., Bowers, A. A. *et al.*, 'The genome-wide distribution of non-B DNA motifs is shaped by operon structure and suggests the transcriptional importance of non-B DNA structures in *Escherichia coli*', *Nucleic Acids Research*, vol. 41, 2013, pp. 5965–77.

Dubarle, D., 'Sens philosophique et portée pratique de la cybernétique', *La Nouvelle Revue Française*, vol. 10, 1953, pp. 60–85.

Dubos, R. J., *The Professor, the Institute and DNA*, New York, Rockefeller University Press, 1976.

Dudai, Y., Jan, Y. N., Byers, D. *et al.*, '*dunce*, a mutant of *Drosophila* deficient in learning', *Proceedings of the National Academy of Sciences USA*, vol. 73, 1976, pp. 1684–8.

Dunn, A. R. and Hassell, J. A., 'A novel method to map transcripts: evidence for homology between an adenovirus mRNA and discrete multiple regions of the viral genome', *Cell*, vol. 12, 1977, pp. 23–36.

Easterling, K., Walter Pitts. *Cabinet*, 5, http://cabinetmagazine.org/issues/5/walterpitts.php, 2001.

Eck, R. V., 'Non-randomness in amino-acid "alleles"', *Nature*, vol. 191, 1961, pp. 1284–5.

Eck, R. V., 'Genetic code: emergence of a symmetrical pattern', *Science*, vol. 140, 1963, pp. 477–81.

Eddy, S. R., 'The C-value paradox, junk DNA and ENCODE', *Current Biology*, vol. 22, 2012, pp. R898–9.

Eddy, S. R., 'The ENCODE project: Missteps overshadowing a success', *Current Biology*, vol. 23, 2013, pp. R259–61.

Eisenhart, C., 'Cybernetics: A new discipline', *Science*, vol. 109, 1949, pp. 397–8.

Elias, P., 'Two famous papers', *IRE Transactions on Information Theory*, vol. 4, 1958, p. 99.

Elias, P., 'Coding and information theory', *Reviews of Modern Physics*, vol. 31, 1959, pp. 221–5.

Elsila, J. E., Glavin, D. P. and Dworkin, J. P., 'Cometary glycine detected in samples returned by Stardust', *Meteoritics and Planetary Science*, vol. 44, 2009, pp. 1323–30.

ENCODE Project Consortium, 'An integrated encyclopedia of DNA elements in the human genome', *Nature*, vol. 489, 2012, pp. 57–74.

Ephrussi, B., Leopold, U., Watson, J. D. and Weigle, J. J., 'Terminology in bacterial genetics', *Nature*, vol. 171, 1953, p. 701.

Ephrussi-Taylor, H., 'Genetic aspects of transformations of pneumococci', *Cold Spring Harbor Symposia on Quantitative Biology*, vol. 16, 1951, pp. 445–56.

Erb, T. J., Kiefer, P., Hattendorf, B. *et al.*, 'GFAJ-1 is an arsenate-resistant, phosphate-dependent organism', *Science*, vol. 337, 2012, pp. 467–70.

Esvelt, K. E., Smidler, A. L., Catteruccia, F. and Church, G. M., 'Concerning RNA-guided gene drives for the alteration of wild populations', *eLife*, 2014, article 03401.

Eyre-Walker, A. and Hurst, L. D., 'The evolution of isochores', *Nature Reviews Genetics*, vol. 2, 2001, pp. 549–55.

Ezkurdia, I., Juan, D., Rodriguez, J. M. *et al.*, 'Multiple evidence strands suggest that there may be as few as 19 000 human protein-coding genes', *Human Molecular Genetics*, vol. 23, 2014, pp. 5866–78.

Fabris, F., 'Shannon Information Theory and molecular biology', *Journal of Interdisciplinary Mathematics*, vol. 12, 2009, pp. 41–87.

Falk, R., *Genetic Analysis: A History of Genetic Thinking*, Cambridge, Cambridge University Press, 2009.

Farzadfard, F. and Lu, T. K., 'Genomically encoded analog memory with precise in vivo DNA writing in living cell populations', *Science*, vol. 346, 2014, article 1256272.

Fechotte, C. and Pritham, E., 'DNA transposons and the evolution of eukaryotic genomes', *Annual Review of Genetics*, vol. 41, 2007, pp. 331–68.

Fedorov A., Suboch G., Bujakov M. and Fedorova L., 'Analysis of nonuniformity in intron phase distribution', *Nucleic Acids Research*, vol. 20, 1992, pp. 2553–7.

Fellermann, H. and Solé, R. V., 'Minimal model of self-replicating nanocells: a physically embodied information-free scenario', *Philosophical Transactions of the Royal Society: B. Biological Sciences*, vol. 362, 2007, pp. 1803–11.

Ferry, G., *Max Perutz and the Secret of Life*, London, Chatto & Windus, 2007.

Forsdyke, D. R., 'Heredity as transmission of information: Butlerian "intelligent design"', *Centaurus*, vol. 48, 2006, pp. 133–148.

Francis, D., Diorio, J., Liu, D. and Meaney, M. J., 'Nongenomic transmission across generations of maternal behavior and stress responses in the rat', *Science*, vol. 286, 1999, pp. 1155–8.

Francis, R. C., *Epigenetics: How Environment Shapes Our Genes*, New York, Norton, 2011.

Fraser, M. J. and Fraser, R. D. B., 'Evidence on the structure of deoxyribonucleic acid from measurements with polarized infra-red radiation', *Nature*, vol. 167, 1951, pp. 761–2.

Freedman, A. H., Gronau, I., Schweizer, R. M. *et al.*, 'Genome sequencing highlights the dynamic early history of dogs', *PLoS Biology*, vol. 10, 2014, article e1004016.

Freeland, S. J. and Hurst, L. D., 'The genetic code is one in a million', *Journal of Molecular Evolution*, vol. 47, 1998, pp. 238–48.

Freeland, S. J., Knight, R. D., Landweber, L. F. and Hurst, L. D., 'Early fixation of an optimal genetic code', *Molecular Biology and Evolution*, vol. 17, 2000, pp. 511–18.

Friedberg, E., *Sydney Brenner: A Biography*, Cold Spring Harbor, Cold Spring Harbor Laboratory Press, 2010.

Friedmann, T. and Roblin, R., 'Gene therapy for human genetic disease?', *Science*, vol. 175, 1972, pp. 949–55.

Fu, Q., Li, H., Moorjani, P. *et al.*, 'Genome sequence of a 45,000-year-old modern human from western Siberia', *Nature*, vol. 514, 2014, pp. 445–9.

Gabor, D., 'Information theory', *The Times Science Review*, February 1953.

Galison, P., 'The ontology of the enemy: Norbert Wiener and the cybernetic vision', *Critical Inquiry*, vol. 21, 1994, pp. 228–66.

Gamow, G., 'Possible relation between deoxyribonucleic acid and protein structure', *Nature*, vol. 173, 1954, p. 318.

Gamow, G., 'Information transfer in the living cell', *Scientific American*, vol. 193 (10), 1955, pp. 70–8.

Gamow, G. and Abelson, P. H., 'The ninth Washington Conference on Theoretical Physics', *Science*, vol. 104, 1946, p. 574.

Gamow, G. and Metropolis, N., 'Numerology of polypeptide chains', *Science*, vol. 120, 1954, pp. 779–80.

Gamow, G., Rich, A. and Yčas, M., 'The problem of information transfer from the nucleic acids to proteins', *Advances in Biological and Medical Physics*, vol. 4, 1957, pp. 23–68.

Gann, A., 'Jacob and Monod: From operons to EvoDevo', *Current Biology*, vol. 20, 2010, pp. R718–23.

Gann, A. and Witkowski, J., 'The lost correspondence of Francis Crick', *Nature*, vol. 467, 2010, pp. 519–24.

Gann, A. and Witkowski, J. (eds), *The Annotated and Illustrated Double Helix*, London, Simon & Schuster, 2012.

García-Sancho, M., 'The rise and fall of the idea of genetic information (1948- 2006)', *Genomics, Society and Policy*, vol. 2, 2007.

García-Sancho, M., 'A new insight into Sanger's development of sequencing: From proteins to DNA, 1943–1977', *Journal of the History of Biology*, vol. 43, 2010, pp. 265–323.

García-Sancho, M., *Biology, Computing and the History of Molecular Sequencing: From Proteins to DNA, 1945–2000*, Basingstoke, Palgrave Macmillan, 2012.

Gardner, R. S., Wahba, A. J., Basilio, C. *et al.*, 'Synthetic polynucleotides and the amino acid code, VII', *Proceedings of the National Academy of Sciences USA*, vol. 48, 1962, pp. 2087–94.

Gasiunas, G., Barrangou, R., Horvath, P. and Siksnys, V., 'Cas9-crRNA ribonucleoprotein complex mediates specific DNA cleavage for adaptive immunity in bacteria', *Proceedings of the National Academy of Sciences USA*, vol. 109, 2012, pp. E2579–86.

Gatlin L. L., 'The information content of DNA', *Journal of Theoretical Biology*, vol. 10, 1966, pp. 281–300.

Gatlin, L. L., 'The information content of DNA. II.', *Journal of Theoretical Biology*, vol. 18, 1968, pp. 181–94.

Gatlin, L. L., *Information Theory and the Living System*, New York, Columbia University Press, 1972.

Gaudillière, J.-P., 'Paris-New York roundtrip: transatlantic crossings and the reconstruction of the biological sciences in post-war France', *Studies in History and Philosophy of Biological and Biomedical Sciences*, vol. 33, 2002, pp. 389–417.

Gayon, J., *Darwinism's Struggle for Survival: Heredity and the Hypothesis of Natural Selection*, Cambridge, Cambridge University Press, 1998.

Gayon, J., 'Hérédité des caractères acquis', in J. Gayon (ed.), *Lamarck, philosophe de la nature*, Paris, Presses Universitaires de France, 2006, pp. 105–63.

Genome of the Netherlands Consortium, 'Whole-genome sequence variation, population structure and demographic history of the Dutch population', *Nature Genetics*, vol. 46, 2014, pp. 818–25.

George, F. H., 'Models in cybernetics', *Symposia of the Society for Experimental Biology*, vol. 14, 1960, pp. 169–98.

George, F. H., *The Brain as a Computer*, London, Pergamon, 1962.

Germain, P.-L., Ratti, E. and Boem, F., 'Junk or functional DNA? ENCODE and the function controversy', *Biology and Philosophy*, vol. 29, 2014, pp. 807–31.

Gerstein, M. B., Bruce, C., Rozowsky, J. S. *et al.*, 'What is a gene, post-ENCODE? History and updated definition', *Genome Research*, vol. 17, 2007, pp. 669–81.

Gibson, D. G., Glass, J. I., Lartigue, C. *et al.*, 'Creation of a bacterial cell controlled by a chemically synthesized genome', *Science*, vol. 329, 2010, pp. 52–6.

Gilbert, W., 'Why genes in pieces', *Nature*, vol. 271, 1978, p. 501.

Gilbert, W., 'The RNA world', *Nature*, vol. 319, 1986, p. 618.

Gilbert, W., 'Towards a paradigm shift in biology', *Nature*, vol. 349, 1991, p. 99.

Gingras, Y., 'Revisiting the 'quiet debut' of the double helix: a bibliometric and methodological note on the 'impact' of scientific publications', *Journal of the History of Biology*, vol. 43, 2010, pp. 159–81.

Glass, B., 'Summary', in W. D. McElroy and B. Glass (eds), *A Symposium on the Chemical Basis of Heredity*, Baltimore, The Johns Hopkins Press, 1957, pp. 757–834.

Gleick, J., *The Information*, London, Fourth Estate, 2011.

Godfrey-Smith, P., 'On the theoretical role of 'genetic coding'', *Philosophy of Science*, vol. 67, 2000a, pp. 26–44.

Godfrey-Smith, P., 'Information, arbitrariness, and selection: Comments on Maynard Smith', *Philosophy of Science*, vol. 67, 2000b, pp. 202–7.

Godfrey-Smith, P., 'Information in biology', in D. Hull and M. Ruse (eds), *The Cambridge Companion to the Philosophy of Biology*, Cambridge, Cambridge University Press, 2007, pp. 103–19.

Godfrey-Smith, P., 'Senders, receivers, and genetic information: comments on Bergstrom and Rosvall', *Biology and Philosophy*, vol. 26, 2011, pp. 177–81.

Goldman, N., Bertone, P., Chen, S. *et al.*, 'Towards practical, high-capacity, low-maintenance information storage in synthesized DNA', *Nature*, vol. 494, 2013, pp. 77–80.

Golomb, S. W., 'Efficient coding for the deoxyribonucleic acid channel', in R. Bellman (ed.), *Proceedings of Symposia in Applied Mathematics XIV: Mathematical Problems in the Biological Sciences*, Providence, RI, American Mathematical Society, 1962a, pp. 87–100.

Golomb, S. W., 'Plausibility of the ribonucleic acid code', *Nature*, vol. 196, 1962b, p. 1228.

Grandy, D. A., *Leo Szilárd: Science as a Mode of Being*, London, University Press of America, 1996.

Graur, D., Zheng, Y., Price, N. *et al.*, 'On the immortality of television sets: "function" in the human genome according to the evolution-free gospel of ENCODE', *Genome Biology and Evolution*, vol. 5, 2013, pp. 578–90.

Green, R. E., Krause, J., Briggs, A. W. *et al.*, 'A draft sequence of the Neandertal genome', *Science*, vol. 328, 2010, pp. 710–22.

Gregory, T. R., 'Coincidence, coevolution, or causation? DNA content, cell size, and the C-value enigma', *Biological Reviews*, vol. 76, 2001, pp. 65–101.

Griffith, F., 'The significance of pneumococcal types', *The Journal of Hygiene*, vol. 27, 1928, pp. 113–59.

Griffiths, P. E., 'Genetic information: A metaphor in search of a theory', *Philosophy of Science*, vol. 68, 2001, pp. 394–412.

Griffiths, P. and Stotz, K., *Genetics and Philosophy: An Introduction*, Cambridge, Cambridge University Press, 2013.

Grmek, M. D. and Fantini, B., 'Rôle du hasard dans la naissance du modèle de l'opéron', *Revue de l'histoire des sciences*, vol. 35, 1982, pp. 193–215.

Gros, F., 'The messenger', in A. Lwoff and A. Ullmann (eds), *Origins of Molecular Biology: A Tribute to Jacques Monod*, London, Academic Press, 1979, pp. 117–24.

Gros, F., Hiatt, H, Gilbert, W. *et al.*, 'Unstable ribonucleic acid revealed by pulse labelling of *Escherichia coli*', *Nature*, vol. 190, 1961, pp. 581–5.

Grunberg-Manago, M., Ortiz, P. J. and Ochoa, S., 'Enzymatic synthesis of nucleic acidlike polynucleotides', *Science*, vol. 122, 1955, pp. 907–10.

Gulland, J. M., 'Some aspects of the chemistry of nucleotides', *Journal of the Chemistry Society*, vol. 1944, 1944, pp. 208–17.

Gulland, J. M., 'The structures of nucleic acids', *Symposia of the Society for Experimental Biology*, vol. 1, 1947a, pp. 1–14.

Gulland, J. M., 'The structures of nucleic acids', *Cold Spring Harbor Symposia in Quantitative Biology*, vol. 12, 1947b, pp. 95–103.

Guo, H., Zhu, P., Yan, L. *et al.*, 'The DNA methylation landscape of human early embryos', *Nature*, vol. 511, 2014, pp. 606–10.

Haddow, A., 'Transformation of cells and viruses', *Nature*, vol. 154, 1944, pp. 194–9.

Hager, T., *Force of Nature: The Life of Linus Pauling*, New York, Simon & Schuster, 1995.

Haldane, J. B. S., 'A physicist looks at biology', *Nature*, vol. 155, 1945, pp. 375–6.

Hall, K., 'William Astbury and the biological significance of nucleic acids, 1938–1951', *Studies in the History and Philosophy of Biological and Biomedical Sciences*, vol. 42, 2011, pp. 119–28.

Hall, K., *The Man in the Monkeynut Coat: William Astbury and the Forgotten Road to the Double Helix*, Oxford, Oxford University Press, 2014.

Halloran, S. M., 'The birth of molecular biology: An essay in the rhetorical criticism of scientific discourse', in R. A. Harris (ed.), *Landmark Essays on Rhetoric of Science: Case Studies*, Mahwah, NJ, Hermagoras Press, 1997, pp. 39–53.

Hammer, M. F., Woerner, A. E., Mendez, F. L. *et al.*, 'Genetic evidence for archaic admixture in Africa', *Proceedings of the National Academy of Sciences USA*, vol. 108, 2011, pp. 15123–8.

Hanawalt, P. C., 'Density matters: The semiconservative replication of DNA', *Proceedings of the National Academy of Sciences USA*, vol. 101, 2004, pp. 17889–94.

Hao, B., Gong, W., Ferguson, T. K. *et al.*, 'A new UAG-encoded residue in the structure of a methanogen methyltransferase', *Science*, vol. 296, 2002, pp. 1462–6.

Hargittai, I., *Candid Science II: Conversations with Famous Biomedical Sciences*, London, Imperial College Press, 2002.

Hartl, D. L. and Orel, V., 'What did Gregor Mendel think he discovered?', *Genetics*, vol. 131, 1992, pp. 245–53.

Hartl, F. U., Bracher, A. and Hayer-Hartl, M., 'Molecular chaperones in protein folding and proteostasis', *Nature*, vol. 475, 2011, pp. 324–32.

Hatfield, D. L. and Gladyshev, V. N., 'How selenium has altered our understanding of the genetic code', *Molecular and Cell Biology*, vol. 22, 2002, pp. 3565–76.

Hayden, E. C., 'Is the $1,000 genome for real?', http://www.nature.com/news/is-the-1–000-genome-for-real-1.14530, 2014.

Heard, E. and Martienssen, R. A., 'Transgenerational epigenetic inheritance: myths and mechanisms', *Cell*, vol. 157, 2014, pp. 95–109.

Heaton, N., 'Interview with Norma Heaton, November 18, 2010 by Jason Gart', http://profiles.nlm.nih.gov/ps/access/JJBCCX.pdf, 2010.

Hegreness, M. and Meselson, M., 'What did Sutton see?: Thirty years of confusion over the chromosomal basis of Mendelism', *Genetics*, vol. 176, 2007, pp. 1939–44.

Heidelberger, M., Kneeland, Jr, Y. and Price, K. M., 'Alphonse Raymond Dochez, 1882–1964', *Biographical Memoir*, Washington DC, National Academy of Sciences, 1971.

Heijmans, B. T., Tobi, E. W., Stein, A. D. *et al.*, 'Persistent epigenetic differences associated with prenatal exposure to famine in humans', *Proceedings of the National Academy of Sciences USA*, vol. 105, 2008, pp. 17046–9.

Heims, S. J., *John von Neumann & Norbert Weiner: From Mathematics to the Technologies of Life and Death*, London, MIT Press, 1980.

Heims, S. J., *The Cybernetics Group*, London, MIT Press, 1991.

Henikoff, S., Keene, M. A., Fechtel, K. and Fristrom, J. W., 'Gene within a gene: nested *Drosophila* genes encode unrelated proteins on opposite DNA strands', *Cell*, vol. 44, 1986, pp. 33–42.

Hershey, A. D., 'Functional differentiation within particles of bacteriophage T2', *Cold Spring Harbor Symposia on Quantitative Biology*, vol. 18, 1953, pp. 135–40.

Hershey, A. D., 'The injection of DNA into cells by phage', in J. Cairns, G. S. Stent and J. D. Watson (eds), *Phage and the Origins of Molecular Biology*, Cold Spring Harbor, Cold Spring Harbor Laboratory of Quantitative Biology, 1966, pp. 100–8.

Hershey, A. D., 'Transcript', in F. W. Stahl (ed.), *We Can Sleep Later: Alfred D. Hershey and the Origins of Molecular Biology*, Cold Spring Harbor, Cold Spring Harbor Laboratory Press, 2000, pp. 105–6.

Hershey, A. D. and Chase, M., 'Independent functions of viral protein and nucleic acid in growth of bacteriophage', *Journal of General Physiology*, vol. 36, 1952, pp. 39–56.

Higham, T., Douka, K., Wood, R. *et al.*, 'The timing and spatiotemporal patterning of Neanderthal disappearance', *Nature*, vol. 512, 2014, pp. 306–9.

Hoagland, M. B., Zamecnik, P. C. and Stephenson, M. L., 'Intermediate reactions in protein biosynthesis', *Biochimica Biophysica Acta*, vol. 24, 1957, pp. 215–16.

Hoagland, M. B., Stephenson, M. L., Scott, J. F. *et al.*, 'A soluble ribonucleic acid intermediate in protein synthesis', *Journal of Biological Chemistry*, vol. 231, 1958, pp. 241–57.

Hodges, A., *Alan Turing: The Enigma*, London, Vintage, 2012.

Hödl, M. and Basler, K., 'Transcription in the absence of histone H3.2 and H3K4 methylation', *Current Biology*, vol. 22, 2012, pp. 2253–7.

Holley, R. W., Apgar, J., Everett, G. A. *et al.*, 'Structure of a ribonucleic acid', *Science*, vol. 147, 1965, pp. 1462–5.

Holliday, R., 'Physics and the origins of molecular biology', *Journal of Genetics*, vol. 85, 2006, pp. 93–7.

Holmes, F. L., *Meselson, Stahl, and the Replication of DNA: A History of 'The Most Beautiful Experiment in Biology'*, London, Yale University Press, 2001.

Holmes, F. L., *Reconceiving the Gene: Seymour Benzer's Adventures in Phage Genetics*, ed. W. C. Summers, London, Yale University Press, 2006.

Holzmüller, W., *Information in Biological Systems: The Role of Macromolecules*, Cambridge, Cambridge University Press, 1984.

Hong, X., Scofield, D. G. and Lynch, M., 'Intron size, abundance, and distribution within untranslated regions of genes', *Molecular Biology and Evolution*, vol. 23, 2006, pp. 2392–404.

Horowitz, N. H., Berg, P., Singer, M. *et al.*, 'A centennial: George W. Beadle, 1903–1989', *Genetics*, vol. 166, 2004, pp. 1–10.

Hotchkiss, R. D., 'Etudes chimiques sur le facteur transformant du pneumocoque', *Colloques Internationaux du Centre National de la Recherche Scientifique*, vol. 8, 1949, pp. 57–65.

Hotchkiss, R. D., 'Oswald T. Avery', *Genetics*, vol. 51, 1965, pp. 1–10.

Hotchkiss, R. D., 'The identification of nucleic acids as genetic determinants', *Annals of the New York Academy of Sciences*, vol. 325, 1979, pp. 321–42.

Hotchkiss, R. D., 'DNA in the decade before the double helix', *Annals of the New York Academy of Sciences*, vol. 758, 1995, pp. 55–73.

Hotchkiss, R. D., 'Growing up into our long genes', in F. W. Stahl (ed.), *We Can Sleep Later: Alfred D. Hershey and the Origins of Molecular Biology*, Cold Spring Harbor, Cold Spring Harbor Laboratory Press, 2000, pp. 33–43.

Hsu, P. D., Lander, E. S. and Zhang, F., 'Development and applications of CRISPR-Cas9 for genome engineering', *Cell*, vol. 157, 2014, pp. 1262–78.

Huerta-Sánchez, E., Jin, X., Asan *et al.*, 'Altitude adaptation in Tibetans caused by introgression of Denisovan-like DNA', *Nature*, vol. 512, 2014, pp. 194–7.

Hunter, G. K., *Vital Forces: The Discovery of the Molecular Basis of Life*, London, Academic Press, 2000.

Hunter, N., 'Prion diseases and the central dogma of molecular biology', *Trends in Microbiology*, vol. 7, 1999, pp. 265–6.

Inglis, J., Sambrook, J. and Witkowski, J. (eds), *Inspiring Science: Jim Watson and the Age of DNA*, Plainview, Cold Spring Harbor Laboratory Press, 2003.

Ingram, V. M., 'A specific chemical difference between the globins of normal human and sickle-cell anaemia haemoglobin', *Nature*, vol. 178, 1956, pp. 792–4.

Ingram, V. M., 'Gene mutations in human haemoglobin: the chemical difference between normal and sickle cell haemoglobin', *Nature*, vol. 180, 1957, pp. 326–8.

Ingram, V. M., 'Sickle-cell anemia hemoglobin: The molecular biology of the first 'molecular disease' – The crucial importance of serendipity', *Genetics*, vol. 167, 2004, pp. 1–7.

Itzkovitz, S., Hodis, E. and Segal, E., 'Overlapping codes within protein-coding sequences', *Genome Research*, vol. 20, 2010, pp. 1582–9.

Ivanova, N. N., Schwientek, P., Tripp, H. J. *et al.*, 'Stop codon reassignments in the wild', *Science*, vol. 344, 2014, pp. 909–13.

Jablonka, E., 'Information: its interpretation, its inheritance and its sharing', *Philosophy of Science*, vol. 69, 2002, pp. 578–605.

Jablonka, E. and Lamb, M. J., 'The evolution of information in the major transitions', *Journal of Theoretical Biology*, vol. 239, 2006, pp. 236–46.

Jackson, D. A., Symons, R. H. and Berg, P., 'Biochemical method for inserting new genetic information into DNA of Simian Virus 40: circular SV40 DNA molecules containing lambda phage genes and the galactose operon of *Escherichia coli*', *Proceedings of the National Academy of Sciences USA*, vol. 69, 1972, pp. 2904–9.

Jacob, F., 'Genetics of the bacterial cell – Nobel Lecture, December 11, 1965', in *Nobel Lectures, Physiology or Medicine 1963–1970*, Amsterdam, Elsevier Publishing Company, 1972, pp. 148–71.

Jacob, F., 'Evolution and tinkering', *Science*, vol. 196, 1977, pp. 1161–6.

Jacob, F., 'The switch', in A. Lwoff and A. Ullmann (eds), *Origins of Molecular Biology: A Tribute to Jacques Monod*, London, Academic Press, 1979, pp. 95–108.

Jacob, F., *The Statue Within*, London, Unwin Hyman, 1988.

Jacob, F., 'The birth of the operon', *Science*, vol. 332, 2011, p. 767.

Jacob, F. and Monod, J., 'Genetic regulatory mechanisms in the synthesis of proteins', *Journal of Molecular Biology*, vol. 3, 1961a, pp. 318–56.

Jacob, F. and Monod, J., 'On the regulation of gene activity', *Cold Spring Harbor Symposia on Quantitative Biology*, vol. 26, 1961b, pp. 193–212.

Jacob, F., Perrin, D., Sanchez, C., and Monod, J., 'L'opéron : groupe de gènes à expression coordonnée par un opérateur', *Comptes rendus hebdomodaires des séances de l'Académie des sciences*, vol. 250, 1960, pp. 1727–9.

Jeong, C., Alkorta-Aranburu, G., Basnyat, B. *et al.*, 'Admixture facilitates genetic adaptations to high altitude in Tibet', *Nature Communications*, vol. 5, 2014, p. 3281.

Jinek, M., Chylinski, K., Fonfara, I. *et al.*, 'A programmable dual-RNA-guided DNA endonuclease in adaptive bacterial immunity', *Science*, vol. 337, 2012, pp. 816–21.

Johnson, A. P., Cleaves, H. J., Dworkin, J. P. *et al.*, 'The Miller volcanic spark discharge experiment', *Science*, vol. 322, 2008, p. 404.

Johnson, J. A., Lu, Y. Y., Van Deventer, J. A. and Tirrell, D. A., 'Residue-specific incorporation of non-canonical amino acids into proteins: recent developments and applications', *Current Opinion in Chemical Biology*, vol. 14, 2010, pp. 774–80.

Joyce, G. F., 'The antiquity of RNA-based evolution', *Nature*, vol. 418, 2002, pp. 214–21.

Joyce, G. F., 'Toward an alternative biology', *Science*, vol. 336, 2012a, pp. 307–8.

Joyce, G. F., 'Bit by bit: The Darwinian basis of life', *PLoS Biology*, vol. 10, 2012b, article e1001323.

Judson, H. F., *The Eighth Day of Creation: Makers of the Revolution in Biology*, Plainview, Cold Spring Harbor Laboratory Press, 1996.

Jukes, T. H., 'Relations between mutations and base sequences in the amino acid code', *Proceedings of the National Academy of Sciences USA*, vol. 48, 1962, pp. 1809–15.

Jukes, T. H., 'Observations on the possible nature of the genetic code', *Biochemical and Biophysical Research Communications*, vol. 10, 1963, pp. 155–9.

Kalmus, H., 'A cybernetical aspect of genetics', *The Journal of Heredity*, vol. 41, 1950, pp. 19–22.

Kalmus, H., 'Analogies of language to life', *Language and Speech*, vol. 5, 1962, pp. 15–25.

Katzman, S., Capra, J. A., Haussler, D. and Pollard, K. S., 'Ongoing GC-biased evolution is widespread in the human genome and enriched near recombination hot spots', *Genome Biology and Evolution*, vol. 3, 2011, pp. 614–26.

Kay, L. E., 'W. M. Stanley's crystallization of the Tobacco Mosaic Virus, 1930–1940', *Isis*, vol. 77, 1986, pp. 450–72.

Kay, L. E., 'Who wrote the book of life? Information and the transformation of molecular biology, 1945–55', *Science in Context*, vol. 8, 1995, pp. 609–34.

Kay, L. E., *Who Wrote the Book of Life? A History of the Genetic Code*, Stanford, Stanford University Press, 2000.

Keller, E. F., *Refiguring Life: Metaphors of Twentieth-Century Biology*, New York, Columbia University Press, 1995.

Keller, E. F., *The Century of the Gene*, Cambridge, MA, Harvard University Press, 2000.

Keller, E. F., *Making Sense of Life: Explaining Biological Development with Models, Metaphors and Machines*, Cambridge, MA, Harvard University Press, 2002.

Kellis, M., Wold, B., Snyder, M. P. *et al.*, 'Defining functional DNA elements in the human genome', *Proceedings of the National Academy of Sciences USA*, vol. 111, 2014, pp. 6131–8.

Keyes, M. E., 'The prion challenge to the 'central dogma' of molecular biology. Part I: Prelude to prions', *Studies in History and Philosophy of Science Part C: Studies in History and Philosophy of Biological and Biomedical Sciences*, vol. 30, 1999a, pp. 1–19.

Keyes, M. E., 'The prion challenge to the 'central dogma' of molecular biology. Part II: The problem with prions', *Studies in History and Philosophy of Science Part C: Studies in History and Philosophy of Biological and Biomedical Sciences*, vol. 30, 1999b, pp. 181–218.

Kim, G., LeBlanc, M. L., Wafula, E. K. *et al.*, 'Genomic-scale exchange of mRNA between a parasitic plant and its hosts', *Science*, vol. 345, 2014, pp. 808–11.

Kimura, M., 'Natural selection as the process of accumulating genetic information in adaptive evolution', *Genetical Research*, vol. 2, 1961, pp. 127–40.

King, G. W., 'Information', *Scientific American*, vol. 187 (9), 1952, pp. 132–48.

Kishida, T., Kubota, S., Shirayama, Y. and Fukami, H., 'The olfactory receptor gene repertoires in secondary-adapted marine vertebrates: evidence for reduction of the functional proportions in cetaceans', *Biology Letters*, vol. 3, 2007, pp. 428–30.

Kjosavik, F., 'Genes, structuring powers and the flow of information in living systems', *Biology and Philosophy*, vol. 29, 2014, pp. 379–94.

Klessig, D. F., 'Two adenovirus mRNAs have a common 5' terminal leader sequence encoded at least 10 kb upstream from their main coding regions', *Cell*, vol. 12, 1977, pp. 9–21.

Klug, A., 'The discovery of the DNA double helix', *Journal of Molecular Biology*, vol. 335, 2004, pp. 3–26.

Kogge, W., 'Script, code, information: How to differentiate analogies in the 'prehistory' of molecular biology', *History and Philosophy of the Life Sciences*, vol. 34, 2012, pp. 603–35.

Kohler, R. E., *Lords of the Fly:* Drosophila *Genetics and the Experimental Life*, Chicago, University of Chicago Press, 1994.

Koonin, E. V. and Martin, W., 'On the origin of genomes and cells within inorganic compartments', *Trends in Genetics*, vol. 21, 2005, pp. 647–54.

Koonin, E. V. and Novozhilov, A. S., 'Origin and evolution of the genetic code: The universal enigma', *IUBMB Life*, vol. 61, 2009, pp. 99–111.

Kresge, N., Simoni, R. D. and Hill, R. L., 'H. Edwin Umbarger's contributions to the discovery of feedback inhibition', *Journal of Biological Chemistry*, vol. 280, 2005, article e49.

Krings, M., Stone, A., Schmitz, R. W. *et al.*, 'Neandertal DNA sequences and the origin of modern humans', *Cell*, vol. 90, 1997, pp. 19–30.

Kryukov, G. V., Castellano, S., Novoselov, S. V. *et al.*, 'Characterization of mammalian selenoproteomes', *Science*, vol. 300, 2003, pp. 1439–43.

Lah, G. J-E., Li, J. S. S. and Millard, S. S., 'Cell-specific alternative splicing of *Drosophila Dscam2* is crucial for proper neuronal wiring', *Neuron*, vol. 83, 2014, pp. 1376–88.

Lajoie, M. J., Rovner, A. J., Goodman, D. B. *et al.*, 'Genomically recoded organisms expand biological functions', *Science*, vol. 342, 2013a, pp. 357–60.

Lajoie, M. J., Kosuri, S., Mosberg, J. A. *et al.*, 'Probing the limits of genetic recoding in essential genes', *Science*, vol. 342, 2013b, pp. 361–3.

Laland, K., Uller, T., Feldman, F. *et al.*, 'Does evolutionary theory need a rethink?', *Nature*, vol. 514, 2014, pp. 161–4.

Lamborg, M. R. and Zamecnik, P. C., 'Amino acid incorporation into protein by extracts of *E. coli*', *Biochimica et Biophysica Acta*, vol. 42, 1960, pp. 206–11.

Lander, E. S., Linton, L. M., Birren, B. *et al.*, 'Initial sequencing and analysis of the human genome', *Nature*, vol. 409, 2001, pp. 860–921.

Lane, N., *Life Ascending: The Ten Great Inventions of Evolution*, London, Profile, 2009.

Lane, N. and Martin, W., 'The energetics of genome complexity', *Nature*, vol. 467, 2010, pp. 929–34.

Lane, N. and Martin, W., 'The origin of membrane bioenergetics', *Cell*, vol. 151, 2012, pp. 1407–16.

Lanni, F., 'Biological validity of amino acid codes deduced with synthetic ribonucleotide polymers', *Proceedings of the National Academy of Sciences USA*, vol. 48, 1962, pp. 1623–30.

Lanouette, W., *Genius in the Shadows. A biography of Leo Szilárd: The Man Behind the Bomb*, Chicago, University of Chicago Press, 1994.

Lanouette, W., 'The science and politics of Leo Szilárd, 1898–1964: evolution, revolution, or subversion?', *Science and Public Policy*, vol. 33, 2006, pp. 613–17.

Lazaris, A., Arcidiacono, S., Huang, Y. *et al.*, 'Spider silk fibers spun from soluble recombinant silk produced in mammalian cells', *Science*, vol. 295, 2002, pp. 472–6.

Lean, O. M., 'Getting the most out of Shannon information', *Biology and Philosophy*, vol. 29, 2014, pp. 395–413.

Lederberg, J., 'Problems in microbial genetics', *Heredity*, vol. 2, 1948, pp. 145–98.

Lederberg, J., 'Cell genetics and hereditary symbiosis', *Physiological Reviews*, vol. 32, 1952, pp. 403–30.

Lederberg, J., 'Discussion', in W. D. McElroy and B. Glass (eds), *A Symposium on the Chemical Basis of Heredity*, Baltimore, The Johns Hopkins Press, 1957, pp. 752–4.

Lekomtsev, S., Kolosov, P., Bidou, L. *et al.*, 'Different modes of stop codon restriction by the *Stylonychia* and *Paramecium* eRF1 translation termination factors', *Proceedings of the National Academy of Sciences USA*, vol. 104, 2007, pp. 10824–9.

Lengyel, P., 'Memories of a senior scientist: On passing the fiftieth anniversary of the beginning of deciphering the genetic code', *Annual Review of Microbiology*, vol. 66, 2012, pp. 27–38.

Lengyel, P., Speyer, J. F. and Ochoa, S., 'Synthetic polynucleotides and the amino acid code', *Proceedings of the National Academy of Sciences USA*, vol. 47, 1961, pp. 1936–42.

Lengyel, P., Speyer, J. F., Basilio, C. and Ochoa, S., 'Synthetic polynucleotides and the amino acid code, III', *Proceedings of the National Academy of Sciences USA*, vol. 48, 1962, pp. 282–4.

Levy, A., 'Information in biology: a fictionalist account', *Noûs*, vol. 45, 2011, pp. 640–57.

Lewin, B., 'A journal of exciting biology', *Cell*, vol. 1, 1974, p. 1.

Lewis, J. B., Anderson, C. W. and Atkins, J. F., 'Further mapping of late adenovirus genes by cell-free translation of RNA selected by hybridization to specific DNA fragments', *Cell*, vol. 12, 1977, pp. 37–44.

Lewis, M., 'A tale of two repressors', *Journal of Molecular Biology*, vol. 409, 2011, pp. 14–27.

Li, G.-W. and Xie, X. S., 'Central dogma at the single-molecule level in living cells', *Nature*, vol. 475, 2011, pp. 308–15.

Li, R., Fan, W., Tian, G. *et al.*, 'The sequence and *de novo* assembly of the giant panda genome', *Nature*, vol. 463, 2010, pp. 311–17.

Lim, Y. W., Cuevas, D. A., Silva, G. G. Z. *et al.*, 'Sequencing at sea: challenges and experiences in Ion Torrent PGM sequencing during the 2013 Southern Line Islands Research Expedition', *PeerJ*, vol. 2, 2014, article e520.

Lin, M. F., Kheradpour, P., Washietl, S. *et al.*, 'Locating protein-coding sequences under selection for additional, overlapping functions in 29 mammalian genomes', *Genome Research*, vol. 21, 2011, pp. 1916–28.

Lincoln, T. A. and Joyce, G. F., 'Self-sustained replication of an RNA enzyme', *Science*, vol. 323, 2009, pp. 1229–32.

Linschitz, H., 'The information content of a bacterial cell', in H. Quastler (ed.), *Essays on the Use of Information Theory in Biology*, Urbana, University of Illinois Press, 1953, pp. 251–62.

Lobanov, A. V., Kryukov, G. V., Hatfield, D. L. and Gladyshev, V. N., 'Is there a twenty third amino acid in the genetic code?', *Trends in Genetics*, vol. 22, 2006, pp. 357–60.

Lockyer, R., *Transmission of Chemosensory Information in* Drosophila melanogaster: *Behavioural Modification and Evolution*, Unpublished PhD thesis, University of Manchester, 2014.

Longo, G., Miquel, P.-A., Sonnenschein, C. and Soto, A. M., 'Is information a proper observable for biological organization?', *Progress in Biophysics and Molecular Biology*, vol. 109, 2012, pp. 108–14.

López-Beltrán, C., 'Forging heredity: From metaphor to cause, a reification story', *Studies in History and Philosophy of Science*, vol. 25, 1994, pp. 211–35.

Lozupone C. A., Knight, R. D. and Landweber, L. F., 'The molecular basis of nuclear genetic code change in ciliates', *Current Biology*, vol. 11, 2001, pp. 65–74.

Lumey, L. H., Stein, A. D., Kahn, H. S. and Romijn, J. A., 'Lipid profiles in middle-aged men and women after famine exposure during gestation: the Dutch Hunger Winter Families Study', *American Journal of Clinical Nutrition*, vol. 89, 2009, pp. 1737–43.

Lwoff, A., 'Essai de conclusion', *Colloques Internationaux du Centre National de la Recherche Scientifique*, vol. 8, 1949, pp. 201–3.

Maaløe, O. and Watson, J. D., 'The transfer of radioactive phosphorus from parental to progeny phage', *Proceedings of the National Academy of Sciences USA*, vol. 37, 1951, pp. 507–13.

Maas, W., 'Leo Szilard: A personal remembrance', *Genetics*, vol. 167, 2004, pp. 555–8.

MacColl, L. R. A., *Fundamental Theory of Servomechanisms*, New York , D. Van Nostrand, 1945.

Maclaurin, J., 'Commentary on "The Transmission Sense of Information" by Carl T. Bergstrom and Martin Rosvall', *Biology and Philosophy*, vol. 26, 2011, pp. 191–4.

Macrae, N., *John von Neumann*, New York, Pantheon Books, 1992.

Maddox, B., *Rosalind Franklin: The Dark Lady of DNA*, London, Harper Collins, 2002.

Maderspacher, F., 'Lysenko rising', *Current Biology*, vol. 20, 2010, pp. R835–7.

Mali, P., Yang, L., Esvelt, K. M. *et al.*, 'RNA-guided human genome engineering via Cas9', *Science*, vol. 339, 2013, pp. 823–6.

Malyshev, D. A., Dhami, K., Lavergne, T. *et al.*, 'A semi-synthetic organism with an expanded genetic alphabet', *Nature*, vol. 509, 2014, pp. 385–8.

Manchester, K. L., 'Did a tragic accident delay the discovery of the double helical structure of DNA?', *Trends in Biochemical Sciences*, vol. 20, 1995, pp. 126–8.

Manchester, K. L., 'Historical opinion: Erwin Chargaff and his 'rules' for the base composition of DNA: why did he fail to see the possibility of complementarity?', *Trends in Biochemical Sciences*, vol. 33, 2008, pp. 65–70.

Mandell, D. J., Lajoie, M. J., Mee, M. T., *et al.*, 'Biocontainment of genetically modified organisms by synthetic protein design', *Nature*, vol. 518, 2015, pp. 55–60.

Manuelidis, L., Yu, Z.-X., Barquero, N. and Mullins, B., 'Cells infected with scrapie and Creutzfeldt–Jakob disease agents produce intracellular 25-nm virus-like particles', *Proceedings of the National Academy of Sciences USA*, vol. 104, 2007, pp. 1965–70.

Marahiel, M. A., 'Working outside the protein-synthesis rules: insights into non-ribosomal peptide synthesis', *Journal of Peptide Science*, vol. 15, 2009, pp. 799–807.

Maraia, R. J. and Iben, J. R., 'Different types of secondary information in the genetic code', *RNA*, vol. 20, 2014, pp. 977–84.

Mardis, E. R., 'Next-generation DNA sequencing methods', *Annual Review of Genomics and Human Genetics*, vol. 9, 2008, pp. 387–402.

Martin, R. G., 'A revisionist view of the breaking of the genetic code', in D. Stetten Jr and W. T. Carrigan (eds), *NIH: An Account of Research in Its Laboratories and Clinics*, Orlando, Academic Press, 1984, pp. 282–95.

Martin, R. G., Matthaei, J. H., Jones, O. W. and Nirenberg, M. W., 'Ribonucleotide composition of the genetic code', *Biochemical and Biophysical Research Communications*, vol. 6, 1961, pp. 410–14.

Martin, W. F., Sousa, F. L. and Lane, N., 'Energy at life's origin', *Science*, vol. 344, 2014, pp. 1092–3.

Marvin, D. A., Spencer, M., Wilkins, M. H. F. and Hamilton, L. D., 'The molecular configuration of deoxyribonucleic acid III. X-ray diffraction study of the C-form of the lithium salt', *Journal of Molecular Biology*, vol. 3, 1961, pp. 547–65.

Masani, P. R., *Norbert Wiener 1894–1964*, Berlin, Birkhaüser Verlag, 1990.

Matthaei, J. H. and Nirenberg, M. W., 'Some characteristics of a cell-free DNAase sensitive system incorporating amino acids into protein.', *Federation Proceedings*, vol. 20, 1960, p. 391.

Matthaei, H. and Nirenberg, M. W., 'The dependence of cell-free protein synthesis in *E. coli* upon RNA prepared from ribosomes', *Biochemical and Biophysical Research Communications*, vol. 4, 1961a, pp. 404–8.

Matthaei, J. H. and Nirenberg, M. W., 'Characteristics and stabilization of DNAase-sensitive protein synthesis in *E. coli* extracts', *Proceedings of the National Academy of Sciences USA*, vol. 47, 1961b, pp. 1580–8.

Matthaei, J. H., Jones, O. W., Martin, R. G. and Nirenberg, M. W., 'Characteristics and composition of RNA coding units', *Proceedings of the National Academy of Sciences USA*, vol. 48, 1962, pp. 666–77.

Mattick, J. S., 'Rocking the foundations of molecular genetics', *Proceedings of the National Academy of Sciences USA*, vol. 109, 2012, pp. 16400–1.

Mavilio, F., 'Gene therapies need new development models', *Nature*, vol. 490, 2012, p. 7.

Maynard Smith, J., 'The idea of information in biology', *The Quarterly Review of Biology*, vol. 74, 1999, pp. 395–400.

Maynard Smith, J., 'The concept of information in biology', *Philosophy of Science*, vol. 67, 2000a, pp. 214–18.

Maynard Smith, J., 'Reply to commentaries', *Philosophy of Science*, vol. 67, 2000b, pp. 177–94.

Maynard Smith, J. and Szathmáry, E., *The Major Transitions in Evolution*, Oxford, Oxford University Press, 1997.

Mayr, E., 'Lamarck revisited', *Journal of the History of Biology*, vol. 5, 1972, pp. 55–94.

Mayr, O., *The Origins of Feedback Control*, London, MIT Press, 1970.

Mazia, D., 'Physiology of the cell nucleus', in E. S. G. Barron (ed.), *Modern Trends in Physiology and Biochemistry*, New York, Academic Press, 1952, pp. 77–122.

McCarty, M., *The Transforming Principle: Discovering that Genes Are Made of DNA*, New York, Norton, 1986.

McCarty, M., 'Some observations on the early history of DNA', unpublished speech to the Institute of Human Virology, http://profiles.nlm.nih.gov/ps/access/CCAAAV.pdf, 1996.

McCarty, M., 'Maclyn McCarty', in I. Hargittai (ed.), *Candid Science II: Conversations with Famous Biomedical Scientists*, London, Imperial College Press, 2002, pp. 16–31.

McCarty, M. and Avery, O. T., 'Studies on the chemical nature of the substance inducing transforming of pneumococcal types. II. Effect of desoxyribonuclease on the biological activity of the trnasforming substance', *Journal of Experimental Medicine*, vol. 83, 1946a, pp. 89–96.

McCarty, M. and Avery, O. T., 'Studies on the chemical nature of the substance inducing transformation of pneumococcal types. III. An improved method for the isolation of the transforming substance and its application to pneumococcus types II, III, and VI', *Journal of Experimental Medicine*, vol. 83, 1946b, pp. 97–104.

McCarty, M., Taylor, H. E. and Avery, O. T., 'Biochemical studies of environmental factors essential in transformation of pneumococcal types', *Cold Spring Harbor Symposia on Quantitative Biology*, vol. 11, 1946, pp. 177–83.

McCutcheon, J. P. and Moran, N. A., 'Extreme genome reduction in symbiotic bacteria', *Nature Reviews Microbiology*, vol. 10, 2011, pp. 13–26.

McElheny, V. K., *Watson and DNA: Making a Scientific Revolution*, Cambridge, MA, Perseus, 2003.

McInerney, J. O., O'Connell, M. J. and Pisani, D., 'The hybrid nature of the Eukaryota and a consilient view of life on Earth', *Nature Reviews Microbiology*, vol. 12, 2014, pp. 449–55.

Medina, E., *Cybernetic Revolutionaries: Technology and Politics in Allende's Chile*, Cambridge, MA, MIT Press, 2011.

Meselson, M. and Stahl, F. W., 'The replication of DNA in *Escherichia coli*', *Proceedings of the National Academy of Sciences USA*, vol. 44, 1958, pp. 671–82.

Mignone, F., Gissi, C., Liuni, S. and Pesole, G., 'Untranslated regions of mRNAs', *Genome Biology*, vol. 3, 2002, article reviews0004-reviews0004.10.

Miller, S., 'A production of amino acids under possible primitive Earth conditions', *Science*, vol. 117, 1953, pp. 528–9.

Mindell, D. A., 'Automation's finest hour: Bell Laboratories' control systems in World War II', *IEEE Control Systems Magazine*, December 1995.

Mindell, D. A., 'Automation's finest hour: Radar and system integration in World War II', in T. P. Hughes and A. Hughes (eds), *Systems, Experts, and Computers: The Systems Approach in Management and Engineering, World War II and After*, London, MIT Press, 2000, pp. 27–56.

Mindell, D. A., *Between Human and Machine: Feedback, Control, and Computing before Cybernetics*, Baltimore, Johns Hopkins University Press, 2002.

Mirsky, A. E., 'Chromosomes and nucleoproteins', *Advances in Enzymology*, vol. 3, 1943, pp. 1–34.

Mirsky, A. E. and Pollister, A. W., 'Chromosin, a desoxyribose nucleoprotein complex of the cell nucleus', *Journal of General Physiology*, vol. 30, 1946, pp. 117–48.

Monod, J., 'Biosynthese eines Enzyms: Information, Induktion, Repression', *Angewandte Chemie*, vol. 71, 1959, pp. 685–708.

Monod, J., 'Foreword', in B. T. Field and G. W. Szilard (eds), *The Collected Works of Leo Szilard: Scientific Papers*, London, MIT Press, 1972a, pp. xv–xvii.

Monod, J., 'From enzymatic adaptation to allosteric transitions: Nobel Lecture, December 11, 1965', in *Nobel Lectures, Physiology or Medicine 1963–1970*, Amsterdam, Elsevier Publishing Company, 1972b, pp. 188–209.

Monod, J. and Cohen-Bazire, G., 'L'effet inhibiteur spécifique des beta-galactosides dans la biosynthèse constitutive de la beta-galactosidase chez E. coli', *Comptes rendus hebdomódaires des séances de l'Académie des sciences*, vol. 236, 1953a, pp. 417–19.

Monod, J. and Cohen-Bazire, G., 'L'effet d'inhibition spécifique dans la biosynthèse de la tryptophane-desmase chez Aeriobacter aerogenes', *Comptes rendus hebdomodaires des séances de l'Académie des sciences*, vol. 236, 1953b, pp. 530–2.

Monod, J. and Jacob, F., 'Teleonomic mechanism in cellular metabolism, growth and differentiation', *Cold Spring Harbor Symposia on Quantitative Biology*, vol. 26, 1961, pp. 389–401.

Montefiore, H., 'Heavenly insemination', *Nature*, vol. 296, 1982, pp. 496–7.

Moore, W., *Schrödinger: Life and Thought*, Cambridge, Cambridge University Press, 1989.

Moran, N. A. and Jarvik, T., 'Lateral transfer of genes from fungi underlies carotenoid production in aphids', *Science*, vol. 328, 2010, pp. 624–7.

Morange, M., 'Schrödinger et la biologie moléculaire', *Fundamenta Scientiae*, vol. 4, 1983, pp. 219–34.

Morange, M., *A History of Molecular Biology*, Harvard, Harvard University Press, 1998.

Morange, M., 'What history tells us. I. The operon model and its legacy', *Journal of Bioscience*, vol. 30, 2005a, pp. 461–4.

Morange, M., 'What history tells us. II. The discovery of chaperone function', *Journal of Bioscience*, vol. 30, 2005b, pp. 313–16.

Morange, M., 'What history tells us. VIII. The progressive construction of a mechanism for prion diseases', *Journal of Bioscience*, vol. 32, 2007a, pp. 223–7.

Morange, M., 'What history tells us. IX. Z-DNA: When nature is not opportunistic', *Journal of Bioscience*, vol. 32, 2007b, pp. 657–61.

Morange, M., 'What history tells us. XIII. Fifty years of the Central Dogma', *Journal of Bioscience*, vol. 33, 2008, pp. 171–5.

Morange, M., 'The scientific legacy of Jacques Monod', *Research in Microbiology*, vol. 161, 2010, pp. 77–81.

Morange, M., 'What history tells us. XXIV. The attempt of Nikolai Koltzoff (Koltsov) to link genetics, embryology and physical chemistry', *Journal of Bioscience*, vol. 36, 2011, pp. 211–14.

Morange, M., 'François Jacob (1920–2013)', *Nature*, vol. 497, 2013, p. 440.

Morgan, T. H., *The Physical Basis of Heredity*, London, Lippincott, 1919.

Morgan, T. H., 'Nobel lecture: The relation of genetics to physiology and medicine', http://www.nobelprize.org/nobel_prizes/medicine/laureates/1933/morgan-lecture.html, 1933.

Morgan, W. T. J., 'Transformation of pneumococcal types', *Nature*, vol. 153, 1944, pp. 763–4.

Morgan, W. T. J., 'Fifth International Congress of Biochemistry', *Nature*, vol. 192, 1961, pp. 1115–16.

Moroz, L. L., Kocot, K. M., Citarella, M. R. *et al.*, 'The ctenophore genome and the evolutionary origins of neural systems', *Nature*, vol. 510, 2014, pp. 109–14.

Morris, K. V. and Mattick, J. S., 'The rise of regulatory RNA', *Nature Reviews Genetics*, vol. 15, 2014, pp. 423–37.

Morwood, M. J., Soejono, R. P., Roberts, R. G. *et al.*, 'Archaeology and age of a new hominin from Flores in eastern Indonesia', *Nature*, vol. 431, 2004, pp. 1087–91.

Mosini, V., 'Proteins, the chaperone function and heredity', *Biology and Philosophy*, vol. 28, 2013, pp. 53–74.

Mueller, J. H., 'The chemistry and metabolism of bacteria', *Annual Review of Biochemistry*, vol. 14, 1945, pp. 733–48.

Muller, H. J., 'Variation due to change in the individual gene', *The American Naturalist*, vol. 56, 1922, pp. 32–50.

Muller, H. J., 'Pilgrim Trust Lecture: The Gene', *Proceedings of the Royal Society of London B*, vol. 134, 1947, pp. 1–37.

Müller-Hill, B., *The Lac Operon: A Short History of a Genetic Paradigm*, New York, de Gruyter, 1996.

Müller-Wille, S. and Rheinberger, H.-J. (eds), *Heredity Produced: At the Crossroads of Biology, Politics, and Culture, 1500–1870*, London, MIT Press, 2007.

Müller-Wille, S. and Rheinberger, H.-J., *A Cultural History of Heredity*, London, University of Chicago Press, 2012.

Neale, D. B., Wegrzyn, J. L., Stevens, K. A. *et al.*, 'Decoding the massive genome of loblolly pine using haploid DNA and novel assembly strategies', *Genome Biology*, vol. 15, 2014, article R59.

Neel, J. V., 'The inheritance of sickle cell anemia', *Science*, vol. 110, 1949, pp. 64–6.

Nelson, V. R., Heaney, J. D., Tesar, P. J. *et al.*, 'Transgenerational epigenetic effects of Apobec1 cytidine deaminase deficiency on testicular germ cell tumor susceptibility and embryonic viability', *Proceedings of the National Academy of Sciences USA*, vol. 109, 2012. pp. E2766–73.

Neves, G., Zucker, J., Daly, M. and Chess, A., 'Stochastic yet biased expression of multiple Dscam splice variants by individual cells', *Nature Genetics*, vol. 36, 2004, pp. 240–6.

Newman, S. A., 'The fall and rise of systems biology: recovering from a half-century gene binge', *GeneWatch*, July–August 2003, pp. 8–12.

Nirenberg, M. W., 'The induction of two similar enzymes by one inducer. A test case for shared genetic information', *Federation Proceedings*, vol. 19, 1960, p. 42.

Nirenberg, M. W., 'The genetic code: II', *Scientific American*, vol. 208 (3), 1963, pp. 80–94.

Nirenberg, M., 'Historical review: Deciphering the genetic code – a personal account', *Trends in Biochemical Sciences*, vol. 29, 2004, pp. 46–54.

Nirenberg, M. W. and Jones, O. W., 'The current status of the RNA code', in H. J. Vogel, V. Bryson and J. O. Lampen (eds), *Informational Macromolecules*, London, Academic Press, 1963, pp. 451–66.

Nirenberg, M. W. and Leder, P., 'RNA codewords and protein synthesis', *Science*, vol. 145, 1964, pp. 1399–407.

Nirenberg, M. W. and Matthaei, J. H., 'The dependence of cell-free protein synthesis in *E. coli* upon naturally occurring or synthetic polyribonucleotides', *Proceedings of the National Academy of Sciences USA*, vol. 47, 1961, pp. 1588–602.

Nirenberg, M. W. and Matthaei, J. H., 'The dependence of cell-free protein synthesis in *E. coli* upon naturally occurring or synthetic template RNA', in V. A. Engelhardt (ed.), *Biological Structure and Function at the Molecular Level. Proceedings of the Fifth International Congress of Biochemistry, Moscow, 10–16 August 1961*, London, Pergamon Press, 1963, pp. 184–9.

Nirenberg, M. W., Jones, O. W., Leder, P. *et al.*, 'On the coding of genetic information', *Cold Spring Harbor Symposia on Quantitative Biology*, vol. 28, 1963, pp. 549–58.

Noble, D., 'Modeling the heart: from genes to cells to the whole organ', *Science*, vol. 295, 2002, pp. 1678–82.

Noble, D., 'Physiology is rocking the foundations of evolutionary biology', *Experimental Physiology*, vol. 98, 2013, pp. 1235–43.

Noble, D. F., *Forces of Production: A Social History of Industrial Automation*, Oxford, Oxford University Press, 1986.

Noller, H., 'Evolution of protein synthesis from an RNA world', *Cold Spring Harbor Perspectives in Biology*, vol. 4, 2012, article a003681.

Northrop, J. H., 'Growth and phage production of lysogenic *D. megatherium*', *Journal of General Physiology*, vol. 34, 1951, pp. 715–35.

Novick, A. and Szilárd, L., 'II. Experiments with the chemostat on the rates of amino acid synthesis in bacteria', in E. J. Boell (ed.), *Dynamics of Growth Processes*, Princeton, Princeton University Press, 1954, pp. 21–32.

O'Brien, J. J. and Campoli-Richards, D. M., 'Acyclovir', *Drugs*, vol. 3, 1989, pp. 233–309.

Ochoa, S., 'Chemical basis of heredity, the genetic code', *Experientia*, vol. 20, 1964, pp. 57–68.

Ochoa, S., 'The pursuit of a hobby', *Annual Review of Biochemistry*, vol. 49, 1980, pp. 1–30.

O'Connell, M. R., Oakes, B. L., Sternberg, S. H. *et al.*, 'Programmable RNA recognition and cleavage by CRISPR/Cas9' *Nature*, vol. 516, 2014, pp. 263–6.

Olby, R., 'Schrödinger's problem: What is life?', *Journal of the History of Biology*, vol. 4, 1971, pp. 119–48.

Olby, R., 'Avery in retrospect', *Nature*, vol. 238, 1972, pp. 295–6.

Olby, R., *The Path to the Double Helix: The Discovery of DNA*, New York, Dover, 1994.

Olby, R., 'Quiet debut for the double helix', *Nature*, vol. 421, 2003, pp. 402–5.

Olby, R., *Francis Crick: Hunter of Life's Secrets*, Plainview, Cold Spring Harbor Laboratory Press, 2009.

Olby, R. and Posner, R., 'An early reference to genetic coding', *Nature*, vol. 215, 1967, p. 556.

Organ, C. L., Schweitzer, M. H., Zheng, W., *et al.*, 'Molecular phylogenetics of *Mastodon* and *Tyrannosaurus rex*', *Science*, vol. 320, 2008, p. 499.

Orlando, L., Ginolhac, A., Zhang, G. *et al.*, 'Recalibrating Equus evolution using the genome sequence of an early Middle Pleistocene horse', *Nature*, vol. 499, 2013, pp. 74 8.

Owens, L., 'Mathematicians at War: Warren Weaver and the Applied Mathematics Panel, 1942–1945', in D. E. Rowe and J. McClearly (eds),

The History of Modern Mathematics Vol II: Institutions and Applications,
London, Academic Press, 1989, pp. 287–305.

Oyama, S., *The Ontogeny of Information: Developmental Systems and
Evolution*, Durham, NC, Duke University Press, 2000.

Oye, K. A., Esvelt, K., Appleton, E. *et al.*, 'Regulating gene drives', *Science*,
vol. 345, 2014, pp. 626–8.

Pääbo, S., *Neanderthal Man: In Search of Lost Genomes*, New York, Basic
Books, 2014.

Palazzo, A. F. and Gregory, T. R., 'The case for junk DNA', *PLoS Genetics*,
vol. 10, 2014, article e1004351.

Pappenheimer, A. M. Jr., 'Whatever happened to Pz?', in A. Lwoff and A.
Ullmann (eds), *Origins of Molecular Biology: A Tribute to Jacques Monod*,
London, Academic Press, 1979, pp. 55–60.

Pardee, A. B., 'Mechanisms for control of enzyme synthesis and enzyme
activity in bacteria', in G. E. W. Wolstenholme and C. M. O'Connor
(eds), *CIBA Foundation Symposium on the Regulation of Cell Metabolism*,
London, Churchill, 1959, pp. 295–304.

Pardee, A. B., 'The PaJaMa experiment', in A. Lwoff and A. Ullmann
(eds), *Origins of Molecular Biology: A Tribute to Jacques Monod*, London,
Academic Press, 1979, pp. 109–16.

Pardee, A. B., 'Molecular basis of gene expression: Origins from the Pajama
experiment', *BioEssays*, vol. 2, 1985, pp. 86–9.

Pardee, A. B., 'PaJaMas in Paris', *Trends in Genetics*, vol. 18, 2002, pp. 585–7.

Pardee, A. B., Jacob, F. and Monod, J., 'Sur l'expression et le rôle des allèles
"inductible" et "constitutif" dans la synthèse de la β-galactosidase
chez des zygotes d'*Escherichia coli*', *Comptes rendus hebdomodaires des
séances de l'Académie des sciences*, vol. 246, 1958, pp. 3125–8.

Pardee, A. B., Jacob, F. and Monod, J., 'The genetic control and cytoplasmic
expression of "inducibility" in the synthesis of β-galactosidase by
E. coli', *Journal of Molecular Biology*, vol. 1, 1959, pp. 165–78.

Paul, N. and Joyce, G. F., 'Minimal self-replicating systems', *Current
Opinion in Chemical Biology*, vol. 8, 2004, pp. 634–9.

Pauling, L., 'Abnormality of hemoglobin molecules in hereditary
hemolytic anemias', *Harvey Lectures*, vol. 12, 1955, pp. 216–41.

Pauling, L., 'Schrödinger's contribution to chemistry and biology', in
C. W. Kilmister (ed.), *Schrödinger: Centenary Celebration of a Polymath*,
Cambridge, Cambridge University Press, 1987, pp. 224–33.

Pauling, L. and Corey, R. B., 'A proposed structure for the nucleic acids',
Proceedings of the National Academy of Sciences USA, vol. 39, 1953,
pp. 84–97.

Pauling, L. and Schomaker, V., 'On a phospho-tri-anhydride formula for the nucleic acids', *Journal of the American Chemical Society*, vol. 74, 1952, pp. 3712–13.

Pauling, L., Itano, H. A., Singer, S. J. and Wells, I. C., 'Sickle cell anemia, a molecular disease', *Science*, vol. 110, 1949, pp. 543–8.

Pearson, H., 'Genetics: What is a gene?', *Nature*, vol. 441, 2006, pp. 398–401.

Penney, D., Wadsworth, C., Fox, G. *et al.*, 'Absence of ancient DNA in sub-fossil insect inclusions preserved in 'anthropocene' Colombian copal', *PLoS ONE*, vol. 8, 2013, article e73150.

Pennisi, E., 'ENCODE project writes eulogy for junk DNA', *Science*, vol. 337, 2012, pp. 1159–61.

Perutz, M., 'DNA helix', *Science*, vol. 164, 1969, pp. 1537–9.

Perutz, M. F., 'Erwin Schrödinger's *What is Life?* and molecular biology', in C. W. Kilmister (ed.), *Schrödinger: Centenary Celebration of a Polymath*, Cambridge, Cambridge University Press, 1987, pp. 234–51.

Petruk, S., Sedkov, Y., Johnston, D. M., Hodgson, J. W. *et al.*, 'TrxG and PcG proteins but not methylated histones remain associated with DNA through replication', *Cell*, vol. 150, 2012, pp. 922–33.

Peyrieras, N. and Morange, M., 'The study of lysogeny at the Pasteur Institute (1950–1960): an epistemologically open system', *Studies in History and Philosophy of Biology and Biomedical Sciences*, vol. 33, 2002, pp. 419–30.

Pfeiffer, J. E., 'Woods Hole in 1949', *Scientific American*, vol. 181 (3), 1949, pp. 13–17.

Philippe, H., Brinkmann, H., Lavrov, D. V. *et al.*, 'Resolving difficult phylogenetic questions: why more sequences are not enough', *PLoS Biology*, vol. 9, 2011, article e1000602.

Piccirilli, J. A., Krauch, T., Moroney, S. E. and Benner, S. A., 'Enzymatic incorporation of a new base pair into DNA and RNA extends the genetic alphabet', *Nature*, vol. 343, 1990, pp. 33–7.

Pichot, A., *Histoire de la notion de gène*, Paris, Flammarion, 1999.

Pickering, A., *The Cybernetic Brain: Sketches of Another Future*, London, University of Chicago Press, 2010.

Pickstone, J. V., *Ways of Knowing: A New History of Science, Technology and Medicine*, Manchester, Manchester University Press, 2001.

Pinheiro, V. B., Taylor, A. I., Cozens, C. *et al.*, 'Synthetic genetic polymers capable of heredity and evolution', *Science*, vol. 336, 2012, pp. 341–4.

Piper, A., 'Light on a dark lady', *Trends in Biochemical Sciences*, vol. 23, 1998, pp. 151–4.

Planer, R. J., 'Replacement of the "genetic program" program', *Biology and Philosophy*, vol. 29, 2014, pp. 33–53.

Poczai, P., Bell, N. and Hyvönen, J., 'Imre Festetics and the Sheep Breeders' Society of Moravia: Mendel's forgotten "research network"', PLoS Biology, vol. 12, 2014, article e1001772.

Pollister, A. W., Hewson, S. and Alfert, M., 'Studies on the desoxypentose nucleic acid content of animal nuclei', Journal of Cellular Physiology, vol. 38 (Suppl. 1), 1951, pp. 101–19.

Pollock, M. R., 'The discovery of DNA: An ironic tale of chance, prejudice and insight', Journal of General Microbiology, vol. 63, 1970, pp. 1–20.

Polyanski, A, A., Hlevnjal, M. and Zagrovic, B., 'Proteome-wide analysis reveals clues of complementary interactions between mRNAs and their cognate proteins as the physicochemical foundation of the genetic code', RNA Biology, vol. 10, 2013, pp. 1248–54.

Pontecorvo, G., 'Genetic formulation of gene structure and gene action', Advances in Enzymology, vol. 13, 1952, pp. 121–49.

Powner, M. W., Gerland, B. and Sutherland, J. D., 'Synthesis of activated pyrimidine ribonucleotides in prebiotically plausible conditions', Nature, vol. 459, 2009, pp. 239–42.

Pringle, P., The Murder of Nikolai Vavilov, New York, Simon & Schuster, 2008.

Pross, A., What is Life? How Chemistry Becomes Biology, Oxford, Oxford University Press, 2012.

Prüfer, K., Racimo, F., Patterson, N. et al., 'The complete genome sequence of a Neanderthal from the Altai Mountains', Nature, vol. 505, 2014, pp. 43–9.

Prusiner, S. B., 'Novel proteinaceous infectious particles cause scrapie', Science, vol. 216, 1982, pp. 136–44.

Prusiner, S. B., 'Prions', Proceedings of the National Academy of Sciences USA, vol. 95, 1998, pp. 13363–83.

Prusiner, S. B. and McCarty, M., 'Discovering DNA encodes heredity and prions are infectious proteins', Annual Review of Genetics, vol. 40, 2006, pp. 25–45.

Ptashne, M., 'François Jacob (1920–2013)', Cell, vol. 253, 2013, pp. 1180–2.

Quastler, H., 'A primer on information theory', in H. P. Yockey, R. L. Platzman and H. Quastler (eds), Symposium on Information Theory in Biology, London, Pergamon, 1958a, pp. 3–49.

Quastler, H., 'The domain of information theory in biology', in H. P. Yockey, R. L. Platzman and H. Quastler (eds), Symposium on Information Theory in Biology, London, Pergamon, 1958b, pp. 187–96.

Quastler, H., 'The status of information theory in biology: a round table discussion', in H. P. Yockey, R. L. Platzman and H. Quastler (eds), Symposium on Information Theory in Biology, London, Pergamon, 1958c, pp. 399–402.

Rabinow, P., *Making PCR: A Story of Biotechnology*, Chicago, University of Chicago Press, 1996.

Radford, E. J., Ito, M., Shi, H. *et al.*, 'In utero undernourishment perturbs the adult sperm methylome and intergenerational metabolism', *Science*, vol. 345, 2014, article 1255903.

Rasmussen, N., *Gene Jockeys: Life Science and the Rise of Biotech Enterprise*, Baltimore, Johns Hopkins University Press, 2014.

Reaves, M. L., Sinha, S., Rabinowitz, J. D. *et al.*, 'Absence of detectable arsenate in DNA from arsenate-grown GFAJ-1 cells', *Science*, vol. 337, 2012, pp. 470–3.

Rechavi, O., Minevich, G. and Hobert, O., 'Transgenerational inheritance of an acquired small RNA-based antiviral response in C. *elegans*', *Cell*, vol. 147, 2011, pp. 1248–56.

Regier, J. C., Shultz, J. W., Zwick, A. *et al.*, 'Arthropod relationships revealed by phylogenomic analysis of nuclear protein-coding sequences', *Nature*, vol. 463, 2010, pp. 1079–83.

Reich, D., Green, R. E., Kircher, M. *et al.*, 'Genetic history of an archaic hominin group from Denisova Cave in Siberia', *Nature*, vol. 468, 2010, pp. 1053–60.

Relyea, R. A., 'New effects of Roundup on amphibians: Predators reduce herbicide mortality; herbicides induce antipredator morphology', *Ecological Applications*, vol. 22, 2012, pp. 634–47.

Remigi, P., Capela, D., Clerissi, C. *et al.*, 'Transient hypermutagenesis accelerates the evolution of legume endosymbionts following horizontal gene transfer', *PLoS Biology*, vol. 12, 2014, article e1001942.

Rheinberger, H.-J., *Towards a History of Epistemic Things: Synthesizing Proteins in the Test Tube*, Stanford, Stanford University Press, 1997.

Rich, A., 'An introduction to intra-cellular information transfer', in R. Bellman (ed.), *Proceedings of Symposia in Applied Mathematics XIV: Mathematical Problems in the Biological Sciences*, Providence, RI, American Mathematical Society, 1962, pp. 205–13.

Rich, A., 'Gamow and the genetic code', in E. Harper, W. C. Parke and D. Anderson (eds), *George Gamow Symposium*, ASP Conference Series, Vol. 129), 1997, pp. 114–22.

Rich, A., 'The excitement of discovery', *Annual Review of Biochemistry*, vol. 73, 2004, pp. 1–37.

Rich, A. and Zhang, S., 'Timeline. Z-DNA, the long road to biological function', *Nature Reviews Genetics*, vol. 4, 2003, pp. 566–72.

Rich, A., Davies, D. R., Crick, F. H. and Watson, J. D., 'The molecular structure of polyadenylic acid', *Journal of Molecular Biology*, vol. 3, 1961, pp. 71–86.

Ridley, M., *Francis Crick: Discoverer of the Genetic Code,* London, Harper Perennial, 2006.

Rietveld, C. A., Esko, T., Davies, G. *et al.*, 'Common genetic variants associated with cognitive performance identified using the proxy-phenotype method', *Proceedings of the National Academy of Sciences USA,* vol. 111, 2014, pp. 13790–4.

RIKEN Genome Exploration Research Group and Genome Science Group (Genome Network Project Core Group) and the FANTOM Consortium, 'Antisense transcription in the mammalian transcriptome', *Science,* vol. 309, 2005, pp. 1564–6.

Ring, K. L. and Cavalcanti, A. R. O., 'Consequences of stop codon reassignment on protein evolution in ciliates with alternative genetic codes', *Molecular Biology and Evolution,* vol. 25, 2007, pp. 179–86.

Rinke, C., Schwientek, P., Sczyrba, A. *et al.*, 'Insights into the phylogeny and coding potential of microbial dark matter', *Nature,* vol. 499, 2013, pp. 431–7.

Roberts, R. B. (ed.), *Microsomal Particles and Protein Synthesis,* London, Pergamon, 1958.

Roberts, R. B., 'Alternative codes and templates', *Proceedings of the National Academy of Sciences USA,* vol. 48, 1962, pp. 897–900.

Robertson, M. P. and Joyce, G. F., 'The origins of the RNA world', *Cold Spring Harbor Perspectives in Biology,* vol. 4, 2012, article a003608.

Roch, A., 'Die Geschichte der Computermaus', *Telepolis,* 28 August 1998. [English translation at http://www-sul.stanford.edu/siliconbase/wip/control.html.]

Roch, A., 'Biopolitics and intuitive algebra in the mathematization of cryptology? A review of Shannon's 'A mathematical theory of cryptography' from 1945', *Cryptologia,* vol. 23, 1999, pp. 261–6.

Roche, J., 'Notice nécrologique: André Boivin (1895–1949)', *Bulletin de la Société de Chimie Biologique,* vol. 31, 1949, pp. 1564–7.

Rogers, E. M., 'Claude Shannon's cryptography research during World War II and the mathematical theory of communication', *Proceedings, IEEE 28th International Carnaham Conference on Security Technology,* 1994, pp. 1–5.

Rogozin, I. B., Carmel, L., Csuros, M. and Koonin, E. V., 'Origin and evolution of spliceosomal introns', *Biology Direct,* vol. 7, 2012, p. 11.

Romiguier, J., Ranwez, V., Douzery, E. J. P. and Galtier, N., 'Contrasting GC-content dynamics across 33 mammalian genomes: Relationship with life-history traits and chromosome sizes', *Genome Research,* vol. 20, 2010, pp. 1001–9.

Rosenblith, W. A., 'Cybernetics, or Control and communication in the animal and the machine by Norbert Wiener', *Annals of the American Academy of Political and Social Science*, vol. 264, 1949, pp. 287–8.

Rosenblueth, A., Wiener, N. and Bigelow, J., 'Behavior, purpose and teleology', *Philosophy of Science*, vol. 10, 1943, pp. 18–24.

Rovner, A. J., Haimovich, A. D., Katz, S. R., *et al.*, 'Recoded organisms engineered to depend on synthetic amino acids', *Nature*, vol. XXX, 2015, pp. 89–93.

Rowen, J. W., Eden, M. and Kahler, H., 'Molecular characteristics of sodium desoxyribonucleate', *Biochimica et Biophysica Acta*, vol. 10, 1953, pp. 89–92.

Rutherford, A., *Creation*, London, Penguin, 2013.

Safaee, N., Noronha, A. M., Rodionov, D. *et al.*, 'Structure of the parallel duplex of poly(A) RNA: evaluation of a 50 year-old prediction', *Angewandte Chemie International Edition*, vol. 52, 2013, pp. 10370–3.

Salas, M., Smith M. A., Stanley, W. M. Jr. *et al.*, 'Direction of reading of the genetic message', *Journal of Biological Chemistry*, vol. 240, 1965, pp. 3988–95.

Sánchez-Silva, R., Villalobo, E., Morin, L. and Torres A., 'A new noncanonical nuclear genetic code: Translation of UAA into glutamate', *Current Biology*, vol. 13, 2003, pp. 442–7.

Sanger F., 'Sequences, sequences, and sequences', *Annual Review of Biochemistry*, vol. 57, 1988, pp. 1–28.

Sanger, F. and Thompson, E. O. P., 'The amino-acid sequence in the glycyl chain of insulin. 1. The investigation of lower peptides from partial hydrolysates', *Biochemical Journal*, vol. 53, 1953a, pp. 353–66.

Sanger, F. and Thompson, E. O. P., 'The amino-acid sequence in the glycyl chain of insulin. 2. The investigation of peptides from enzymic hydrolysates', *Biochemical Journal*, vol. 53, 1953b, pp. 366–74.

Sanger, F. and Tuppy, H., 'The amino-acid sequence in the phenylalanyl chain of insulin. I. The investigation of lower peptides from partial hydrolysates', *Biochemical Journal*, vol. 49, 1951a, pp. 463–81.

Sanger, F. and Tuppy, H., 'The amino-acid sequence in the phenylalanyl chain of insulin. II. The investigation of peptides from enzymic hydrolysates', *Biochemical Journal*, vol. 49, 1951b, pp. 481–90.

Sanger, F., Nicklen, S. and Coulson, A. R., 'DNA sequencing with chain-terminating inhibitors', *Proceedings of the National Academy of Sciences USA*, vol. 74, 1977, pp. 5463–7.

Sanger, F., Coulson, A. R., Friedmann, T. *et al.*, 'The nucleotide sequence of bacteriophage phiX174', *Journal of Molecular Biology*, vol. 125, 1978, pp. 225–46.

Sankararaman, S., Mallick, S., Dannemann, M. *et al.*, 'The genomic landscape of Neanderthal ancestry in present-day humans', *Nature*, vol. 507, 2014, pp. 354–7.

Sapp, J. *The New Foundations of Evolution: On the Tree of Life*, Oxford, Oxford University Press, 2009.

Sarkar, S., '*What is Life?* revisited', *BioScience*, vol. 41, 1991, pp. 631–4.

Sarkar, S., 'Biological information: a skeptical look at some central dogmas of molecular biology', in S. Sarkar (ed.), *The Philosophy and History of Molecular Biology: New Perspectives*, Dordrecht, Kluwer, 1996a, pp. 187–231.

Sarkar, S., 'Decoding "coding" – information and DNA', *BioScience*, vol. 46, 1996b, pp. 857–64.

Sarkar, S., 'Information in genetics and developmental biology: Comments on Maynard Smith', *Philosophy of Science*, vol. 67, 2000, pp. 208–13.

Sarkar, S., 'Erwin Schrödinger's excursus on genetics', in O. Harman and M. R. Dietrich (eds), *Outsider Scientists: Routes to Innovation in Biology*, Chicago, Chicago University Press, 2013, pp. 93–109.

Sasidharan, R. and Gerstein, M., 'Genomics: Protein fossils live on as RNA', *Nature*, vol. 453, 2008, pp. 729–31.

Sayre, A., *Rosalind Franklin and DNA*, London, Norton, 1975.

Schaffner, K., 'Logic of discovery and justification in regulatory genetics', *Studies in History and Philosophy of Science*, vol. 4, 1974, pp. 349–85.

Schlatter, M. and Aizawa, K., 'Walter Pitts and "A Logical Calculus"', *Synthese*, vol. 162, 2008, pp. 235–50.

Schmucker D., Clemens, J. C., Shu, H. *et al.*, '*Drosophila* Dscam is an axon guidance receptor exhibiting extraordinary molecular diversity', *Cell*, vol. 101, 2000, pp. 671–84.

Schneider, T. D., Stormo, G. D., Gold, L. and Ehrenfeucht, A., 'Information content of binding sites on nucleotide sequences', *Journal of Molecular Biology*, vol. 188, 1986, pp. 415–31.

Schrödinger, E., *What is Life?*, Cambridge, Cambridge University Press, 2000.

Schultz, J., 'Aspects of the relation between genes and development in *Drosophila*', *The American Naturalist*, vol. 69, 1935, pp. 30–54.

Schultz, J., 'The evidence of the nucleoprotein nature of the gene', *Cold Spring Harbor Symposia on Quantitative Biology*, vol. 9, 1941, pp. 55–65.

Schultz, J., 'The nature of heterochromatin', *Cold Spring Harbor Symposia on Quantitative Biology*, vol. 12, 1947, pp. 179–91.

Schuster, S. C., 'Next-generation sequencing transforms today's biology', *Nature Methods*, vol. 5, 2008, pp. 16–18.

Schwartz, D., 'Speculations on gene action and protein specificity', *Proceedings of the National Academy of Sciences USA*, vol. 41, 1955, pp. 300–7.

Schwartz, J., *In Pursuit of the Gene: From Darwin to DNA*, Cambridge, MA, Harvard University Press, 2008.

Shama, L. N. S. and Wegner, K. W., 'Grandparental effects in marine sticklebacks: transgenerational plasticity across multiple generations', *Journal of Evolutionary Biology*, vol. 27, 2014, pp. 2297–307.

Shannon, C. E., *An Algebra for Theoretical Genetics*, unpublished PhD thesis, Massachusetts Institute of Technology, 1940.

Shannon, C. E., 'A mathematical theory of communication', *The Bell System Technical Journal*, vol. 227, 1948a, pp. 379–423.

Shannon, C. E., 'A mathematical theory of communication', *The Bell System Technical Journal*, vol. 227, 1948b, pp. 623–56.

Shannon, C. E. and Weaver, W., *The Mathematical Theory of Communication*, Urbana, University of Illinois Press, 1949.

Shapiro, B. and Hofreiter, M., 'A paleogenomic perspective on evolution and gene function: new insights from ancient DNA', *Science*, vol. 343, 2014, article 1236573.

Shapiro, J. A., 'Revisiting the Central Dogma in the twenty-first century', *Annals of the New York Academy of Sciences*, vol. 1178, 2009, pp. 6–28.

shCherbak, V. I. and Makukov, M. A., 'The "Wow! signal" of the terrestrial genetic code', *Icarus*, vol. 224, 2013, pp. 228–42.

Shea, N., 'What's transmitted? Inherited information', *Biology and Philosophy*, vol. 26, 2011, pp. 183–9.

Shen, P. S., Park, J., Qin, Y., *et al.*, 'Rqc2p and 60S ribosomal subunits mediate mRNA-independent elongation of nascent chains', *Science*, vol. 347, 2015, pp. 75–8.

Sheridan, C., 'First CRISPR-Cas patent opens race to stake out intellectual property', *Nature Biotechnology*, vol. 32, 2014, pp. 599–601.

Shine, I. and Wrobel, S., *Thomas Hunt Morgan: Pioneer of Genetics*, Lexington, University of Kentucky Press, 1976.

Shreeve, J., *The Genome War*, New York, Knopf, 2004.

Singer, M., 'Leon Heppel and the early days of RNA biochemistry', *Journal of Biological Chemistry*, vol. 278, 2003, pp. 47351–6.

Slater, E. C., '1961 Moscow, USSR: The Fifth International Congress of Biochemistry', *IUBMB Life*, vol. 55, 2003, pp. 185–7.

Sloan, P. R. and Fogel, B., *Creating a Physical Biology. The Three-Man Paper and Early Molecular Biology*, London, University of Chicago Press, 2011.

Smith, Z. D., Chan, M. M., Humm, K. C. *et al.*, 'DNA methylation dynamics of the human preimplantation embryo', *Nature*, vol. 511, 2014, pp. 611–15.

Sofyer, V. N., 'The consequences of political dictatorship for Russian science', *Nature Reviews Genetics*, vol. 2, 2001, pp. 723–9.

Söll, D., Ohtsuka, E., Jones, D. S. *et al.*, 'Studies on polynucleotides, XLIX. Stimulation of the binding of aminoacyl-sRNAs to ribosomes by ribotrinucleotides and a survey of codon assignments for 20 amino acids', *Proceedings of the National Academy of Sciences USA*, vol. 54, 1965, pp. 1378–85.

Sonneborn, T. M., 'Molecular biology of the gene', *Science*, vol. 150, 1965, p. 1282.

Spetner, L. M., 'Information transmission in evolution', *IEEE Transactions on Information Theory*, vol. 14, 1968, pp. 3–6.

Speyer, J. F., Lengyel, P., Basilio, C. and Ochoa, S., 'Synthetic polynucleotides and the amino acid code, II', *Proceedings of the National Academy of Sciences USA*, vol. 48, 1962a, pp. 63–8.

Speyer, J. F., Lengyel, P., Basilio, C. and Ochoa, S., 'Synthetic polynucleotides and the amino acid code, IV', *Proceedings of the National Academy of Sciences USA*, vol. 48, 1962b, pp. 441–8.

Speyer, J. F., Lengyel, P., Basilio, C. *et al.*, 'Synthetic polynucleotides and the amino acid code', *Cold Spring Harbor Symposia on Quantitative Biology*, vol. 28, 1963, pp. 559–68.

Spiegelman, S., 'Nuclear and cytoplasmic factors controlling enzymatic constitution', *Cold Spring Harbor Symposia on Quantitative Biology*, vol. 11, 1946, pp. 256–77.

Spiegelman, S. and Landman, O. E., 'Genetics of microorganisms', *Annual Review of Microbiology*, vol. 8, 1954, pp. 181–236.

Srinivasan, G., James, C. M. and Krzycki, J. A., 'Pyrrolysine encoded by UAG in Archaea: charging of a UAG-decoding specialized tRNA', *Science*, vol. 296, 2002, pp. 1459–62.

Stacey, M., 'Bacterial nucleic acids and nucleoproteins', *Symposia of the Society for Experimental Biology*, vol. 1, 1947, pp. 86–100.

Stamhuis, I. H., Meijer, O. G. and Zevenhuizen, E. J. A., 'Hugo de Vries on heredity, 1889–1903. Statistics, Mendelian laws, pangenes, mutations', *Isis*, vol. 90, 1999, pp. 238–67.

Stanley, W. M., 'Isolation of a crystalline protein possessing the properties of tobacco-mosaic virus', *Science*, vol. 81, 1935, pp. 644–5.

Stanley, W. M., 'The "undiscovered" discovery', *Archives of Environmental Health*, vol. 21, 1970, pp. 256–62.

Starck, S. R., Jiang, V., Pavon-Eternod, M., *et al.*, 'Leucine-tRNA initiates at CUG start codons for protein synthesis and presentation by MHC class I', *Science*, vol. 336, 2012, pp. 1719–23.

Stedman, E. and Stedman, E., 'The function of deoxyribose-nucleic acid in the cell nucleus', *Symposia of the Society for Experimental Biology*, vol. 1, 1947, pp. 232–51.

Stegmann, U. E., 'The arbitrariness of the genetic code', *Biology and Philosophy*, vol. 19, 2004, pp. 205–22.

Stegmann, U. E., 'Genetic information as instructional content', *Philosophy of Science*, vol. 72, 2005, pp. 425–43.

Stegmann, U. E., 'DNA, inference, and information', *British Journal for the Philosophy of Science*, vol. 60, 2009, pp. 1–17.

Stegmann, U. E., 'Varieties of parity', *Biology and Philosophy*, vol. 27, 2012, pp. 903–18.

Stegmann, U. E., 'On the transmission sense of information', *Biology and Philosophy*, vol. 28, 2013, pp. 141–4.

Stegmann, U. E., "Genetic coding' reconsidered: an analysis of actual usage', *British Journal for the Philosophy of Science*, 2014a, in press.

Stegmann, U. E., 'Causal control and genetic causation', *Noûs*, vol. 48, 2014b, pp. 450–65.

Stent, G. S., 'Genetic transcription', *Proceedings of the Royal Society of London Series B*, vol. 164, 1966, pp. 181–97.

Stent, G. S., 'That was the molecular biology that was', *Science*, vol. 160, 1968a, pp. 390–5.

Stent, G. S., 'Reply to Lamanna', *Science*, vol. 160, 1968b, p. 1398.

Stent, G. S., *Molecular Genetics: An Introductory Narrative*, San Francisco, Freeman, 1971.

Stent, G. S., 'Prematurity and uniqueness in scientific discovery', *Scientific American*, vol. 227 (12), 1972, pp. 84–93.

Stergachis, A. B., Haugen, E., Shafer. A. *et al.*, 'Exonic transcription factor binding directs codon choice and affects protein evolution', *Science*, vol. 342, 2013, pp. 1367–72.

Stern, K. G., 'Nucleoproteins and gene structure', *Yale Journal of Biology and Medicine*, vol. 19, 1947, pp. 937–49.

Stevens, H., *Life Out of Sequence: A Data-Driven History of Bioinformatics*, London, University of Chicago Press, 2013.

Strasser, B. J., 'A world in one dimension: Linus Pauling, Francis Crick and the Central Dogma of molecular biology', *History and Philosophy of the Life Sciences*, vol. 28, 2006, pp. 491–512.

Stretton, A. O. W., 'The first sequence: Fred Sanger and insulin', *Genetics*, vol. 162, 2002, pp. 527–32.

Sturtevant, A. H., *A History of Genetics*, London, Harper, 1965.

Suárez-Díaz, E., 'The long and winding road of molecular data in phylogenetic analysis', *Journal of the History of Biology*, vol. 47, 2014, pp. 443–78.

Subak-Sharpe, H., Bürk, R. R., Crawford, L. V. *et al.*, 'An approach to evolutionary relationships of mammalian DNA viruses through analysis of the pattern of nearest neighbor base sequences', *Cold Spring Harbor Symposia on Quantitative Biology*, vol. 31, 1966, pp. 737–48.

Sueoka, N., Marmur, J. and Doty, P., 'II. Dependence of the density of deoxyribonucleic acids on guanine-cytosine content', *Nature*, vol. 183, 1959, pp. 1429–31.

Sulston, J. and Ferry, G., *The Common Thread: Science, Politics, Ethics and the Human Genome*, London, Bantam, 2002.

Summers, W. C., 'How bacteriophage came to be used by the phage group', *Journal of the History of Biology*, vol. 26, 1993, pp. 255–67.

Supattapone, S. and Miller, M. B., 'Cofactor involvement in prion propagation', in W.-Q. Zou and P. Gambetti (eds), *Prions and Diseases: Volume 1, Physiology and Pathophysiology*, New York, Springer, 2013, pp. 93–105.

Šustar, P., 'Crick's notion of genetic information and the 'central dogma' of molecular biology', *British Journal for the Philosophy of Science*, vol. 58, 2007, pp. 13–24.

Sutton, W. S., 'On the morphology of the chromosome group in *Brachystola magna*', *Biological Bulletin*, vol. 4, 1902, pp. 24–39.

Sutton, W. S., 'The chromosomes in heredity', *Biological Bulletin*, vol. 4, 1903, pp. 231–51.

Switzer, C., Moroney, S. E. and Benner, S. A., 'Enzymatic incorporation of a new base pair into DNA and RNA', *Journal of the American Chemical Society*, vol. 111, 1989, pp. 8322–3.

Symonds, N., 'What is Life?: Schrödinger's influence in biology', *Quarterly Review of Biology*, vol. 61, 1986, pp. 221–6.

Symonds, N., 'Reminiscence', in F. W. Stahl (ed.), *We Can Sleep Later: Alfred D. Hershey and the Origins of Molecular Biology*, Cold Spring Harbor, Cold Spring Harbor Laboratory Press, 2000, pp. 91–4.

Szilárd, L., 'The control of the formation of specific proteins in bacteria and in animal cells', *Proceedings of the National Academy of Sciences USA*, vol. 46, 1960, pp. 277–92.

Tatum, E. L., 'Chairman's Remarks', in H. J. Vogel, B. Vernon and J. O. Lampen (eds), *Informational Macromolecules*, London, Academic Press, 1963, pp. 175–6.

Tatum, E. L. and Beadle, G. W., 'Genetic control of biochemical reactions in *Neurospora*: An 'aminobenzoicless' mutant', *Proceedings of the National Academy of Sciences USA*, vol. 28, 1942, pp. 234–43.

Taylor, H. E., 'Nouvelles transformations induites spécifiquement chez le pneumocoque', *Colloques Internationaux du Centre National de la Recherche Scientifique*, vol. 8, 1949, pp. 45–55.

Temin, H. M., 'The protovirus hypothesis: Speculations on the significance of RNA-directed DNA synthesis for normal development and for carcinogenesis', *Journal of the National Cancer Institute*, vol. 46, 1971, pp. iii–vii.

Theobald, D. L., 'A formal test of the theory of universal common ancestry' *Nature*, vol. 465, 2010, pp. 219–22.

Thieffry, D., 'Contributions of the "Rouge-Cloître group" to the notion of "messenger RNA"', *History and Philosophy of the Life Sciences*, vol. 19, 1997, pp. 89–113.

Thieffry, D. and Burian, R. M., 'Jean Brachet's alternative scheme for protein synthesis', *Trends in Biochemical Sciences*, vol. 21, 1996, pp. 115–17.

Thomas, C. A., 'The genetic organization of chromosomes', *Annual Review of Genetics*, vol. 5, 1971, pp. 237–56.

Thomas, R., 'Molecular genetics under an embryologist's microscope: Jean Brachet, 1909–1988', *Genetics*, vol. 131, 1992, pp. 515–18.

Tissières, A., Schlessinger, D. and Gros, F., 'Amino acid incorporation into proteins by *Escherichia coli* ribosomes', *Proceedings of the National Academy of Sciences USA*, vol. 46, 1960, pp. 1450–63.

Toomey, D., *Weird Life: The Search for Life That Is Very, Very Different from Our Own*, London, Norton, 2013.

Triclot, M., 'Norbert Wiener's politics and the history of cybernetics', in M. Kokowski (ed.), *The Global and the Local: The History of Science and the Cultural Integration of Europe. Proceedings of the Second International Conference of the European Society for the History of Science (Cracow, Poland, 6–9 September 2006)*, Warsaw, Press of the Polish Academy of Arts and Sciences, 2007.

Triclot, M., *Le Moment cybernétique : La constitution de la notion d'information*, Paris, Champ Vallon, 2008.

Troland, L. T., 'Biological enigmas and the theory of enzyme action', *The American Naturalist*, vol. 51, 1917, pp. 321–50.

Tsugita, A., 'The proteins of mutants of TMV: composition and structure of chemically evoked mutants of TMV RNA', *Journal of Molecular Biology*, vol. 5, 1962, pp. 284–92.

Ullmann, A., 'Jacques Monod, 1910–1976: his life, his work and his commitments', *Research in Microbiology*, vol. 161, 2010, pp. 68–73.

Umbarger, H. E., 'Evidence for a negative-feedback mechanism on the biosynthesis of isoleucine', *Science*, vol. 123, 1956, p. 848.

van Dam, L. and Levitt, M. H., 'BII nucleotides in the B and C forms of natural-sequence polymeric DNA: A new model for the C form of DNA', *Journal of Molecular Biology*, vol. 304, 2000, pp. 541–61.

van Noorden, R., Maher, B. and Nuzzo, R., 'The top 100 papers', *Nature*, vol. 514, 2014, pp. 550–3.

Varmus, H., *The Art and Politics of Science*, New York, Norton, 2009.

Venter, J. C., *A Life Decoded. My Genome: My Life*, London, Penguin, 2007.

Venter, J. C., *Life at the Speed of Light*, London, Little, Brown, 2013.

Venter, J. C., Adams, M. D., Myers, E. W. *et al.*, 'The sequence of the human genome', *Science*, vol. 291, 2001, pp. 1304–51.

Vernot, B. and Akey, J. M., 'Resurrecting surviving Neandertal lineages from modern human genomes', *Science*, vol. 343, 2014, pp. 1017–21.

Vogel, H., 'Repressed and induced enzyme formation: a unified hypothesis', *Proceedings of the National Academy of Sciences USA*, vol. 43, 1957, pp. 491–6.

Vogel, H. J., Vernon, B. and Lampen, J. O. (eds), *Informational Macromolecules*, London, Academic Press, 1963.

von Neumann, J., 'The general and logical theory of automata', in L. A. Jeffress (ed.), *Cerebral Mechanisms in Behavior. The Hixon Symposium*, New York, Wiley, 1951, pp. 1–31.

von Neumann, J., 'Letter to Norbert Wiener from John von Neumann', *Proceedings of Symposia in Applied Mathematics*, vol. 52, 1997, pp. 506–12.

von Schwerin, A., 'Medical physicists, biology, and the physiology of the cell (1920–1940)', in L. Campos and A. von Schwerin (eds), *Making Mutations: Objects, Practices, Contexts, Preprint 393*, Berlin, Max Planck Institute for the History of Science, 2010, pp. 231–58.

Waddington, C. H., 'Some European contributions to the prehistory of molecular biology', *Nature*, vol. 221, 1969, pp. 318–21.

Wahba, A. J., Basilio, C., Speyer, J. F. *et al.*, 'Synthetic polynucleotides and the amino acid code, VI', *Proceedings of the National Academy of Sciences USA*, vol. 48, 1962, pp. 1683–6.

Wahba, A. J., Gardner, R. S., Basilio, C. *et al.*, 'Synthetic polynucleotides and the amino acid code, VIII', *Proceedings of the National Academy of Sciences USA*, vol. 49, 1963a, pp. 116–22.

Wahba, A. J., Miller, R. S., Basilio, C. *et al.*, 'Synthetic polynucleotides and the amino acid code, IX', *Proceedings of the National Academy of Sciences USA*, vol. 49, 1963b, pp. 880–5.

Wahlsten, D., Bachmanov, A., Finn, D. A. and Crabbe, J. C., 'Stability of inbred mouse strain differences in behavior and brain size between laboratories and across decades', *Proceedings of the National Academy of Sciences USA*, vol. 103, 2006, pp. 16364–9.

Wain, H. M., Bruford, E. A., Lovering, R. C. *et al.*, 'Guidelines for human gene nomenclature', *Genomics*, vol. 79, 2002, pp. 464–70.

Wall, R., 'Overlapping genetic codes', *Nature*, vol. 193, 1962, pp. 1268–70.

Walker, F. O., 'Huntington's disease', *The Lancet*, vol. 369, 2007, pp. 218–28.

Wang, A. H., Quigley, G. J., Kolpak, F. J. *et al.*, 'Molecular structure of a left-handed double helical DNA fragment at atomic resolution', *Nature*, vol. 282, 1979, pp. 680–6.

Watanabe, T., Zhong, G., Russell, C. A. *et al.*, 'Circulating avian influenza viruses closely related to the 1918 virus have pandemic potential', *Cell Host Microbe*, vol. 15, 2014, pp. 692–705.

Watson, J. D., *Molecular Biology of the Gene*, New York, Benjamin, 1965.

Watson, J. D., *The Double Helix: A Personal Account of the Discovery of the Structure of DNA*, London, Weidenfeld & Nicolson, 1968.

Watson, J. D., *Genes, Girls and Gamow*, Oxford, Oxford University Press, 2001.

Watson, J. D. and Berry, A., *DNA: The Secret of Life*, New York, Knopf, 2003.

Watson, J. D. and Crick, F. H. C., 'A structure for deoxyribose nucleic acid', *Nature*, vol. 171, 1953a, pp. 737–8.

Watson, J. D. and Crick, F. H. C., 'Genetical implications of the structure of deoxyribose nucleic acid', *Nature*, vol. 171, 1953b, pp. 964–7.

Watson, J. D. and Crick, F. H. C., 'The structure of DNA', *Cold Spring Harbor Symposia on Quantitative Biology*, vol. 18, 1953c, pp. 123–31.

Watson, J. D. and Maaløe, O., 'Nucleic acid transfer from parental to progeny bacteriophage', *Biochimica et Biophysica Acta*, vol. 10, 1953, pp. 432–42.

Weaver, W., 'The mathematics of communication', *Scientific American*, vol. 181 (7), 1949, pp. 11–15.

White, M. A., Myers, C. A., Corbo, J. C. and Cohen, B. A., 'Massively parallel in vivo enhancer assay reveals that highly local features determine the cis-regulatory function of ChIP-seq peaks', *Proceedings of the National Academy of Sciences USA*, vol. 110, 2013, pp. 11952–7.

Wiener, N., 'Time, communication and the nervous system', *Annals of the New York Academy of Science*, vol. 50, 1948a, pp. 197–220.

Wiener, N., *Cybernetics: or, Control and Communication in the Animal and the Machine*, New York, Technology Press, 1948b.

Wiener, N., *Extrapolation, Interpolation, and Smoothing of Stationary Time Series: With Engineering Applications*, Boston, MIT Press, 1949.

Wiener, N., *The Human Use of Human Beings: Cybernetics and Society*, Boston, Houghton Mifflin, 1950.

Wiener, N., *I am a Mathematician*, London, Gollancz, 1956.

Wilkins, M. H. F., 'The molecular configuration of nucleic acids: Nobel Lecture, December 11, 1962', in *Nobel Lectures Physiology or Medicine 1942–1962*, Elsevier Publishing Company, Amsterdam, 1964.

Wilkins, M., *The Third Man of the Double Helix: An Autobiography*, Oxford, Oxford University Press, 2003.

Wilkins, M. H. F., Gosling, R. G. and Seeds, W. E., 'Physical studies of nucleic acid: an extensible molecule?', *Nature*, vol. 167, 1951, pp. 759–60.

Williams, T. A., Foster, P. G., Cox, C. J. and Embley, T. M., 'An archaeal origin of eukaryotes supports only two primary domains of life', *Nature*, vol. 504, 2013, pp. 231–6.

Witkowski, J. A., 'The discovery of split genes', *Trends in Biochemical Sciences*, vol. 13, 1988, pp. 110–13.

Witkowski, J. (ed.), *The Inside Story: DNA to RNA to Protein*, Cold Spring Harbor, Cold Spring Harbor Laboratory Press, 2005.

Woese, C., 'Nature of the biological code', *Nature*, vol. 194, 1962, pp. 1114–15.

Woese, C., 'On the evolution of the genetic code', *Proceedings of the National Academy of Sciences USA*, vol. 54, 1965, pp. 1546–52.

Woese, C. R., *The Genetic Code*, London, Harper & Row, 1967.

Woese, C. R. and Fox, G. E., 'Phylogenetic structure of the prokaryotic domain: the primary kingdoms', *Proceedings of the National Academy of Sciences USA*, vol. 74, 1977a, pp. 5088–90.

Woese, C. R. and Fox, G. E., 'The concept of cellular evolution', *Journal of Molecular Evolution*, vol. 10, 1977b, pp. 1–6.

Woese, C. R., Dugre, D. H., Saxinger, W. C. and Dugre, S. A., 'The molecular basis of the genetic code', *Proceedings of the National Academy of Sciences USA*, vol. 55, 1966, pp. 966–74.

Wolfe, A. D., 'The Cold War context of the Golden Jubilee, or, why we think of Mendel as the father of genetics', *Journal of the History of Biology*, vol. 45, 2012, pp. 389–414.

Wolfe-Simon, F., Davies, P. C. W. and Anbar, A. D., 'Did nature also choose arsenic?', *International Journal of Astrobiology*, vol. 8, 2009, pp. 69–74.

Wolfe-Simon, F., Blum, J. S., Kulp, T. R. *et al.*, 'A bacterium that can grow by using arsenic instead of phosphorus', *Science*, vol. 332, 2011, pp. 1163–6.

Wood, A. R., Esko, R., Yang, Y. *et al.*, 'Defining the role of common variation in the genomic and biological architecture of adult human height', *Nature Genetics*, vol. 46, 2014, pp. 294–8.

Wood, R. J. and Orel, V., *Genetic Prehistory in Selective Breeding: A Prelude to Mendel*, Oxford, Oxford University Press, 2001.

Worley, K. C. and Gibbs, R. A., 'Decoding a national treasure', *Nature*, vol. 463, 2010, pp. 303–4.

Wright, S., *Molecular Politics: Developing American and British Regulatory Policy for Genetic Engineering, 1972–1982*, London, University of Chicago Press, 1994.

Wrinch, D. M., 'The molecular structure of chromosomes' *Protoplasma*, vol. 25, 1936, pp. 550–69.

Wyatt, H. V., 'When does information become knowledge?', *Nature*, vol. 235, 1972, pp. 86–9.

Wyatt, H. V., 'How history has blended', *Nature*, vol. 249, 1974, pp. 803–5.

Wyatt, H. V., 'Knowledge and prematurity: the journey from transformation to DNA', *Perspectives in Biology and Medicine*, vol. 18, 1975, pp. 149–56.

Yáñez-Cuna, J. O., Kvon, E. Z. and Stark, A., 'Deciphering the transcriptional *cis*-regulatory code', *Trends in Genetics*, vol. 29, 2013, pp. 11–22.

Yang, Z., Chen, F., Alvarado, J. B. and Benner, S. A., 'Amplification, mutation, and sequencing of a six-letter synthetic genetic system', *Journal of the American Chemical Society*, vol. 133, 2011, pp. 15105–12.

Yaniv, M., 'The 50th anniversary of the publication of the operon theory in the Journal of Molecular Biology: Past, present and future', *Journal of Molecular Biology*, vol. 409, 2011, pp. 1–6.

Yanofsky, C., 'Establishing the triplet nature of the genetic code', *Cell*, vol. 128, 2007, pp. 815–18.

Yanofsky, C., Carlton, C. C., Guest, J. R. *et al.*, 'On the colinearity of gene structure and protein structure', *Proceedings of the National Academy of Sciences USA*, vol. 51, 1964, pp. 262–72.

Yarus, M., *Life from an RNA World: The Ancestor Within*, Harvard, Harvard University Press, 2010.

Yarus, M., Caporaso, J. G. and Knight, R., 'Origins of the genetic code: The escaped triplet theory', *Annual Review of Biochemistry*, vol. 74, 2005, pp. 179–98.

Yates, R. A. and Pardee, A. B., 'Control of pyrimidine biosynthesis in *Escherichia coli* by a feed-back mechanism', *Journal of Biological Chemistry*, vol. 221, 1956, pp. 757–70.

Yčas, M., 'The protein text', in H. P. Yockey, R. L. Platzman and H. Quastler (eds), *Symposium on Information Theory in Biology*, London, Pergamon, 1958, pp. 70–100.

Yčas, M., *The Biological Code*, London, North-Holland, 1969.

Yčas, M. and Vincent, W. S., 'A ribonucleic acid fraction from yeast related in composition to desoxyribonucleic acid', *Proceedings of the National Academy of Sciences USA*, vol. 46, 1960, pp. 804–11.

Yockey, H. P., 'Some introductory ideas concerning the application of information theory in biology', in H. P. Yockey, R. L. Platzman and H. Quastler (eds), *Symposium on Information Theory in Biology*, London, Pergamon, 1958, pp. 50–9.

Yockey, H. P., *Information Theory and Molecular Biology*, Cambridge, Cambridge University Press, 1992.

Yockey, H. P., Platzman, R. L. and Quastler, H. (eds), *Symposium on Information Theory in Biology*, London, Pergamon, 1958.

Yong, E., 'The unique merger that made you (and ewe, and yew)', *Nautilus*, 10, http://nautil.us/issue/10/mergers--acquisitions/the-unique-merger-that-made-you-and-ewe-and-yew, 2014.

Young, J. Z., 'Memory, heredity and information', in J. Huxley, A. C. Hardy and E. B. Ford (eds), *Evolution as a Process*, London, Allen & Unwin, 1954, pp. 281–99.

Yoxen, E. J., 'Where does Schroedinger's *What is Life?* belong in the history of molecular biology?', *History of Science*, vol. 17, 1979, pp. 17–52.

Yu, A., Lepère, G., Jay, F. *et al.*, 'Dynamics and biological relevance of DNA demethylation in *Arabidopsis* antibacterial defense', *Proceedings of the National Academy of Sciences USA*, vol. 110, 2013, pp. 2389–94.

Zagorski, N., 'Profile of Alec J. Jeffreys', *Proceedings of the National Academy of Sciences USA*, vol. 26, 2006, pp. 8918–20.

Zamecnik, P., 'From protein synthesis to genetic insertion', *Annual Review of Biochemistry*, vol. 74, 2005, pp. 1–28.

Zamecnik, P. C. and Keller, E. B., 'Relation between phosphate energy donors and incorporation of labeled amino acids into proteins', *Journal of Biological Chemistry*, vol. 209, 1954, pp. 337–54.

Zamenhof, S., 'Properties of the transforming principle', in W. D. McElroy and B. Glass (eds), *A Symposium on the Chemical Basis of Heredity*, Baltimore, The Johns Hopkins Press, 1957, pp. 351–72.

Zubay, G., 'A possible mechanism for the initial transfer of the genetic code from deoxyribonucleic acid to ribonucleic acid', *Nature*, vol. 182, 1958, pp. 112–13.

Zubay, G. and Quastler, H., 'An RNA-protein code based on replacement data', *Proceedings of the National Academy of Sciences USA*, vol. 48, 1962, pp. 461–71.

NOTES

Chapter 1

1. Wood and Orel (2001), p. 258; see also Cobb (2006a), Poczai *et al.* (2014).
2. López-Beltrán (1994), Müller-Wille and Rheinberger (2007, 2012).
3. Harvey basically shrugged his shoulders and gave up (Cobb, 2006b).
4. Cobb (2006a).
5. For Mendel's work and its implications, see Bowler (1989), Gayon (1998), Hartl and Orel (1992). For critical accounts of the way in which Mendel's work has been interpreted and used, see Brannigan (1979) and Wolfe (2012).
6. There are many historical accounts of twentieth-century genetics, for example Carlson (1966, 1981, 2004), Hunter (2000), Pichot (1999), Schwartz (2008), Sturtevant (1965). For conceptual aspects see the articles in Beurton, Falk and Rheinberger (2000) as well as Falk (2009) and Müller-Wille and Rheinberger (2012). For the changing views of de Vries, see Stamhuis, Meijer and Zevenhuizen (1999).
7. Sutton (1902), p. 39. See Crow and Crow (2002).
8. Sutton (1903), p. 236.
9. Hegreness and Meselson (2007).
10. Boveri (1904), cited in Crow and Crow (2002).
11. Pichot (1999), p. 111.
12. Shine and Wrobel (1976).
13. Carlson (1981), Kohler (1994), Sturtevant (1965).
14. Morgan (1933).
15. All details from Carlson (2004).

16. Morgan (1919), p. 246.
17. Morgan (1933), p. 316.
18. von Schwerin (2010).
19. For a translation of the Three-Man Paper, and discussions of its significance by historians and philosophers, see Sloan and Fogel (2011).
20. Sloan and Fogel (2011), p. 257.
21. Sofyer (2001), Morange (2011). Koltsov's name can also be transliterated as Koltzoff. For a discussion of Koltsov's contribution to ideas of messages and codes, see Kogge (2012).
22. Olby (1994), Sofyer (2001).
23. Muller (1922), p. 37; Troland (1917).
24. Quoted in Pollock (1970), p. 13.
25. Haldane (1945), Morange (2011).
26. Pringle (2008).
27. Olby (1994), pp. 73–96.
28. Caspersson *et al.* (1935), p. 369.
29. Stanley (1935).
30. Muller (1922).
31. Cairns *et al.* (1966), Summers (1993).
32. Kay (1986).
33. All quotes from Wrinch (1936).
34. Schultz (1935), p. 30.
35. Beadle and Tatum (1941).
36. For example Troland (1917).
37. Horowitz *et al.* (2004), p. 4.
38. Tatum and Beadle (1942), p. 240.
39. Berg and Singer (2003), pp. 171–86.
40. *Time*, 5 April 1943; *The Irish Press*, 6 February 1943.
41. *The Irish Press*, 13 and 16 February 1943.
42. Moore (1989).
43. 5, 12 and 19 February 1943. Moore (1989), p. 35.
44. Using different calculations, Schrödinger at one point suggested that a gene was composed of a few million atoms, at another '1,000 and possibly much less'. Schrödinger (2000), p. 46.
45. Schrödinger (2000), p. 20.
46. Schrödinger (2000), p. 21.
47. Schrödinger (2000), p. 22.
48. Schrödinger (2000), p. 62.
49. Olby and Posner (1967).
50. In 1999, Joshua Lederberg argued that Schrödinger did not really mean that the genetic material was 'aperiodic', but rather that it had

'elements of crystallinity' or was 'near-crystal' (Dromanraju, 1999, p. 1074).
51. *The Irish Press*, 6 and 16 February 1943.
52. *The Kerryman*, 22 January 1944.
53. In 1945 Schrödinger had a brief correspondence with the geneticist J. B. S. Haldane over the genetics of hornless cattle (Crow, 1992).
54. Yoxen (1979), p. 45, note 9; Olby (1971), p. 122.
55. Pauling (1987), p. 229.
56. Perutz (1987), p. 243; Waddington (1969), p. 321.
57. Wilkins (2003), p. 84; Crick (1988), p. 18; Inglis *et al.* (2003), p. 3.
58. For example, Morange (1983), Symonds (1986), Kay (2000), Sarkar (2013).

Chapter 2
1. Administrative Framework of OSRD (1948).
2. Conway and Siegelman (2005), p. 199.
3. Mindell (1995), p. 91. See also Bennett (1994), Masani (1990) and Mindell (2000, 2002).
4. Mindell (1995), p. 92; Owens (1989). Wiener's grant was the smallest awarded by D-2.
5. In 2013, the then occupant of the room, Dr Bjorn Poonen, kindly sent me photos of his office, looking pretty much as it must have done in 1940, with the exception of the floor. The room was due to be completely remodelled in a few months. Dr Poonen is the Claude Shannon Professor of Mathematics at MIT.
6. Wiener (1956), p. 249.
7. Rosenblueth *et al.* (1943).
8. Mindell (1995), p. 95.
9. Kay (2000), p. 83.
10. On Pitts, see Easterling (2001) and Schlatter and Aizawa (2008). Easterling (2001) begins: 'There are no biographies of Walter Pitts, and any honest discussion of him resists conventional biography.'
11. Conway and Siegelman (2005), p. 134.
12. Galison (1994).
13. It has been argued that there is a direct engineering and conceptual link between this device and the computer mouse (Roch, 1998).
14. Kay (2000), p. 81.
15. Wiener (1949), p. 2.
16. Shannon (1940); Roch (1999), p. 265.
17. Rogers (1994).
18. Hodges (2012), p. 251.
19. Conway and Siegelman (2005), p. 126.

20. For a comparison of the relatively minor editorial differences between the 1945 original and the two published articles (Shannon 1948a, b), see Roch (1999).
21. Conway and Siegelman (2005), p. 146.
22. Macrae (1992), p. 242; Heims (1980), pp. 192–9.
23. Galison (1994), p. 253; Triclot (2007, 2008).
24. Conway and Siegelman (2005), p. 155.
25. Wiener (1948a).
26. von Neumann (1997).
27. Carlson (1981), pp. 307 and 310.
28. http://encyclopedia.gwu.edu/index.php?title=Theoretical_Physics_Conference,_1946. A more sober, and less interesting, summary was provided by Gamow and Abelson (1946).

Chapter 3

1. Judson (1996), p. 44.
2. Burnet (1968), p. 81.
3. Heidelberger *et al.* (1971).
4. Griffith (1928).
5. Hotchkiss (1965), p. 5.
6. Dobzhansky (1941), pp. 48–9.
7. Dobzhansky (1941), pp. 49–50.
8. Biscoe *et al.* (1936).
9. McCarty (1986), p. 104. McCarty's memoir is the main source for much of the detail of life in the Avery laboratory. Some of Avery's lab books, along with reports, articles and letters, can be found at http://profiles.nlm.nih.gov/ps/retrieve/Collection/CID/CC.
10. McCarty (1986), p. 127. Transformation as studied in the Avery lab in fact involved two different types of pneumococcus, the Type II R form and the Type III S form. For the sake of simplicity, I have referred only to the R and S characteristics.
11. McCarty (2002), p. 25.
12. Report of the Director of the Hospital to the Corporation of the Rockefeller Institute for Medical Research, 19 April 1941. Rockefeller Archive Center. http://profiles.nlm.nih.gov/ps/retrieve/ResourceMetadata/CCAANJ.
13. Schultz (1941), p. 56.
14. Mirsky (1943), p. 19.
15. Report of the Director of the Hospital to the Corporation of the Rockefeller Institute for Medical Research, 17 April 1943, pp. 151–2. Rockefeller Archive Center. http://profiles.nlm.nih.gov/ps/retrieve/ResourceMetadata/CCAADS.

16. Letter from Roy Avery to Wendell Stanley, 26 January 1970. University of California, Berkeley. Bancroft Library. Wendell M. Stanley Papers, Box 4, Folder 7. http://profiles.nlm.nih.gov/ps/retrieve/ResourceMetadata/CCAAHG.
17. The scientific part of the letter is reproduced in Dubos (1976), pp. 216–20. A full transcript and links to a scanned version of the letter can be found at http://profiles.nlm.nih.gov/ps/retrieve/ResourceMetadata/CCBDBF#transcript.
18. McCarty (1986), p. 168.
19. Avery *et al.* (1944), p. 155.
20. McCarty (1986), p. 168.
21. McCarty (1986), p. 195.
22. McCarty and Avery (1946a, b).
23. McCarty and Avery (1946a), p. 94.
24. McCarty and Avery (1946a), p. 95.
25. Morgan (1944), p. 764; Haddow (1944), p. 196.
26. Anonymous (1944), p. 329.
27. Bearn (1996), p. 552.
28. Muller (1947), p. 22.
29. Mueller (1945), p. 734.
30. Lederberg diary entry, 20 January 1945. http://profiles.nlm.nih.gov/ps/access/CCAAAB.pdf.
31. Boivin *et al.* (1945a), p. 648.
32. Judson (1996), p. 44. Salvador Luria showed Boivin's paper to Avery and then lunched with Avery's group.

Chapter 4
1. Chargaff and Vischer (1948).
2. These and subsequent quotes are from Hall (2011), p. 124.
3. Astbury (1947), p. 69. The diffraction pattern produced by a helix had yet to be described – this was Francis Crick's PhD work, which helped give him the insight into the problem of DNA structure (Cochran *et al.*, 1952). See comment 3 at http://paulingblog.wordpress.com/2009/07/09/the-x-ray-crystallography-that-propelled-the-race-for-dna-astburys-pictures-vs-franklins-photo-51/.
4. This and subsequent quotes are from Gulland (1947a), pp. 3–4.
5. Stacey (1947), p. 96.
6. Stedman and Stedman (1947), p. 244.
7. McCarty, Taylor and Avery (1946), p. 177.
8. Spiegelman (1946), p. 274. This was Cohen's contribution to the discussion of Spiegelman's paper.

9. Mirsky and Pollister (1946), pp. 134–5. For Mirsky's life and work, see Cohen (1998).
10. Muller (1947), pp. 22–3.
11. Cohen (1947) seems to have been the English-language pioneer in this respect. Soon the abbreviation was everywhere, and it has now passed into the English language.
12. Boivin *et al.* (1945a, b), Boivin (1947).
13. Boivin (1947), pp. 12–13. He also thought that changes acquired during the organism's life might be stored in RNA molecules.
14. Boivin (1947), p. 16. Mirsky's comments can be found on the same page.
15. Chargaff (1947), p. 32.
16. Gulland (1947b), p. 97.
17. Gulland (1947b), p. 102. It is not clear whether Gulland would have pursued this work had he lived – he had recently taken up a post in industry. On what Gulland's role might have been, see Manchester (1995).
18. Stedman and Stedman (1947), p. 235.
19. Schultz (1947), p. 221.
20. Gulland (1944), Thieffry (1997), Thieffry and Burian (1996).
21. Lederberg (1948), p. 182.
22. Dubos (1976), p. 159.
23. Bohlin (2009) conducted interviews with Swedish scientists who were involved with nucleic acids research in the 1950s and explored why Avery was not given the prize. An English translation of Bohlin's Swedish-language paper would be most welcome. Avery's contribution was eventually acknowledged when a crater on the Moon was named after him. Mendel, Schrödinger, Szilárd, von Neumann and Wiener have all been similarly honoured.
24. *New York Times*, 21 February 1955.
25. *New York Times*, 23 January 1949.
26. Taylor (1949).
27. Hotchkiss (1949).
28. Lwoff (1949), p. 202.
29. Delbrück (1949), Hotchkiss (1979), p. 330.
30. Boivin *et al.* (1949), p. 67.
31. Boivin *et al.* (1949), p. 75.
32. Olby (1994), p. 201.
33. There is no biography of Boivin. Roche (1949) wrote a brief obituary.
34. Mazia (1952), p. 109.
35. Mazia (1952), p. 114.
36. Pollister *et al.* (1951), p. 115.

37. Chargaff (1951).
38. This and subsequent quotes are from Ephrussi-Taylor (1951), pp. 445–8.
39. Anonymous (1980), p. 25.
40. Judson (1996), p. 41.
41. Deichmann (2004, 2008), Olby (1972), Pollock (1970), Wyatt (1972, 1975). For a survey of the long-running debate over this issue, see Cobb (2014). For a pugnacious defence of Avery's work by a participant, see Hotchkiss (1979).
42. Stent (1972). Stent was prompted to write this piece because in an earlier article on the history of molecular genetics he had not mentioned the name of Avery and had been criticised for this (Stent, 1968a, b).
43. Stanley (1970), p. 262.
44. Judson (1996), p. 41.
45. Judson (1996), p. 43.
46. Hershey (2000), p. 105.
47. Hotchkiss (2000), p. 36.
48. Northrop (1951), p. 732.
49. Letter of 16 November 1951, Hershey (1966), p. 102.
50. Anderson (1966), p. 76.
51. Creager (2009).
52. http://library.cshl.edu/oralhistory/interview/cshl/memories/szybalski-martha-chase/.
53. Hershey and Chase (1952); for a discussion of the impact of this paper and a detailed analysis of all its experimental steps, see Wyatt (1974). Only the key experiments are described here. A similar experiment was carried out by Watson and Maaløe (1953).
54. Hershey and Chase (1952), p. 56.
55. Symonds (2000), p. 93.
56. Hershey (1953).
57. Hershey (1966), p. 106.
58. Stern (1947). The only people to have been interested in Stern's models, rather than his biochemical procedures, are historians from 1970 onwards.
59. Caspersson and Schultz (1939), Brachet (1942), Caspersson (1947).
60. Boivin and Vendrely (1947).
61. Caldwell and Hinshelwood (1950).
62. Dounce (1952).
63. Dounce (1952).

Chapter 5

1. *Business Week*, 15 February 1949.
2. Conway and Siegelman (2005), p. 183.
3. Wiener (1956), pp. 315–17.
4. *'Cybernétique'* had been used by the French physicist Ampère in 1845 to describe the science of civil government; Wiener's meaning was both far more broad and far more precise. De Latil (1953), pp. 23–4. In a lecture given in 1950, Wiener implied that he had been inspired to use the Greek word for 'steersman' by the use of negative feedback in power steering on ships; see http://www.wnyc.org/story/ men-machines-and-the-world-about-them/.
5. *New York Times*, 10 April 1949.
6. Wiener (1948b), pp. 27–9.
7. Quotes in this paragraph are from Wiener (1948b), pp. 11, 58 and 132.
8. In the middle of the 1940s, feedback was mathematically formalised by Wiener and his group, as well as by Hans Sartorius in Germany and a Bell Labs mathematician, Le Roy MacColl. MacColl (1945), Mayr (1970), Bennett (1996).
9. *New York Times*, 19 December 1948.
10. Rosenblith (1949), p. 187.
11. Eisenhart (1949); Brillouin (1949), p. 566; Brillouin (1956).
12. Pfeiffer (1949), p. 16.
13. *Le Monde*, 28 December 1948. Wiener was sufficiently impressed (or flattered) by Dubarle's article to reproduce substantial parts of it (Wiener, 1950). See also Dubarle (1953).
14. Wiener (1956), p. 331.
15. Shannon and Weaver (1949), p. 31.
16. Shannon and Weaver (1949), p. 8.
17. Shannon and Weaver (1949), p. 17.
18. As Weaver put it in an article in *Scientific American*: 'it is most significant that an entropy-like expression appears in communication theory as a measure of information' (Weaver, 1949, p. 12).
19. Kay (2000), p. 94.
20. Kay (2000), p. 101.
21. All material in this paragraph is from von Neumann (1951), pp. 28–31.
22. Kay (1995), p. 623.
23. Kay (2000), pp. 118–19.
24. Anonymous (1950), pp. 193–4.
25. http://www.bbc.co.uk/radio4/features/the-reith-lectures/ transcripts/1948/#y1950.
26. de Latil (1953, 1957), Colnort-Bodet (1954).
27. King (1952), Gabor (1953).

28. Keller (1995), p. 92.

29. Wiener (1950), p. 16.

30. *Times Literary Supplement*, 20 July 1951.

31. Wiener (1950), p. 15.

32. Wiener (1950), p. 110.

33. Gleick (2011), p. 232.

34. Kay (1995), p. 624.

35. See the letters from Lederberg to Quastler in the Lederberg papers, http://profiles.nlm.nih.gov/ps/retrieve/Collection/CID/BB. Letters are dated 3 May 1951 (reference BBARI) and 16 May 1951 (BBAFAL). For Lederberg's 1993 view of this episode, see his handwritten comments to Lily E. Kay, on a copy of the 3 May 1951 letter (BBAFAK).

36. Linschitz (1953), p. 251.

37. Dancoff and Quastler (1953), pp. 269–70.

38. Apter and Wolpert (1965), pp. 249–50.

39. Macrae (1992).

40. Conway and Siegelman (2005).

41. The suggestion that heredity involves a kind of memory was first put forward by Ewald Hering and Samuel Butler in the 1870s (Forsdyke, 2006).

42. Kalmus (1950, 1962).

43. Watson (2001), p. 12.

44. Lederberg (1952). One of Lederberg's coinages – 'plasmid' – is still used to describe small extrachromosomal bacterial DNA molecules.

45. Ephrussi *et al.* (1953).

46. Biologists Spiegelman and Landman (1954), Cavalli-Sforza (1957) and Thomas (1992) all showed they knew it was a joke. Historians and philosophers Kay (1995, 2000), Keller (1995) and Sarkar (1991) all took it seriously.

47. Kay (1995), p. 627.

Chapter 6

1. Maddox (2002), Wilkins (2003).

2. The main books used in this chapter are Crick (1988), Ferry (2007), Gann and Witkowski (2012), Hager (1995), Inglis *et al.* (2003), Judson (1996), Maddox (2002), McElheny (2003), Olby (1994, 2009), Ridley (2006), Sayre (1975), Watson (1968, 2001), Wilkins (2003). The most potent account, which has framed all others, is Jim Watson's *The Double Helix* (Watson 1968, Gann and Witkowski 2012). For a collection of articles covering this period and afterwards, see Witkowski (2005).

3. Daly *et al.* (1950), p. 506.

4. Chargaff (1951), p. 44. See also Chargaff (1950) and Manchester (2008).

5. Daly *et al.* (1950).
6. Chargaff *et al.* (1951), p. 229.
7. Chargaff (1950).
8. Creeth *et al.* (1947), p. 1141.
9. Wilkins (1964).
10. Wilkins (2003), p. 121; Attar (2013).
11. Attar (2013), p. 5.
12. Olby (1994), p. 355.
13. This was the meeting in Naples that excited Jim Watson so much.
14. Wilkins *et al.* (1951).
15. Fraser and Fraser (1951).
16. There were seven articles by Pauling on the α-helix in the May 1951 issue of the *Proceedings of the National Academy of Sciences*.
17. Cochran and Crick (1952), Cochran, Crick and Vand (1952).
18. Judson (1996), p. 95.
19. Maddox (2002), p. 160.
20. Gann and Witkowski (2012), p. 11. For Franklin, see Maddox (2002), Piper (1998), Sayre (1975).
21. Maddox (2002), p. 151.
22. Wilkins (2003), Attar (2013).
23. Olby (1994), pp. 338–9.
24. Maaløe and Watson (1951).
25. Perutz recalled this moment shortly before he died, in his last letter to Watson: Inglis *et al.* (2003), p. 73.
26. Gann and Witkowski (2012), p. 43.
27. Judson (1996), p. 88.
28. Olby (1994), p. 354.
29. Maddox (2002), p. 149.
30. Maddox (2002), p. 154.
31. Judson (1996), p. 104.
32. Klug (2004).
33. Gann and Witkowski (2010), p. 524.
34. Judson (1996), p. 117.
35. Chargaff (1978), p. 101.
36. Judson (1996), p. 120.
37. Cochran and Crick (1952), Cochran *et al.* (1952).
38. Davies (1990), Hall (2011, 2014), Olby (1994).
39. Pauling to Tinker http://osulibrary.oregonstate.edu/specialcollections/coll/pauling/dna/corr/corr410.17-lp-tinker-19520506–02.html. Pauling eventually published a brief article describing Ronwin's model as 'extraordinary. Deserves no serious consideration' (Pauling and Schomaker, 1952).

40. Rowen *et al.* (1953), p. 90.
41. Interview with Gosling, 2013. http://www.nature.com/nature/podcast/index-gosling-2013–04–20.html.
42. Judson (1996), p. 131.
43. Olby (1994), pp. 376–7.
44. Pauling and Corey (1953).
45. Interview with Gosling, 2013. http://www.nature.com/nature/podcast/index-gosling-2013–04–20.html.
46. Gann and Witkowski (2012), p. 181.
47. Wilkins (2003), p. 224; Maddox (2002), p. 196.
48. Judson (1996), p. 142.
49. Judson (1996), p. 132; Maddox (2002), p. 190.
50. Maddox (2002), pp. 201–2.
51. A few days later, the iconic picture of Watson and Crick, with the model of DNA, was taken by Antony Barington Brown, although the photo was not used at the time. De Chadarevian (2003).
52. Wilkins (2003), pp. 212–14.
53. Olby (1994), p. 422. For contrasting views on the impact of the double helix paper, see Olby (2003) and Gingras (2010). There were two contemporary press accounts of the discovery: 'Why you are you: nearer secret of life', which appeared in the London-based *News Chronicle* (15 May 1953), and 'Clue to chemistry of heredity found', which appeared in the *New York Times* (13 June 1953).
54. Creager and Morgan (2008).
55. Perutz (1969).
56. Donohue (1978), p. 135.
57. Watson and Crick (1953a), p. 737.

Chapter 7

1. Judson (1996), p. 153.
2. Watson and Crick (1953b).
3. They did not use the term 'double helix' until a year later (Crick and Watson, 1954).
4. Watson (2001), p. 11. For a discussion of the language used by Watson and Crick and its significance, see Halloran (1997); for a post-modern exploration of the rhetoric of molecular biology, see Doyle (1997).
5. Olby (1994), p. 421.
6. Wilkins (2003), p. 224.
7. http://www.webofstories.com/play/francis.crick/84.
8. The nuclear historian Alex Wellerstein kindly sent me 120 pages of FBI documents relating to Gamow that he obtained through a freedom of information request. In 1951 the FBI concluded that Gamow was 'not

the type of individual who would possess deep-rooted convictions of loyalty to any government.' US Federal Bureau of Investigation, George Gamow FBI file (116-HQ-12246), via Freedom of Information Act Request 1227772–0.

9. Watson (2001), p. 125. 'Combinatorix' presumably refers to the branch of mathematics known as 'combinatorics'.

10. Watson (2001), p. 24.

11. Crick (1966a); Crick (1988), pp. 92–3; Watson (2001), pp. 46–7. Crick (1988) recalls the 'Tompkins' article as being the *Nature* paper; this contradicts his earlier account.

12. Gamow (1954).

13. Olby (2009), p. 221.

14. Crick (1958), p. 140.

15. Watson and Crick (1953c), p. 127.

16. Crick recalls discussing Gamow's diamond model and his list of 20 amino acids with Watson in Cambridge after the receipt of Gamow's first letter (Crick, 1988, p. 91). This must be an error: Gamow's letter does not contain the diamond model and makes no mention of amino acids at all.

17. See, for example, Gamow's letter to Yčas of 2 July 1954, http://www.loc.gov/exhibits/treasures/images/125.6as.jpg and http://www.loc.gov/exhibits/treasures/images/125.6bs.jpg.

18. Watson (2001), Brenner (2001), Crick (1988), among many others.

19. Watson, letter to Crick, 10 February 1955, p. 2. http://profiles.nlm.nih.gov/ps/access/SCBBJL.pdf.

20. Judson (1996), p. 264.

21. Watson's tie can be seen on the cover of Watson (2001). Gamow can be seen wearing his tie in the photo in the plate section of this book.

22. http://www.webofstories.com/play/francis.crick/84.

23. Judson (1996), pp. 307–12.

24. Kay (2000), pp. 141–2.

25. Rich (1997), p. 122.

26. For example, Dounce *et al.* (1955).

27. Gamow and Metropolis (1954), Gamow *et al.* (1957).

28. Crick (1955), p. 1.

29. Sanger and Tuppy (1951a, b), Sanger and Thompson (1953a, b), Stretton (2002).

30. Crick (1955), p. 4.

31. Crick (1955), pp. 5–6.

32. Crick (1955), p. 17.

33. Schwartz (1955).

34. Gamow *et al.* (1957).

35. Judson (1996), p. 282.
36. Brenner (1956), p. 3.
37. Brenner (2001), p. 55; Friedberg (2010), p. 82.
38. Brenner (1957), Gamow (1955), Gamow *et al.* (1957).
39. Judson (1996), pp. 282 and 299.
40. Judson (1996), p. 299.
41. Olby (2009), p. 263.
42. Neel (1949).
43. Pauling *et al.* (1949), Hager (1995).
44. Allison (2004).
45. Ingram (2004).
46. Pauling (1955), p. 222.
47. Ingram (1956), p. 794.
48. *The Times*, 1 September 1956.
49. Ingram (1957).
50. *The Times*, 23 August 1957.
51. Morange (1998), pp. 130–1; Strasser (2006).

Chapter 8

1. Olby (2009), p. 247; Crick (1988), p. 108.
2. Jacob (1988), pp. 287–8.
3. Judson (1996), p. 335.
4. Crick (1957, 1958). Crick (1957) has been cited less than 20 times.
5. Crick (1958), p. 144, pp. 138–9.
6. Crick (1958), p. 144.
7. Glass (1957), p. 757.
8. Zamenhof (1957), p. 354. For Zamenhof's early acceptance of Avery's findings, see Zamenhof's 28 February 1978 letter to Joshua Lederberg. http://profiles.nlm.nih.gov/ps/access/CCAALE.pdf.
9. Beadle (1957), p. 5.
10. Crick (1958), p. 145.
11. Crick (1958), p. 144.
12. Crick (1958), p. 144.
13. Crick (1958), p. 152.
14. Chargaff (1957), pp. 521, 526.
15. Burnet (1956), p. 25.
16. Roberts (1958), p. viii. Roberts's explanation for the change of vocabulary was as follows: during the conference, entitled 'Microsomal particles and protein synthesis', 'a semantic difficulty became apparent' as different people used the term microsome to mean very different things. Roberts wrote: 'During the meeting the

word "ribosome" was suggested; this seems a very satisfactory name, and it has a pleasant sound.'

17. Zamecnik and Keller (1954).
18. Hoagland *et al.* (1957).
19. Hoagland *et al.* (1958).
20. Crick (1958), pp. 143–4. See also Rich (1962).
21. Crick (1957), pp. 198–200.
22. Crick (1970), p. 562.
23. http://profiles.nlm.nih.gov/ps/access/SCBBFT.pdf.
24. Crick (1988), p. 109.
25. Crick (1970), p. 562.
26. Olby (2009), p. 253; Morange (1998), pp. 169–70.
27. Judson (1996), p. 333.
28. Burnet (1968), Davis (2013).
29. Burnet (1956), pp. 170–1.
30. Burnet (1956), p. 171.
31. Watson (1965).
32. Morange (1998), pp. 172–3.
33. Crick (1988), p. 110.
34. Crick (1958), p. 142.
35. Organ *et al.* (2008).
36. Burnet (1956), p. 21.
37. Burnet (1956), p. 22.
38. The book that contains the papers from the meeting has the more enticing title *Symposium on Information Theory in Biology* (Yockey *et al.*, 1958).
39. Yockey (1958), p. 51.
40. Yockey (1958), p. 52.
41. Yčas (1958), p. 94.
42. Bar-Hillel (1953).
43. Augenstine (1958), p. 112.
44. Quastler (1958a), p. 41.
45. Augenstine (1958), p. 115.
46. Quastler (1958b), p. 190.
47. Quastler (1958b), p. 188.
48. Quastler (1958c), p. 399.
49. Young (1954), p. 281.
50. Correspondence between Lederberg and von Neumann, 10 March 1955 – 16 September 1955. http://profiles.nlm.nih.gov/ps/retrieve/Series/2722.
51. Burnet (1956), pp. 164–5.
52. Young (1954), pp. 284–5.

53. George (1960), p. 190.
54. Elias (1958).
55. Elias (1959), p. 225.
56. Heims (1991). For an example of this approach, see George (1962).
57. Quoted in Kay (2000), p. 125.
58. Kay (2000), p. 115.
59. Quoted in Kay (2000), p. 126.
60. Quastler (1958c), p. 402.

Chapter 9

1. For biographical studies of Monod, his science and his politics, see Carroll (2013), Debré (1996), Morange (2010) and Ullmann (2010).
2. Quoted by Kay (2000), p. 200, from an original draft. In the published version, Monod removed the fruit-merchant reference (Monod, 1972a).
3. Grandy (1996), Lanouette (1994, 2006), Maas (2004).
4. Monod (1972a), p. xv.
5. Jacob (1988), p. 293.
6. Pappenheimer (1979); Yates and Pardee (1956), p. 770.
7. For the anti-Lysenko explanation of this change, see Carroll (2013), Morange (1998) and above all Kay (2000), pp. 201–3.
8. Novick and Szilárd (1954), p. 21.
9. Cohn *et al.* (1953a), Monod and Cohen-Bazire (1953a, b), Pardee (1959).
10. Yates and Pardee (1956).
11. Umbarger (1956), p. 848. See also Kresge *et al.* (2005).
12. Morange (2013).
13. Grmek and Fantini (1982), p. 204.
14. Pardee *et al.* (1958, 1959), Jacob (1979), Pardee (1979). For the role of US links in work at the Institut Pasteur, see Burian and Gayon (1999) and Gaudillière (2002).
15. Pardee (1985, 2002).
16. In Belgium, Chantrenne and Jeener were developing similar ideas – see Thieffry (1997).
17. Pardee *et al.* (1959).
18. Szilárd (1960), Schaffner (1974).
19. Monod (1972b), p. 199.
20. Monod (1972b), p. 199.
21. Summarised in Szilárd (1960). For Maas's only published speculations on the matter, made in a conference discussion in September 1957 and published in 1958, see Schaffner (1974), p. 361. See also Maas (2004).
22. Schaffner (1974), p. 360; Maas (2004).
23. Vogel (1957).
24. Pardee *et al.* (1958); Schaffner (1974), p. 374.

25. Jacob (1988), p. 298. The fiftieth anniversary of the operon led to a flurry of papers putting the discovery into historical perspective, for example Beckwith (2011), Gann (2010), Lewis (2011) and Yaniv (2011). For Jacob's work on phage, see Peyrieras and Morange (2002).
26. Grmek and Fantini (1982), p. 209.
27. Jacob (1988), p. 308.
28. Kay (2000), p. 217.
29. Jacob (1988), p. 304. As Lily Kay has noted, in his 1965 Nobel Lecture, Jacob used the more neutral analogy of doors in a house, each controlled 'by a little radio receiver' (Jacob, 1972, p. 154).
30. Monod (1959).
31. Pardee *et al.* (1959).
32. Pontecorvo (1952).
33. Lederberg (1957), p. 753.
34. Benzer (1957), p. 70. For the impact of Benzer's work on a young researcher, see Holliday (2006).
35. Benzer (1966).
36. Benzer (1957), p. 90.
37. Benzer (1959, 1961).
38. Beadle (1957), p. 129. For a full discussion of Benzer's work on the rII region and its implications, see Holmes (2006).
39. Meselson and Stahl (1958). See also: Holmes (2001), Davis (2004) and Hanawalt (2004).
40. Delbrück and Stent (1957).
41. Judson (1996), p. 416.
42. Crick (1988), p. 119. See also Brenner (2001), pp. 73–87.
43. Jacob (1988), p. 312.
44. Judson (1996), p. 419.
45. Jacob (1988), pp. 313–14.
46. Yčas and Vincent (1960).
47. Brenner *et al.* (1961), Gros *et al.* (1961). For Gros's view of the race to identify the messenger, see Gros (1979). Watson's telegram to Brenner asking him to delay publication until his group's paper was ready can be seen at http://libgallery.cshl.edu/items/show/66514.
48. Jacob and Monod (1961a, b), Monod and Jacob (1961).
49. Jacob and Monod (1961a), p. 318.
50. Jacob and Monod (1961a), p. 334.
51. Jacob and Monod (1961a), p. 344. They had first used the term operon in the previous year, in a French publication (Jacob *et al.*, 1960). The precise functioning of the *Lac* operon would not be fully understood for some years. See Müller-Hill (1996) for a detailed insider's account. For an overview of the operon and its legacy, see Morange (2005a).

52. Jacob and Monod (1961a), p. 354.
53. Jacob and Monod (1961a), p. 354. Jacob could not recall any conscious reference to Schrödinger (Morange, 1998, p. 295, note 36).
54. Monod and Jacob (1961), p. 401.
55. Jacob (2011).
56. Ptashne (2013), p. 1181.
57. Brenner (1961); Monod and Jacob (1961), p. 393.

Chapter 10
1. Crick (1959).
2. Belozersky and Spirin (1958), Sueoka *et al.* (1959).
3. Golomb (1962a), p. 100.
4. Rheinberger (1997), p. 213.
5. There is a large amount of material covering Nirenberg's career in the Modern Manuscripts Collection, History of Medicine Division, National Library of Medicine at Bethesda. This includes over 40 volumes of diaries and notebooks – a future researcher's goldmine. The diary covering this period is D9 IXA, 1960 Sep–[1961] May, to be found in Box 22, Folder 44, Marshall W. Nirenberg Papers, 1937–2003. The Nirenberg diary entries quoted here are all taken from Kay (2000). A tiny proportion of the papers can be found online at http://profiles.nlm.nih.gov/JJ/.
6. Kay (2000), p. 240.
7. Hoagland *et al.* (1957, 1958). For a thorough exploration of Zamecnik's work and its conceptual implications, see Rheinberger (1997).
8. Grunberg-Manago *et al.* (1955). Ochoa shared the 1959 Nobel Prize with his ex-student Arthur Kornberg, who in 1956 had isolated the enzyme that enables DNA molecules to copy themselves. There was no prize for the co-discoverer of polynucleotide phosphorylase, the French biochemist Marianne Grunberg-Manago.
9. Singer (2003).
10. Kay (2000), p. 241.
11. Nirenberg (1960).
12. Lamborg and Zamecnik (1960); Rheinberger (1997), pp. 208–21.
13. Kay (2000), p. 246.
14. Tissières *et al.* (1960).
15. Kay (2000), p. 246.
16. Nirenberg (1963), p. 84.
17. Kay (2000), p. 247.
18. Rheinberger (1997), p. 213.
19. Crick *et al.* (1957), p. 420.
20. Crick (1958), p. 160.

21. Kay (2000), p. 248.
22. Matthaei and Nirenberg (1960).
23. Matthaei and Nirenberg (1961a).
24. Matthaei and Nirenberg (1961a), pp. 405–6.
25. Matthaei and Nirenberg (1961a), p. 407.
26. Kay (2000), p. 249.
27. Judson (1996), pp. 458–9.
28. Judson (1996), p. 460.
29. Judson (1996), p. 462.
30. Dr Jerry Hurwitz, e-mail to the author, 9 April 2014.
31. Lengyel (2012).
32. Hargittai (2002), p. 140.
33. Matthaei and Nirenberg (1961b), p. 1587. One of the controls later caused much confusion: to show that acidity was not involved in the DNase effect, they added several compounds, including polyadenylic acid, none of which affected protein synthesis. Polyadenylic acid is better known as poly(A); this 'negative control' later led some competitors to unfairly cast doubt on whether they had intended to get an effect with poly(U). See Kay (2000), pp. 249–50; Rheinberger (1997), p. 210.
34. Nirenberg and Matthaei (1961), p. 1601.
35. When interviewed by Judson in the 1970s, Nirenberg seemed unaware of these key papers (Judson, 1996, p. 462).
36. Anonymous (1961).
37. Anonymous (1961), Morgan (1961), Slater (2003). See also the informal photos of the Congress in the collections of Watson and Brenner, held at Cold Spring Harbor: http://libgallery.cshl.edu/items/show/51693 and http://libgallery.cshl.edu/items/show/52212.
38. The manuscript version of this talk, with Nirenberg's handwritten edits, can be found at http://profiles.nlm.nih.gov/ps/access/JJBBKB. pdf. For the published version see Nirenberg and Matthaei (1963).
39. Nirenberg (2004) recalled the size of the audience as '~35' (p. 49).
40. Hargittai (2002), p. 137.
41. Judson (1996), pp. 463–4.
42. Watson (2001), p. 265.
43. Watson and Berry (2003), p. 76; Crick et al. (1961), p. 1232; Nirenberg (2004), p. 49.
44. Judson (1996), p. 464.
45. Interview with Nirenberg by Ruth Harris, 1995–1996. http://history. nih.gov/archives/downloads/Nirenberg%20oral%20history%20 Chap%203a-%20%20Recognition%20Moscow,%20MIT.pdf
46. Judson (1996), p. 464.

47. Dr Jerry Hurwitz, e-mail to the author, 9 April 2014.
48. Varmus (2009), p. 24.
49. Letter from Lengyel to Ochoa, 19 August 1961, in Supplemental Material, Lengyel (2012).
50. Hurwitz recalled, 'I was impressed with the data presented by Nirenberg in Moscow but puzzled by the properties of the product (which were due to my own lack of information)' (e-mail to the author, 9 April 2014).
51. Judson (1996), p. 464.
52. Lipmann to Crick, 27 November 1961. http://profiles.nlm.nih.gov/ps/retrieve/ResourceMetadata/SCBBBV.
53. Kay (2000), p. 255.
54. Judson (1996), p. 465.
55. Judson (1996), pp. 464–5.
56. Stent (1971) described the experiment as follows: 'One day, Nirenberg added artificially synthesised polyuridylic acid to this reaction mixture instead of natural mRNA and obtained a most surprising result' (p. 528). See also Brenner (2001), p. 99.
57. Woese (1967), p. 53, note 1; Nirenberg (2004), p. 50. According to Woese, Beljanski's results were 'uninterpretable'. Nirenberg claims that Tissière had also tried and failed to get poly(A) to work, but Tissière said that although poly(A) was sitting in a freezer in the lab next door, he never thought to use it. He described his lack of initiative as 'idiotic' (Judson, 1996, p. 465).
58. Zamecnik (2005).
59. Nirenberg (2004), p. 50.
60. Nirenberg (2004), p. 49.
61. Judson (1996), p. 465.
62. Ochoa (1980), p. 20. Lengyel (2012), p. 32, uses very similar words.
63. Judson (1996), p. 469.
64. Nirenberg (2004), p. 49.
65. Martin (1984), p. 293.
66. Nirenberg (2004), p. 50.
67. Transcript of BBC talk 'Cracking the genetic code' by Crick, 22 January 1962. http://profiles.nlm.nih.gov/ps/access/SCBBFX.pdf.
68. Crick *et al.* (1961), p. 1229. For an appreciation of this paper, see Yanofsky (2007).
69. Judson (1996), p. 467.
70. Crick *et al.* (1961), p. 1227.
71. Crick *et al.* (1961), p. 1231.
72. Crick *et al.* (1961), p. 1232.
73. Anonymous (1962), p. 19.

74. Transcript of BBC talk 'Cracking the genetic code' by Crick, 22 January 1962. http://profiles.nlm.nih.gov/ps/access/SCBBFX.pdf.
75. Crick (1962), p. 16.

Chapter 11
1. In chronological order: Lengyel *et al.* (1961, 1962), Basilio *et al.* (1962), Gardner *et al.* (1962), Speyer *et al.* (1962a, b), Wahba *et al.* (1962, 1963a, 1963b).
2. Letter from Tomkins to Nirenberg, 25 October 1961. http://profiles.nlm.nih.gov/ps/access/JJBCBB.pdf.
3. Lengyel *et al.* (1961), p. 1941.
4. Martin *et al.* (1961).
5. Lengyel (2012), p. 35.
6. *The Sunday Times*, 31 December 1961. According to the paper, Crick was 'leader of the Cambridge team which discovered code', while Ochoa's colleague Speyer was 'a British biologist'.
7. 'I have stressed that it is your discovery which was the real breakthrough.' Crick to Nirenberg, 4 January 1962. http://profiles.nlm.nih.gov/ps/access/JJBBFL.pdf.
8. Nirenberg to Crick, 15 January 1962. http://profiles.nlm.nih.gov/ps/access/JJBBFJ.pdf.
9. Speyer *et al.* (1962a, b).
10. When each batch of these synthetic RNAs was created, the nucleotides were assembled in a different, random way, producing slightly different molecular sequences, and making it difficult to compare studies that claimed to be looking at the same nucleotide ratios (Matthaei *et al.*, 1962, p. 671).
11. Martin *et al.* (1961), Speyer *et al.* (1962a, b).
12. Crick to Ochoa, 21 September 1962. http://profiles.nlm.nih.gov/ps/access/SCBBSY.pdf.
13. Speyer *et al.* (1962b), p. 443.
14. Speyer *et al.* (1962b), p. 445.
15. Matthaei *et al.* (1962), p. 674.
16. Matthaei *et al.* (1962). See also Crick to Nirenberg, 29 January 1962. http://profiles.nlm.nih.gov/ps/access/JJBBGN.pdf.
17. Crick (1966a), p. 6.
18. Eck (1961). See also Jukes (1962, 1963), Lanni (1962), Wall (1962) and Woese (1962).
19. Tsugita (1962).
20. Crick (1963a), p. 170.
21. Ageno (1962).
22. Woese (1962).

23. Roberts (1962).
24. Eck (1963).
25. Zubay and Quastler (1962).
26. Golomb (1962b).
27. Bretscher and Grunberg-Manago (1962), Gardner *et al.* (1962).
28. Gardner *et al.* (1962).
29. Chantrenne (1963), p. 30.
30. Couffignal (1965), p. 182.
31. Couffignal (1965), p. 78.
32. Chantrenne (1963), p. 27. The term was also used at the meeting by André Lwoff (Couffignal, 1965, p. 176).
33. Ochoa (1964), pp. 4, 3.
34. Tatum (1963), p. 175.
35. Vogel *et al.* (1963), p. 517.
36. Nirenberg and Jones (1963), p. 461.
37. Vogel *et al.* (1963), p. 503.
38. Vogel *et al.* (1963), pp. 517–18.
39. Crick (1963a), pp. 177, 180.
40. Crick (1963a), p. 182.
41. Crick (1963a), p. 212. The original draft of the article contains some even sharper formulations (http://libgallery.cshl.edu/items/show/52223). Crick wrote to Ochoa apologising for criticising his work 'in certain ways' (Crick to Ochoa, 21 September 1962; http://profiles.nlm.nih.gov/ps/access/SCBBGY.pdf). A few months later Crick published a less combative version of these arguments in *Science*, but repeated his crushing statement that the experimental evidence for establishing a codon 'falls short of proof in almost all cases' (Crick, 1963b, p. 463).
42. Crick (1963a), p. 198.
43. Crick (1963a), p. 202.
44. Crick (1966a), p. 5.
45. Crick (1963a), pp. 213–14.
46. A number of Watson's colleagues, including the joker Seymour Benzer, sent Watson a congratulatory telegram that concluded 'PLEASE DONT REFUSE' http://libgallery.cshl.edu/items/show/46113.
47. Nirenberg *et al.* (1963).
48. Nirenberg *et al.* (1963), p. 557.
49. Speyer *et al.* (1963).
50. Ochoa (1964).
51. Nirenberg and Leder (1964).
52. Heaton (2010), pp. 26, 31.
53. Heaton (2010), pp. 29–30.

54. Crick (1966b), p. 554.
55. Söll *et al.* (1965).
56. Salas et al. (1965). For a contemporaneous review, see Stent (1966).
57. Clark and Marcker (1966).
58. Friedberg (2010), p. 150.
59. Yanofsky *et al.* (1964).
60. Brenner *et al.* (1967).
61. Subak-Sharpe *et al.* (1966), Woese *et al.* (1966).
62. Crick (1966a), p. 3.
63. Jacob (1977).
64. http://profiles.nlm.nih.gov/ps/access/JJBCCQ_.jpg.
65. Stent (1968a).
66. Yčas (1969), p. 284.
67. Jacob (2011).
68. Cairns (1966).

Chapter 12
1. Monod and Jacob (1961), p. 393.
2. Crick (1970), p. 561.
3. Lewin (1974).
4. Chow *et al.* (1977), Klessig (1977), Dunn and Hassell (1977), Lewis *et al.* (1977), Berk and Sharp (1977), Berget *et al.* (1977).
5. Gilbert (1978), Crick (1979), Witkowski (1988).
6. Gilbert (1978).
7. Boyce *et al.* (1991), Fedorov *et al.* (1992), Hong *et al.* (2006).
8. Henikoff *et al.* (1986).
9. Crick (1979).
10. Burnet (1956), p. 22.
11. Crick (1959).
12. Rogozin *et al.* (2012).
13. Schmucker *et al.* (2000), Neves *et al.* (2004).
14. Lah *et al.* (2014).
15. Barrell *et al.* (1979).
16. Sapp (2009).
17. Archibald (2014).
18. Lane and Martin (2010), McInerney *et al.* (2014), Yong (2014).
19. Sánchez-Silva *et al.* (2003).
20. Hatfield and Gladyshev (2002), Srinivasan *et al.* (2002), Hao *et al.* (2002), Lobanov *et al.* (2006).
21. Kryukov *et al.* (2003).
22. Berry *et al.* (1993).
23. Rinke *et al.* (2013), Ivanova *et al.* (2014).

24. Starck et al. (2012).
25. Cavalcanti and Landweber (2004).
26. Lozupone *et al.* (2001), Lekomtsev *et al.* (2007).
27. Ring and Cavalcanti (2007).
28. Lajoie *et al.* (2013a, b).
29. Lane (2009).
30. Theobald (2010).
31. Holley *et al.* (1965).
32. Sanger (1988), p. 22. See García-Sancho (2010) for how Sanger's sequencing strategy changed as he moved from proteins to DNA.
33. Sanger *et al.* (1977), van Noorden *et al.* (2014).
34. Sanger *et al.* (1978).
35. Rabinow (1996). For Mullis's account of how he came up with the method see http://www.nobelprize.org/nobel_prizes/chemistry/laureates/1993/mullis-lecture.html.
36. Chien *et al.* (1976).
37. Zagorski (2006).
38. Botstein *et al.* (1980).
39. García-Sancho (2012).
40. Davies (2002), Sulston and Ferry (2002), Shreeve (2004), Ashburner (2006), Venter (2007).
41. The transcript of the ceremony, which includes some creepy banter between Clinton and Blair, can be found here: http://transcripts.cnn.com/TRANSCRIPTS/0006/26/bn.01.html.
42. Lander *et al.* (2001), Venter *et al.* (2001).
43. *The Guardian*, 13 June 2013.
44. *The Guardian*, 5 September 2014. Bizarrely, the court allowed patenting of genes only in material that has been removed from the human body. How else would the sequence be determined?
45. Mardis (2008), Schuster (2008).
46. Li *et al.* (2010), Worley and Gibbs (2010).
47. For microbes there is *Genome Announcements*, http://genomea.asm.org/.
48. Lim *et al.* (2014).
49. Hayden (2014).
50. *MIT Technological Review*, September 2014. http://www.technologyreview.com/news/531091/emtech-illumina-says-228000-human-genomes-will-be-sequenced-this-year/.
51. Genome of the Netherlands consortium (2014).
52. http://cancergenome.nih.gov.
53. Callaway (2014b).
54. Philippe *et al.* (2011).

55. http://www.genomesonline.org. For how to sequence a genome, see http://sciblogs.co.nz/tuataragenome/2013/06/25/first-find-your-tuatara-or-how-to-sequence-a-genome/.
56. Neale *et al.* (2014).
57. Bennett and Moran (2013).
58. McCutcheon and Moran (2011).
59. Woese and Fox (1977a, b). For an analysis of Woese's work and its implications, see Sapp (2009).
60. Williams *et al.* (2013), McInerney *et al.* (2014).
61. Axelsson *et al.* (2013), Freedman *et al.* (2014).
62. Moroz *et al.* (2014).
63. Regier *et al.* (2010).
64. Crick (1958), p. 142. For historical analyses of the significance of this approach, see Stevens (2013) and Suárez-Díaz (2014).
65. http://timetree.org. The app is also called Timetree.
66. Orlando *et al.* (2013).
67. Penney *et al.* (2013).
68. Shapiro and Hofreiter (2014).
69. Pääbo (2014).
70. Krings *et al.* (1997), Green *et al.* (2010).
71. Fu *et al.* (2014).
72. Higham *et al.* (2014).
73. Sankararaman *et al.* (2014), Vernot and Akey (2014).
74. Prüfer *et al.* (2014).
75. Reich *et al.* (2010).
76. Jeong *et al.* (2014).
77. Huerta-Sánchez *et al.* (2014).
78. Hammer *et al.* (2011), Callaway (2014a), Prüfer *et al.* (2014).
79. *New York Times*, 3 June 2003.
80. Ezkurdia *et al.* (2014).
81. Britten and Davidson (1969).
82. Morange (2008).
83. Morris and Mattick (2014).
84. RIKEN *et al.* (2005).
85. Corden *et al.* (1980).
86. Keller (2000), Gerstein *et al.* (2007).
87. See Chapter 9.
88. Pearson (2006).
89. Coyne (2000), Wain *et al.* (2002).
90. Kishida *et al.* (2007).
91. Doolittle and Sapienza (1980).
92. Fechotte and Pritham (2007).

93. Sasidharan and Gerstein (2008).
94. Cornelis *et al.* (2014).
95. Thomas (1971), Gregory (2001).
96. Palazzo and Gregory (2014).
97. http://judgestarling.tumblr.com/post/64504735261/the-origin-of-junk-dna-a-historical-whodunnit.
98. ENCODE Project Consortium (2012). See http://www.nature.com/encode/ and http://blogs.discovermagazine.com/notrocketscience/2012/09/05/encode-the-rough-guide-to-the-human-genome. Birney's own view: http://genomeinformatician.blogspot.co.uk/2012/09/encode-my-own-thoughts.html.
99. Pennisi (2012); *New York Times*, 5 September 2012; *The Guardian*, 5 September 2012.
100. Eddy (2012). Probably the most outspoken critic has been Dan Graur: http://judgestarling.tumblr.com, Graur *et al.* (2013), Bhattacharjee (2014). See also Doolittle *et al.* (2014) and Larry Moran's blog, http://sandwalk.blogspot.co.uk. Germain *et al.* (2014) are supportive of ENCODE's approach from a philosophical point of view.
101. ENCODE Project Consortium (2012), p. 57.
102. Eddy (2013).
103. White *et al.* (2013).
104. http://thefinchandpea.com/2013/07/17/using-a-null-hypothesis-to-find-function-in-the-genome/.
105. Kellis *et al.* (2014).

Chapter 13
1. Crick (1957), pp. 198–200.
2. Anonymous (1970), p. 1198.
3. Watson (1965), Keyes (1999a). In the final chapter of his book, Watson discussed the possibility that cancer might involve exceptions to the central dogma; this was noted at the time, but seems to have been forgotten since (Sonneborn, 1965).
4. Crick (1970), p. 562.
5. Temin (1971), p. iv.
6. Crick (1970), p. 562. In a letter to Temin, Crick said that his classification of the various kinds of information transfer, including the apparent special exceptions, was 'tentative, and may need revision from time to time'. Crick to Temin, 3 August 1970. http://profiles.nlm.nih.gov/ps/access/SCBBMG.pdf.
7. Keyes (1999b), Morange (2007a).
8. Prusiner (1982).

9.   Hunter (1999), Prusiner and McCarty (2006). For a more nuanced view, see Morange (2007a).
10. Prusiner (1998).
11. Manuelidis *et al.* (2007).
12. Supattapone and Miller (2013).
13. Bremer *et al.* (2010).
14. Keyes (1999b).
15. Morange (2008).
16. Denenberg and Rosenberg (1967).
17. Lockyer (2014).
18. Shama and Wegner (2014).
19. Arai *et al.* (2009).
20. Francis *et al.* (1999), Carey (2011).
21. Burkeman (2010), Danchin (2013), Jablonka and Lamb (2006), Noble (2013) and Shapiro (2009) all consider that we are in the midst of a revolution. Maderspacher (2010) is not so convinced. For a point-by-point rebuttal of Noble's views see http://whyevolutionistrue. wordpress.com/2013/08/25/famous-physiologist-embarrasses-himself-by-claiming-that-the-modern-theory-of-evolution-is-in-tatters/
22. See Chapter 11.
23. Guo *et al.* (2014), Smith *et al.* (2014).
24. Radford *et al.* (2014).
25. Hüdl and Basler (2012).
26. Petruk *et al.* (2012).
27. Carey (2011), pp. 217–19.
28. Nelson *et al.* (2012), Mattick (2012).
29. Lumey *et al.* (2009), Heijmans *et al.* (2008).
30. Daxinger and Whitelaw (2012).
31. Francis (2011).
32. Rechavi *et al.* (2011).
33. Heard and Martienssen (2014), Yu *et al.* (2013).
34. Cortijo *et al.* (2014), Bond and Baulcombe (2014, 2015).
35. Heard and Martienssen (2014).
36. For a recent presentation of these two views, see Laland *et al.* (2014).
37. Gayon (2006).
38. For a fascinating exploration of Lamarck's ideas in their historical context, and the importance of his thinking for the acceptance of the idea of evolution, see Mayr (1972).
39. Gayon (1998).
40. See, for example, http://www.technologyreview.com/news/411880/a-comeback-for-lamarckian-evolution.

41. Li and Xie (2011).
42. Crick (1958), p. 144.
43. Hartl *et al.* (2011), Morange (2005b).
44. Anonymous (1997).
45. Crick to Temin, 3 August 1970. http://profiles.nlm.nih.gov/ps/access/SCBBMG.pdf. See also Strasser (2006).
46. Marahiel (2009).
47. Shen *et al.* (2015).
48. Mosini (2013).
49. Morange (2005b).

Chapter 14
1. Gibson *et al.* (2010).
2. For Venter's account of this work, see Venter (2013). The existence of the code was announced here: http://www.jcvi.org/cms/press/press-releases/full-text/article/first-self-replicating-synthetic-bacterial-cell-constructed-by-j-craig-venter-institute-researcher/. For a clear explanation of the code, and details of how it was cracked, see https://genomevolution.org/wiki/index.php/Mycoplasma_mycoides_JCVI-syn1.0_Decoded.
3. *New York Times*, 18 November 2013.
4. For an excellent overview of all aspects of biotechnology, which covers the subject in far more detail than space allows here, see Rutherford (2013).
5. Readers of a certain age may recall the satirical 1983 hit song by Orchestral Manoeuvres in the Dark entitled 'Genetic engineering' (http://www.youtube.com/watch?v=OddgsPyCJmU). According to the *Oxford English Dictionary*, 'biotechnology' was first used in 1921, by the US Department of Agriculture.
6. Lazaris *et al.* (2002); *The Guardian*, 14 January 2012; Rutherford (2013).
7. Source: USDA Economic Research Service. http://www.ers.usda.gov/data-products/adoption-of-genetically-engineered-crops-in-the-us.aspx.
8. Relyea (2012), Annett *et al.* (2014).
9. Moran and Jarvik (2010), Boto (2014).
10. Remigi *et al.* (2014).
11. Kim *et al.* (2014).
12. Goldman *et al.* (2013). In the 1980s, a series of sequences were introduced into *E. coli* bacteria that encoded simple Venus-like icons as part of an art project called Microvenus (Davis, 1996).
13. Church *et al.* (2012); *The New Yorker*, 24 November 2014.
14. Farzadfard and Lu (2014).

15. Schneider's web site is delightfully retro: http://users.fred.net/tds/leftdna/.
16. Marvin *et al.* (1961), van Dam and Levitt (2000).
17. Wang *et al.* (1979), Morange (2007b).
18. Morange (2007b).
19. Rich (2004). Bizarro is the name of Superman's alter ego, a fractured mirror image for whom all of Superman's moral code is reversed.
20. *New York Times*, 29 June 1999; Rich and Zhang (2003).
21. Du *et al.* (2013).
22. Rich *et al.* (1961).
23. Safaee *et al.* (2013).
24. Pinheiro *et al.* (2012).
25. Joyce (2012a, b).
26. Wolfe-Simon *et al.* (2011). The article was accompanied by a series of critical comments and an explanation from the editor of *Science*, Bruce Alberts.
27. Reaves *et al.* (2012), Erb *et al.* (2012).
28. Cleland and Copley (2005), Wolfe-Simon *et al.* (2009). The title of the Wolfe-Simon *et al.* article is 'Did nature also choose arsenic?', thereby inadvertently extending to scientific articles Betteridge's law of headlines, which states that 'Any headline which ends in a question mark can be answered by the word no.'
29. Davies *et al.* (2009), Toomey (2013).
30. Davis and Chin (2013). For a description of other approaches, see Johnson *et al.* (2010).
31. Switzer *et al.* (1989), Piccirilli *et al.* (1990).
32. Yang *et al.* (2011).
33. For example, the drug acyclovir – O'Brien and Campoli-Richards (1989).
34. Malyshev *et al.* (2014).
35. Jackson *et al.* (1972).
36. Berg *et al.* (1974).
37. Berg *et al.* (1975), Berg (2008). See also Morange (1998), Rasmussen (2014), Rutherford (2013) and above all Wright (1994). Some researchers had already begun to circumvent the voluntary moratorium, such was the pressure to make use of these new techniques (Comfort, 2014; Rasmussen, 2014).
38. Watanabe *et al.* (2014).
39. *The Guardian*, 11 June 2014. For an online discussion of this experiment, with useful points on both sides, see http://whyevolutionistrue.wordpress.com/2014/06/12/mad-scientists-or-is-there-any-justification-for-trying-to-recreate-a-deadly-virus/.

40. http://news.sciencemag.org/biology/2014/10/researchers-rail-against-moratorium-risky-virus-experiments.
41. Friedmann and Roblin (1972).
42. Mavilio (2012).
43. 'Gene therapy's big comeback', *Forbes Magazine*, 14 April 2014.
44. Jinek *et al.* (2012). This publication was followed weeks later by a similar announcement from a group based in France, Lithuania and the US (Gasiunas *et al.*, 2012).
45. Cong *et al.* (2013), Mali *et al.* (2013).
46. *New York Times*, 4 March 2014. For a more technical summary, see Hsu *et al.* (2014).
47. *The Independent*, 7 November 2013.
48. Sheridan (2014), http://www.technologyreview.com/view/526726/broad-institute-gets-patent-on-revolutionary-gene-editing-method.
49. O'Connell *et al.* (2014).
50. Esvelt *et al.* (2014).
51. Oye *et al.* (2014).
52. Mandel *et al.* (2015), Rover *et al.* (2015).
53. Berg (2008), pp. 290–1.

Chapter 15

1. Miller (1953), Bada and Lazcano (2000).
2. Elsila *et al.* (2009). In 2008, after Miller's death, researchers examined vials left over from a similar experiment he had carried out in the 1950s. Using modern techniques, they discovered that the levels of amino acids produced by the experiment were even higher than those originally reported (Johnson *et al.*, 2008).
3. Lane and Martin (2012), Martin *et al.* (2014). See also Koonin and Martin (2005), Fellermann and Solé (2007).
4. Baaske *et al.* (2007).
5. For a summary of molecular evidence supporting this view, see Di Giulio (2013a).
6. Crick (1981), pp. 15–16. *Nature* mischievously asked the Bishop of Birmingham to review the book; the Bishop countered Crick's directed panspermia hypothesis by outlining the evidence for a purely terrestrial origin of life, while inevitably leaving the door open for 'divine providence', for which there is arguably even less evidence than for space aliens (Montefiore, 1982).
7. See shCherbak and Makukov (2013) for a suggestion that the genetic code contains an intelligent signature.
8. Gilbert (1986).
9. Joyce (2002), Paul and Joyce (2004), Robertson and Joyce (2012).

10. Pross (2012), p. 63.
11. A handful of scientists are not convinced there was such a thing as the RNA world. See, for example, Caetano-Anollés and Seufferheld (2013).
12. Hotchkiss (1995). This brilliant insight had little consequence.
13. Powner *et al.* (2009). However, as Adam Rutherford points out, John Sutherland's research on the spontaneous synthesis of RNA bases found that the yield of uracil was increased in the presence of ultraviolet radiation, hinting that the synthesis of early RNA may have taken place close to the surface (Rutherford, 2013, p. 96).
14. Lincoln and Joyce (2009).
15. Yarus (2010), p. 97.
16. Noller (2012).
17. Yarus (2010), p. 179.
18. There is a massive literature on this topic. What follows is based mainly on Koonin and Novozhilov (2009), Yarus (2010) and Rutherford (2013).
19. See, for example, Woese (1965).
20. Polyanski *et al.* (2013).
21. Crick (1968).
22. Freeland and Hurst (1998), Freeland *et al.* (2000).
23. Yarus *et al.* (2005).
24. Koonin and Novozhilov (2009), p. 108. See the special issue of *Journal of Molecular Evolution* in 2013, which contained four papers, each outlining a different explanation (Di Giulio, 2013b).
25. Behura and Severson (2013).
26. Cannarozzi *et al.* (2010).
27. Bernardi (2000).
28. Eyre-Walker and Hurst (2001).
29. Romiguier *et al.* (2010), Katzman *et al.* (2011).
30. Stergachis *et al.* (2013).
31. http://www.washington.edu/news/2013/12/12/scientists-discover-double-meaning-in-genetic-code/.
32. For example, Birnbaum *et al.* (2012) and Lin *et al.* (2011).
33. People were also annoyed by the use of the neologism 'duon' to describe the codons that have both coding and transcription factor binding functions (I predict this coinage will not have a long life). For examples of spontaneous and more considered responses, see https://twitter.com/edyong209/status/411283930294534144 and http://pasteursquadrant.wordpress.com/2013/12/14/on-duons-and-cargo-cult-science/.
34. Itzkovitz *et al.* (2010).
35. Mignone *et al.* (2002).

36. Yáñez-Cuna *et al.* (2013).
37. Maraia and Iben (2014).
38. Apter and Wolpert (1965). For a discussion of the role of metaphor in science in general, with a particular emphasis on chemistry, see Brown (2003).
39. For example, Fabris (2009), Gatlin (1966, 1968, 1972), Holzmüller (1984), Lean (2014), Longo *et al.* (2012), Schneider *et al.* (1986), Spetner (1968), Yockey (1992).
40. Maynard Smith and Szathmáry (1997).
41. Maynard Smith (1999, 2000a, b). Other contributors to the debates include: Bergstrom and Rosvall (2011a, b), Collier (2008), Garcìa-Sancho (2007) Godfrey-Smith (2000a, b, 2007, 2011), Griffiths (2001), Kjosavik (2014), Kogge (2012), Levy (2011), Maclaurin (2011), Sarkar (1996a, b, 2000, 2013), Shea (2011), Stegmann (2004, 2005, 2009, 2012, 2013, 2014a, b), Šustar (2007). I am grateful to Ulrich Stegmann for his comments on this section on the philosophy of genetic information; however, grumpy philosophers and others should address their criticisms to me, not him.
42. Maynard Smith (2000a), p. 190. This point was first made by Kimura (1961).
43. Maynard Smith (2000a), p. 190.
44. For example, Jablonka (2002).
45. Maynard Smith (2000a), p. 193.
46. Griffiths (2001), Stegmann (2014b).
47. For a critique of the idea of the gene as a program, see Planer (2014).
48. Sarkar (1996a), p. 107.
49. Sarkar (2000).
50. Godfrey-Smith (2011), p. 180.
51. Sarkar (1996b), p. 863.
52. Dudai *et al.* (1976).
53. http://flybase.org/reports/FBgn0000479.html.
54. For example, Sarkar (1996a), Keller (1995, 2000, 2002)
55. Crick (1958), p. 144, pp. 138–9.
56. For example, Oyama (2000). For a clear introduction to this view, and to other aspects of philosophy and genetics, see Griffiths and Stotz (2013).
57. For example, Commoner (1968), de Lorenzo (2014), Noble (2013), Shapiro (2009).
58. Walker (2007).
59. Crabbe *et al.* (1999).
60. Wahlsten *et al.* (2006).
61. Rietveld *et al.* (2014).

62. Maynard Smith (2000a, b).
63. Noble (2002), Newman (2003).
64. For example, Cosentino and Bates (2012).
65. Cobb (2011). For the fate of the British branch of cybernetics, see Pickering (2010). Medina (2011) provides a fascinating exploration of how in the early 1970s applied cybernetics was used in the doomed attempt to peacefully transform the Chilean economy under the socialist President, Salvador Allende.

Conclusion

1. Pickstone (2001).
2. Gilbert (1991).
3. Wood *et al.* (2014).

LIST OF ILLUSTRATIONS

INDEX

BC	7/15

```
CTAATGGAGTGGGAAAAATGGAAAATTAAGAGGGGACAGTGAATACTTGAGGAAACTAAGCTA
AAGCTAAACTTGTGGCATATATCCATTTTATTAATCTCTCACACAATGCATATTAGGGATTGA
TCCGAATTGATCTGCTATAGATTAGTGGAATCAATTTTTTTTTTTATAATTAAAAGCAGAAAT
ATATGTGTTATGTGTCATTCTAATGTGCGTTTTTCTCTGTTTCTTTGCAGGTACGTTTATTCT
CGAACATCCAATTAGGTGTGACCAAATGTATGTATCTACGAATCGTACGAACTTATATATATG
TCAGGCTTTTATATACACTCCAAGTCCGATCCAAATCCCCCAACGTCGTGTGTTAAACAAATT
AGTTTAATTGGCAGTGAGCGGGTAACTACTTTGGTGACTGACTGGTTTGAACGGTGGCAAATT
TATAAAGTTCCAAATGAAATTATAGCCATTGAATTAACTAATTGCATTTGCCGTGGCCCCCCC
CAAAAGGGGTTTACGGCATACACATTCAGTACATACATGTGACAAACAGTACAAACAGGACAAA
ATGCTGGACTTCCCCCAATTTCCCCCCTTTCCGTTGGTTAATCAAGTATAATCAATGGCTGTG
TTCTTAGGGATGATTACATTCTGTGGACTGCGGTCAAGGAATTTGAGCACATTCCTGCGTGCA
AACATTTCGATTAATTAGCCGTTGTTCATTGTGCGCAAATTAGAAATATATCTTCAAATTACC
GAATAAATACCGTGGAAAGTGTGACACAATCAAATTTCGAGCAAATTGTGCTGCTTTTTGCAG
TTCAGTTGCTACGTTATGATGTTTAATATCTCCAGCCCTTTTATCCCTGAGTCCTTCGCATTA
TGGTAGTTCCTTTAATGCGCTTAACATCTCCTGAGCATCATGAGCTGCTGTTCCTTGAGCGCA
TGTGTGCCCGATTGTGGTCGGGATTCGGGATTCGGGATCTGGGAGAGCACAAAACCCTCTCGA
GTAGGCAAATTTTTGAAAGCATTTGAATACTACTAAAAGAAAAAAAACCTAATTCAATTGGTA
GAAAATATGTATCAATTTCAAGCACAACAACTTGGGAGCAGTACTTTTCCCACTATAAAGCAT
TTTTTTTTTTCACAACTGTTTGATTTGGCTTAAAATTCTTAACAAAGCGAATATTTTATTGAA
ATTTATTTTCTGCGTGTATTCTCCACACATGCATCGAGGGCGTTGTTAATGATGCGCAGTTCG
AATCTTGAGTTTTGGGATTGAACACAGAACAGGAATAAAAAAAAGTGGCTAGAAAAGCAGCGA
AATAGCTTAGCTTAGCTGGCGAGCGGATGTGAGGAAAAACCCCCGTGAAAACTGGCGAAAATG
CCTTAAGCCTTTGCCATGGCAATTTGTTTCGCAACAGTTGAACAAATTTAATGCAATAAGCTA
CGCAAAAAATTGTAATTTTTATCTGTTTAATATGTATGTTATATTGAATCACATATGTAATAT
CCAACACTTGTTTCAATTAAGTGCTGCAAATAGATTGGTGTGTCCACACTTTTATCATAGACT
CATTTTATTTCGCCAAGTGTTTATATAGCGCTGCCGCCATAGTAACGCATCGAATCTGTGTCT
TGTGGCCCAAGGATATTTCGTTTAACTTTTTCTGATTGAAATATTCCATTCTATGAACGAGAT
TTTTGCGTTTGCGTTTGCCTTGGCCAAAGTTACACAGCCGGCTCGTTTCTAATGGTAAGCTAC
TAACAATTTAATTTACACATAACTCAGCACTAAGCTTCAACTAAGCCACAAAATGTGCATCTTG
GCCAAGTGTCAACAAGAGGAACACATAGGGGTTAAGGAGGCAATGGGCCCTCAAGGGGTTAAA
GGCTAAAAATATTTTCCTATTGCAATTTAAAATGTTTACTTTGTGGGCAAAGCGGGCGCGCGG
CGTAATAAAATGTATTTGCCATTTTGCAATAAAATGGAAATTCGTTGGAAAATGGAAATTCAT
AAGAGCTAAAATGGATGACAACGAGTGCTGATTGTTGTTCAAGTGTCTTTCCTACATATATAT
GTATATGTATGTACATATGTTAGAACACATACCCATTTAAGTGGTAACCACTCACAAATTGAAAC
AGAAATATGAGCACAAAAGTTGTGATATATTGAAATAACACAATGGCTTGCACCTTTTTGCCA
ATTTTTTGGTTTAACAAGAGCTCGTAACTTATAGGTTTTCGTGTGTTTAAATTTAATTTGTA
TATTTGTGCGTCGTATTTCACTGTGATGATACCTCCCTCATCAATGTCCTCATTCGTCAATCA
TTGTTAATTCAACTAAAGCAGTTTGCCAAACGAATAAAATAAAATGATGTGCGAAGTAAATTT
GGCGGTTGAAATTCGCATATAAATCAGAATAATTTACAACTGTTTTTGGAAAACTTAAGTATT
GGTGGGTGTGTGGGTTCAGCATTATAAATATACACACATACAAATTGATCGTAAATCTTGACC
TTTTTTTTTTTTTGCCTGCCCTTGTCCGATTTGCATTTCCCTTGGGCATAAATTTCGATTTTCCA
TCTTCTTTGATCATTCCAATTTCAACACGAATTAGCATTTGGCAACTTAATTAAAACGATTTTT
CATTTTCTTTTTTTCATAATTTTTGGGACACTTGTGCAACAAGTTGACGGCTGCCGGTGACGCA
AAATTTATGATTACCGCAAAAACGAAAAACCGAATTTCAGAAATTATGAATTTTATTTATGCC
TACCCAAGTTTGCTGTTCGTCGAAATTTATTGGAAAAATGTAAATGAAACATAAATATAAGGC
CATGTGTGCGGTGCGCTGGGTCTTTAGTTTTTCGCTTGCCATCTCTGAATATTTTCCTTTGCC
TCTTGAGGGTTTTTTTTTTCACTCTTAAACCATTTTCAATTGCATCGTTAAGAGCTTTGAGGTC
TGCTTAGCCTTTCGCTATATGGACGAGATGCCATATGTATTTATATATATATATATATATATAT
TTAAAGACTTGGAGATGACAGCCAGTGATGAGATCAATTGTGGGTATTGCAAAAATAAAACGTT
GGAACAAACATCATTTCAAAAGTATGCTCAATGCTTAAAAGAATAATGTTAATCTTGATTAAT
TTTGAAATTAGTTATCAGACTTGGGTGTATTTATGTGTGAGTCACAACTGCTTAAACTGGATGA
TTTATATTTGCCGCCGAAATGACATCTCTAATGCCAATATGCACCCGCAAGATAGTTTTCGCA
TTTTTTTTCAGCGGAAGGCTGACAGAGTGGAAGCCAAGTTAAAATGCAATTGAAAAGCCATTTA
TCACTGAAAGGGAAAAAAGAGTGCCTTTAGGGAAGTTTAAGGGCAGATATCCAAATGCTTTTC
```